Advances in Stereotactic and Functional Neurosurgery 7

Proceedings of the 7th Meeting
of the European Society for Stereotactic
and Functional Neurosurgery,
Birmingham 1986

Edited by
J. Gybels, E. R. Hitchcock, B. Meyerson,
Ch. Ostertag, G. F. Rossi

Acta Neurochirurgica
Supplementum 39

Springer-Verlag Wien New York

Professor Jan Gybels
Department of Neurology and Neurosurgery, Leuven, Belgium

Professor Edward R. Hitchcock
The Midland Centre for Neurosurgery and Neurology, West Midland, U.K.

Professor Björn Meyerson
Department of Neurosurgery, Karolinska Hospital, Stockholm, Sweden

Professor Christoph Ostertag
Abteilung für Stereotaktische Neurochirurgie, Neurochirurgische Universitäts-Klinik, Homburg/Saar,
Federal Republic of Germany

Professor Gian Franco Rossi
Istituto di Neurochirurgia, Università Cattolica del Sacro Cuore, Rome, Italy

With 121 Figures

Library of Congress Cataloging-in-Publication Data. European Society for Stereotactic and Functional Neurosurgery. Meeting (7th: 1986: Birmingham, West Midlands, England) Advances in stereotactic and functional neurosurgery 7. (Acta neurochirurgica. Supplementum, ISSN 0065-1419: 39). 1. Stereoencephalotomy— Congresses. I. Gybels, J. II. Title. III. Series. [DNLM: 1. Brain Neoplasms—surgery—congresses. 2. Movement Disorders—surgery—congresses. 3. Neurosurgery— congresses. 4. Pain—surgery—congresses. 5. Stereotaxic Technics—congresses. W1 AC8661 no. 39/WL368 E89 1986] RD594.E85 1986. 617'.481. 87-17286

ISSN 0065-1419 (Acta Neurochirurgica/Suppl.)
ISSN 0720-7972 (Advances in Stereotactic and Functional Neurosurgery)
ISBN-13: 978-3-7091-8911-5   e-ISBN-13: 978-3-7091-8909-2
DOI: 10.1007/978-3-7091-8909-2

Preface

These collected papers represent only a small part of the large amount of new work in the field of stereotaxy. The number of contributions to the Birmingham meeting was such that only selected papers, chosen as representative of advances in the field, could be printed.

These papers present the most up to date accounts of major advances in stereotactic imaging and the renewed interest in the stereotactic treatment of movement disorders.

Contents

B. Spasticity

Section III. Pain and Miscellaneous

Section I

Stereotactic Imaging, Tumours and Haematomas

Acta Neurochirurgica, Suppl. 39, 3–6 (1987)

Transformation Modes in Computerized Human Thalamic Brain Mapping

H. G. Lipinski*, A. Struppler, and **P. Birk**

Department of Neurology and Clinical Neurophysiology, Technical University Munich, Federal Republic of Germany

Summary

An exact transfer of data intraoperatively gathered in thalamic nuclei to an anatomical atlas requires an efficient mathematical transformation mode. Three different kinds of transformation modes were analyzed: First, the AC PC distance was used as a parameter to correlate data with the atlas coordinate system. Second, the influence of the patients 3rd ventricle widths on the transformation procedure. Third, a transformation mode was performed based on "noise"-data, registered when the electrode penetrated patient's thalamus. This method was also used to combine the atlas and CT images.

Keywords: Brain mapping; thalamus; stereotaxy; thalamotomy.

Introduction

Electrical stimulation in the target area was performed routinely during localization in stereotaxis thalamotomy for movement disorders[1, 4]. The atlas of Schaltenbrand and Wahren[3] served as a reference for physiological data as well as CT images. The physiological data was transferred to the atlas according to the stereotaxic coordinates of their respective stimulation and recording sites[2]. For exact transfer mathematical calculations must be performed on a computer.

Methods and Materials

The upper and lower thalamic border was identified by recording spontaneous neural noise which changes when the electrode enters (*cf.*, Fig. 1, *1*) or leaves (*cf.*, Fig. 1, *3*) the thalamic nucleus (VL). All the collected data of stimulation sites and the noise data were related to a rectangular (cartesian) coordinate system (CCS). This system

based on the AC PC plane had its origin at 2/3 AC PC. The frontal direction was the (positive) S-axis, the laterality to the left was defined as (positive) L axis and the cranial direction was the (positive) H axis (*cf.*, Fig. 1).

Schaltenbrand and Wahren's atlas[3] was completely digitized to match the data with thalamic nuclei. A computer program was created to digitize the atlas with the aid of a videocamera and the data was stored in computer memory. The atlas coordinate system was defined as having the same origin and same orientation as CCS. Three different kinds of transformation modes were used to correlate the recorded data with the atlas. The efficacy of these modes were tested by matching the sites of noise data with the respective outlines of atlas-thalamus. The atlas also served as a reference for CT images. Hence, the patient's ventriculogram referenced to the raw data and the stereotaxic atlas was digitized and correlated with the CT scan. All calculations were done with the aid of DEC's LSI 11 computer in combination with SIGNUM's image module IM 512. SIEMENS Somatom DR 3 was used for CT scanning.

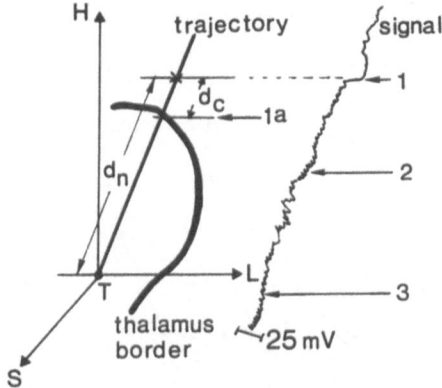

Fig. 1. Spatial distance dc was used to adjust patient's coordinates to the atlas system (*S* frontal, *L* lateral, *H* cranial, *T* origin of the coordinate system). dc was calculated as a "distance vector" from atlas section point of the trajectory (*1 a*) to the coordinates at which noise change occurred (*1*). (*2 noise registered inside of the thalamic nucleus, 3 change of noise when electrode is passing the thalamic-subthalamic border line*). Signal was filtered (1 Hz, lowpass) and rectified

* Dr. Hans-Gerd Lipinski, Department of Neurology and Clinical Neurophysiology, Technical University Munich, Moehlstrasse 28, D-8000 Munich 80, Federal Republic of Germany.

Results

The AC PC distance of the patient is an important parameter to relate the individual coordinate system (CCS) with the atlas coordinate system (ACS). The AC

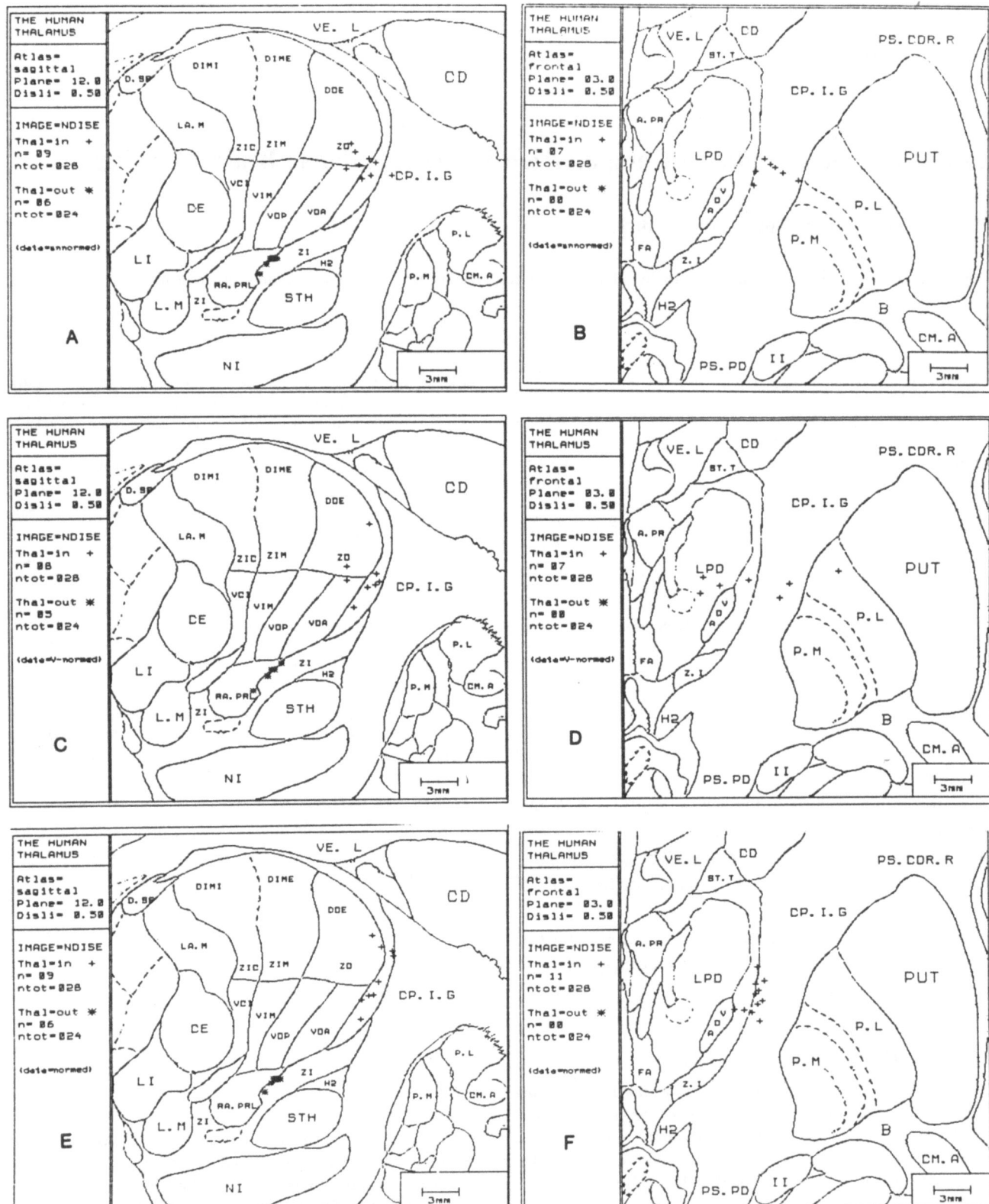

Fig. 2. Efficacies of AC PC-transformation mode (A, B), adjustment of ventricle widths (C, D) and of noise transformation mode (E, F) were tested by matching the sites of noise data with the outlines of atlas thalamus

PC distances of our latest 28 patients ranged from 24 up to 31 mm (mean value = 27.2 mm, SD = 1.8 mm, Wilks = 0.55, n = 28). Generally, the quotient Q = AC PCa/AC PCp (AC PCa = AC PC distance of the atlas, AC PCp = patient's AC PC distance) was used to transfer any point of CCS to ACS. The accuracy of this "classical" transformation mode[5, 6] was tested by transferring noise data to the atlas. The result is demonstrated in Figs. 2 A and B. The "electrophysiological" thalamic boundaries did not coincide with the thalamus border of the atlas, neither in the saggital plane (*cf.*, Fig. 2 A) nor in the frontal plane (*cf.*, Fig. 2 B) (penetration of the thalamus = +, passing of the thalamic-subthalamic border = ★).

The individual widths of the 3rd ventricle of the 28 patients ranged from 4 up to 20 mm (mean value = 7.5 mm, SD = 3.5 mm, Wilks = 0.899, n = 28). The scatter of ventricle widths was eliminated by creating a "standard ventricle width" which corresponded to the ventricular mean value of 7.5 mm. After this calculation an AC PC-transformation mode was used to transfer those "ventricle adjusted" noise data to the atlas. The effect of the transformation mode is shown in Figs. 2 C and D. The discrepancy between thalamus border in the atlas and the location of noise data was enlarged, in the lateral (*cf.*, Fig. 2 D), as well as in the sagittal direction (*cf.*, Fig. 2 C).

Both tested transformation modes were not sufficient for transferring data to the atlas of Schaltenbrand and Wahren. Therefore, noise itself was used to create a transformation mode. The distances on from the centre of the individual coordinate system (*cf.*, Fig. 1, *T*) up to the coordinates where noise change occurred (*cf.*, Fig. 1, *I*) ranged from 10 up to 20 mm (mean value = 15.0 mm, SD = 2.0 mm, Wilks = 0.891, n = 28). The coordinates corresponding to the section of the trajectory with the thalamic (VL) border of the atlas were calculated (*cf.*, Fig. 1, *1 a*). The spatial distance dc between that section point and the coordinates of thalamic border according to noise was used for the adjustment of patient's coordinate system to the atlas coordinate system. The efficacy of that "noise transformation mode" is demonstrated in Figs. 2 D and E. Noise data matched the thalamic boundaries of the atlas with the smallest discrepancy in the sagittal planes (*cf.*, Fig. 2 E) as well as in the frontal planes (*cf.*, Fig. 2 F).

Additionally, a transformation mode was created which permitted combination of the CT scan of the individual thalamus with the atlas. For this purpose the length of midsagittal line of the 3rd ventricle was calculated from the CT image (*cf.*, Fig. 3 A). The value was correlated to the (digitized) ventriculogram. The PC in the CT scan was reconstructed by this procedure (*cf.*, Fig. 3 B) and gantry tilt angle in reference to AC PC plane was calculated. Furthermore, the origin of the atlas coordinate system could be identified from Fig. 3 B. When these procedures were done, horizontal

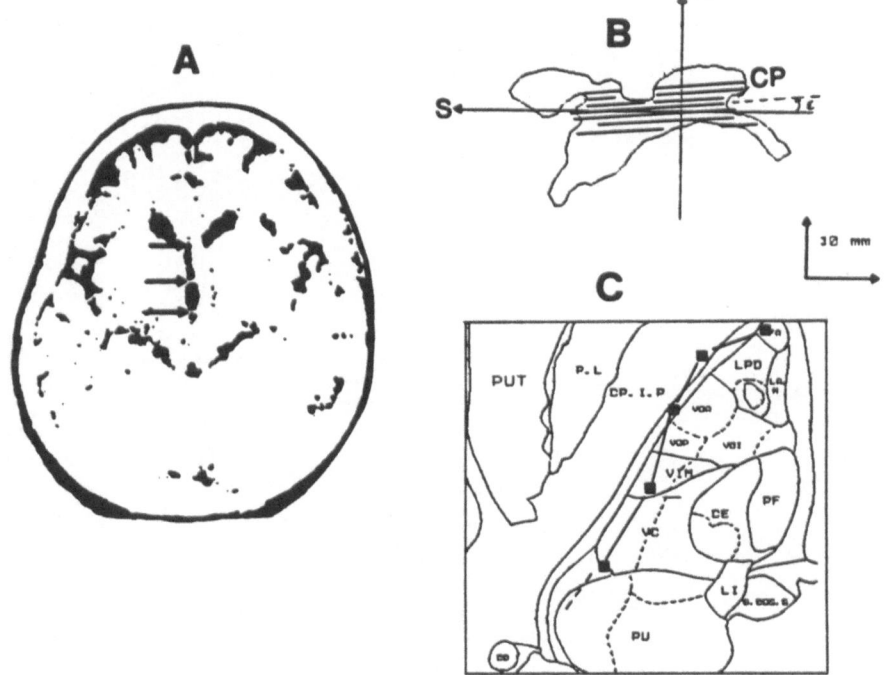

Fig. 3. Combination of Schaltenbrand-Wahren's atlas (horizontal plane 2.0) and reconstruction of lateral thalamic border from (filtered) CT image. From a CT scan (A) PC was reconstructed and transfered to the (digitized) ventriculogram (B). Outlines of the thalamus in CT image were combined with the atlas (C)

outlines of the thalamus in the CT image were transfered to the atlas. Fig. 3 C demonstrates the outlines of the individual thalamus in relation to the atlas plane 2.0.

Discussion

Using AC PC-transformation mode atlas thalamic border did not match with the sites of noise-data. Thus, the lateral and height extensions of the thalamus did not exactly correlate with the AC PC distance. The individual widths of the 3rd ventricles (Wv) ranged from 4 to 20 mm. Therefore, an adjustment of Wv to the atlas was done. Any individual value of Wv was replaced by the mean ventricle width before transformation of noise data started. As a result the sites of transformed noise data were even more deviated from the atlas thalamic boundaries. Thus, ventricle widths did not correlate with the (lateral) thalamic extension. Finally, noise itself was used to create a special transformation mode. A complex computer program permitted the calculation of the coordinates through the trajectory passing through the thalamic boundaries of the atlas. Using this method noise data could be adjusted to the atlas border of thalamus (VL) not perfectly but with smaller deviation than before. Thus, "noise transformation mode" should be used to transfer data to an atlas.

Actually, stereotaxic surgery is combined with CT scanning. A CT scan could be transferred to the atlas, if the AC PC would be identified in the CT image. Unfortunately, routine reconstructions of the midventricular plane are mostly insufficient to identify AC PC. Thus, the CT image must be correlated with the (digitized) ventriculogram. Using this approximation method the outlines of the indiviudal thalamus can be transferred to the atlas. The latest result of our calculations indicates that this approximation method permits the combination of atlas and CT scan, but for high accuracy it has to be improved.

References

1. Andy OJ (1966) Sensory motor responses from the diencephalon. J Neurosurg 24: 612–620
2. Ohye C, Narabayashi H (1979) Physiological study of presumed ventralis intermedius neurons in the human thalamus. J Neurosurg 50: 290–297
3. Schaltenbrand G, Wahren W (1977) Atlas for stereotaxy of the human brain. Thieme, Stuttgart
4. Taren J, Guiot G, Derome P, Trigo C (1969) Thalamic target localization in stereotaxic surgery: A comparison of the accuracy of radiologic and electrophysiologic methods. Conf Neurol 31: 116–122
5. Tasker RR, Hawrylyshyn P, Organ LW (1978) Computerized graphic display of physiological data collected during human stereotactic surgery. Appl Neurophysiol 41: 183–187
6. Velasco F, Monia-Negro P, Bertrand C, Hardy J (1972) Further definition of the subthalamic target for the arrest of tremor. J Neurosurg 36: 184–191

Acta Neurochirurgica, Suppl. 39, 7–9 (1987)

Stereotactic Computer Graphic System with Brain Maps

F. Giunta*, G. Marini, and **M. Bertossi**

Department of Neurosurgery, University of Brescia, Italy

Summary

We have developed a sterotactic computer graphic system with brain maps that runs on a personal computer. This system consists of three parts: 1. firmware for dizitizing radiological films with a TV camera or scanner (when digital image is not directly obtained from floppy disks); 2. software for introducing or processing brain maps by matching them with CT or NMR images; 3. hardware.

Our system is designed to: 1. recognize the relevant frame points and calculate targets, their volume and surgical instrument trajectory; 2. match between brain maps of an ideal brain and patient's CT or NMR images with visible and invisible pathology; 3. monitor during surgery the position of the surgical instrument or target modification.

The whole procedure and processed images are stored in a data-base for further study.

Keywords: Stereotaxy, brain mapping; computer graphics.

Introduction

About fifteen years ago the computer was introduced to help stereotactic neurosurgery[2–10, 12, 13].

Main frame or minicomputers were used in all cases, but these systems are more complex and much more expensive than the stereoactic apparatus to which they have to be connected. Previous studies are very interesting but are often not transferable from the research center where they have been developed for the operating theatre.

We have developed a stereotactic computer graphic system with brain maps that runs on a personal computer.

Description of the Method

Brain maps are digitized from a fixed human brain, embedded in paraffin and cut with a microtome. Each cerebral structure is a record in a data-base file, which contains graphic and alphanumeric data. A TV camera digitizes CT or NMR images (when they are not directly read from a floppy disk) and X-ray films taken in operating theatre then reads relevant points from the frame for target calculation. There is a match between the patient's brain dimensions and computerized brain maps.

With these data, targets or target-areas are calculated and serial biopsies are planned. The database is on line for storage of CT or NMR characteristic of the target points or some clinical or instrumental alphanumeric data.

During the stereotactic procedure, a new image can be obtained from X-ray film and the program provides its digitalization and recalculates the target point or surgical instrument position.

The system consists of three parts: 1. firmware for digitizing radiological films with a TV camera, when digital images are not directly obtained from floppy disks; 2. software for introducing or processing brain maps by matching them with CT or NMR images; 3. hardware compatible with a personal computer.

Our system is designed to: 1. recognize the relevant frame points and calculate targets, their volume and surgical instrument trajectory; 2. match between brain maps of an ideal brain (Fig. 1) and patient's CT or NMR images with visible or invisible pathology; 3. measure length, angles, areas and volumes; 4. monitor during surgery the position of the surgical instrument

---

* Filippo Giunta, M.D., Neurochirurgia, Spedali Civili, I-25100 Brescia, Italy.

or target modification. The whole procedure images are stored in a data-base for further study.

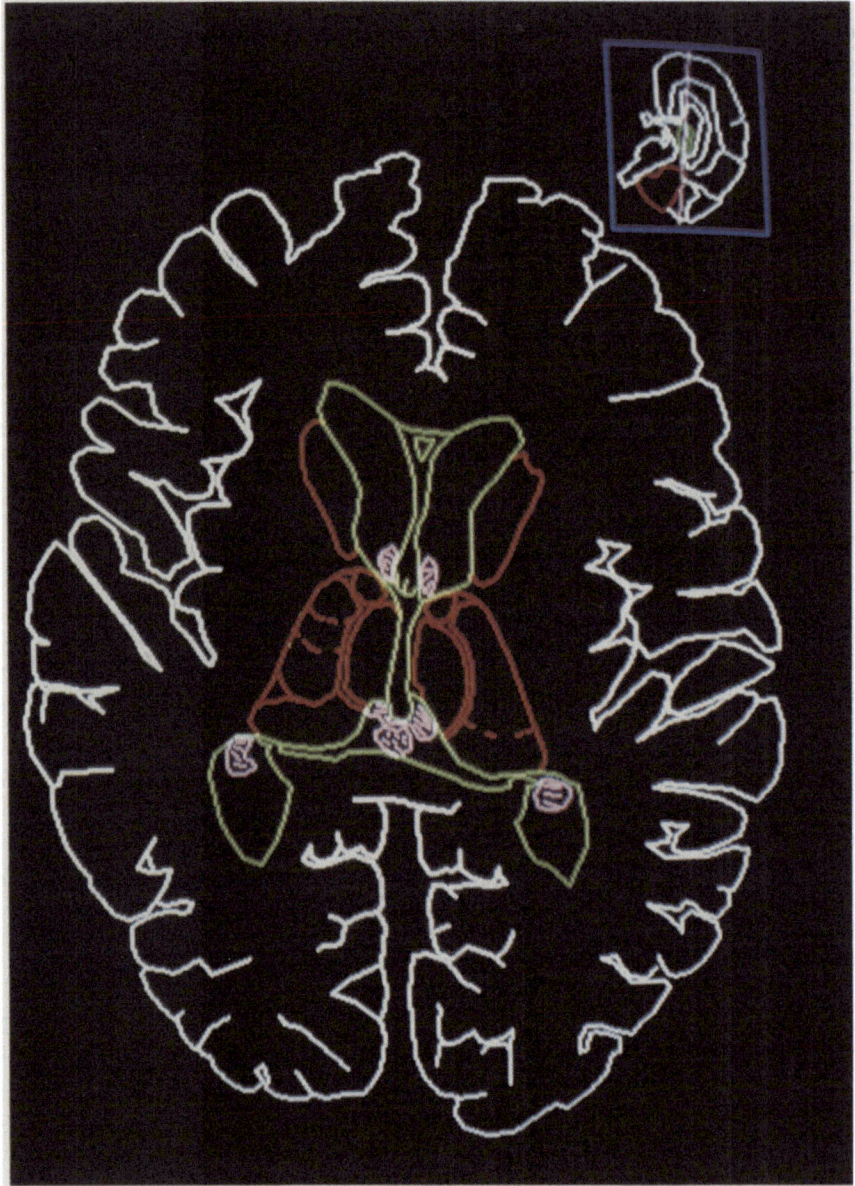

Fig. 1. Digitized brain map

References

1. Afshar F, Dikes E, Watkins ES (1983) Three-dimensional stereotactic anatomy of the human trigeminal nerve nuclear complex. Appl Neurophysiol 46: 147–153
2. Barcia-Salorio JL, Martinez-Corrillo JA (1969) Calculation of the target point by means of an analogue field plotter. In: Third Symposium on Parkinson Disease, Livingstone, London
3. Bertrand G, Olivier A, Thompson CJ (1974) Computer display of stereotaxic brain maps and probe tracts. Acta Neurochir (Wien) [Suppl] 21: 235–343
4. Dawson BH, Derwin E, Heywood OB (1969) The development of a mechanical analog for directing and tracking the electrode during stereotaxic operations. Technical note. J Neurosurg 31: 361–366
5. Derwin E, Heywood OB, Crossley TR, Dawson BH (1974) The use of a small digital computer for stereotactic surgery. Acta Neurochir (Wien) [Suppl] 21: 254–272
6. Garibotto G, Giorgi C, Cerchiari U (1985) Stereoscopic analysis of vascular and neuroanatomical data with application in functional stereotactic surgery. Computer aided radiology (Berlin), pp 146–148

7. Giorgi C, Garibotto G, Cerchiari U, Broggi G, Franzini A (1984) The use of highly detailed three-dimensional neuroanatomical images in the placing of cerebral stimulating electrodes for pain. Acta Neurochir (Wien) [Suppl] 33: 527–528

8. Hardy RT L, Kock J, Lassiter A (1983) Computer graphics with computerized tomography for functional neurosurgery. Appl Neurophysiol 46: 217–226

9. Kelly PJ, Kall B, Goerss S (1984) Functional stereotactic surgery utilizing CT data and computer generated stereotactic atlas. Acta Neurochir (Wien) [Suppl] 33: 577–583

10. Peluso F, Gybels J (1974) Computer calculation of two-target trajectory with "centre of area target" stereotaxic equipment. Acta Neurochir (Wien) [Suppl] 21: 173–180

11. Peters TM, Olivier A, Bernard G (1983) The role of computed tomographic digital radiographic techniques in stereotactic procedures for electrode implantation and mapping, and lesions localization. Appl Neurophysiol 46: 200–205

12. Suetens PJ, Gybels J, Jansen P, Oosterlink A, Haegemans A, Dierkx P (1984) A global 3-D image of the blood vessels, tumor and simulated electrode. Acta Neurochir (Wien) [Suppl] 33: 225–232

13. Towsend HRA (1974) Towards a three dimensional brain model stored in a computer. Acta Neurochir (Wien) [Suppl] 21: 265–266

Acta Neurochirurgica, Suppl. 39, 10–12 (1987)

An Intraoperative Interactive Method to Monitor Stereotactic Functional Procedures

C. Giorgi*, U. Cerchiari[1], G. Broggi, N. Contardi[1], P. Birk[2], and A. Struppler[2]

Division of Neurosurgery, Istituto Neurologico, Milano, Italy, [1] Department of Physics, Istituto dei Tumori, Milano, Italy, [2] Neurologische Klinik, Technische Universität, Munich, Federal Republic of Germany

Summary

A computer graphic technique is presented, which makes it possible to handle neurophysiological and neuroanatomical data collected during functional stereotactic procedures.

Keywords: Computer graphics; stereotaxy, brain mapping.

Introduction

In a stereotactic functional procedure a transposition of the atlas anatomy on to the patient's reference system is performed. Next, within the selected nucleus the process of functional localization is completed with the acquisition of neurophysiological observations, recorded along the stereotactic trajectory. Different authors have dealt with the problem of functional localization within the anatomical reference system[1, 2, 4-6]. We propose a method that allows collection and retrieval of functional data during the procedure, using a graphic data base.

Description of the Method

The system that we have developed is based upon computer graphic technology and consists of two subunits. The first is an interactive program that employs a graphic tablet to enter a patient's radiological data. These can be biplanar ventriculograms, or CT or MR images, performed under stereotactic con-

ditions: they relate the anatomical and stereotactic reference points. Doing this the programme performs "scaling" of the stored models to the individual case. The adequacy of this scaling depends upon the radiological information available. A "menu" of possible interactions is provided; its most relevant feature refers to the possibility of anatomical and functional data retrieval (Fig. 1). From the resulting images, which are oriented within the anatomical frame or reference, the

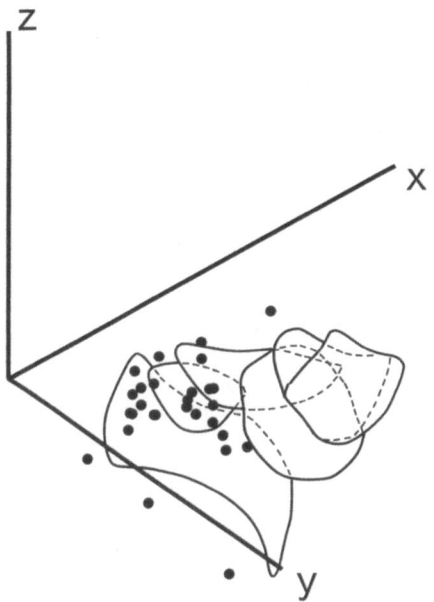

Fig. 1. Drawing from computer monitor output, not suitable for reproduction. Black dots represent functional observations; outlines indicate the position of Voa-Vop complex, digitized from the Schaltenbrand and Wahren atlas. The anatomic reference system has its origin at the midpoint of CA-CP line, Z being the vertical axis, X and Y lying respectively in the sagittal and frontal planes

* Cesare Giorgio, M.D., Division of Neurosurgery, Istituto Neurologico, 11, via Celoria, I-20133 Milano, Italy.

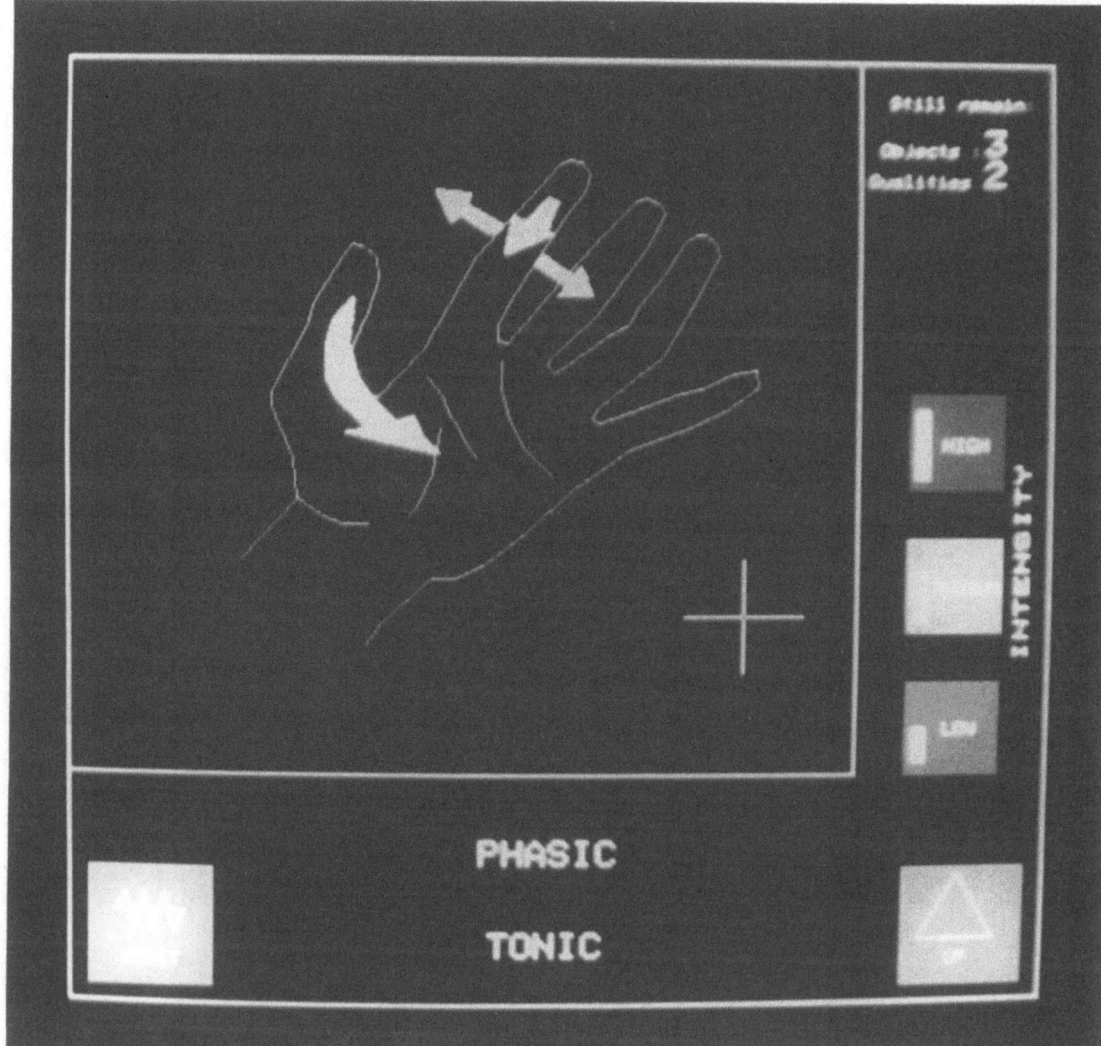

Fig. 2. An example of a data acquisition mask. Finger movements are symbolized by arrows, that can be selected with touch-screen device. Modalities of the observed movements are entered by touching the corresponding areas on the screen

surgeon can select the most suitable trajectory, converted into stereotactic coordinates. The outlines of anatomical structures are derived from the Schaltenbrand and Wahren atlas[3], while the graphics symbols which represent functional information are retrieved from the second program of the system, *i.e.* the graphic data base. This consists of symbols shown to the surgeon during the procedure, during acquisition of neurophysiological observations. The symbols summarize all major phenomena related to central or peripheral stimulation procedures. Their graphic modality speeds up and standardizes the process of acquisition (Fig. 2). The selection is operated via a touch pen device. The corresponding modalities, together with alphanumeric data on the patient, the parameters of central or peripheral stimulation, and the description of

sensory phenomena, are delivered to the data base. From the data base, information can be retrieved in any possible combination or range of parameters and shown, after appropriate scaling, on the anatomical frame of reference of a given case. Graphic symbols and colors enhance the data. The system is written in Fortran 77 and runs on a Digital PDP 11 23, with a Tesak EGP 414 graphic controller. It can be readily adapted to any suitable graphic workstation. Test trials are in progress at the Istituto Neurologico of Milano and at the Neurology Clinic of the Technische Universität in Munich. The wide variety of configurations permitted by the system's architecture allows its use in multicentric protocols.

This work has been supported by Italian Research Council grant no 85.00391.11.

References

1. Emers R, Tasker RR (1975) The human somesthesic thalamus with maps for physiologic target localization during stereotactic neurosurgery. Raven Press, New York
2. Hardy TL (1975) Computer display of the electrophysiological topography of the diencephalon during stereotactic surgery. MSc Thesis McGill University
3. Shaltenbrant G, Wahren W (1977) Atlas for stereotaxy of the human brain, II Ed. Thieme, Stuttgart
4. Tasker RR, Rowe IH, Hawrylyshyn P, Organ LW (1976) Computer mapping of brainstem sensory centers in man. J Neurosurg 44: 458–464
5. Tasker RR, Hawrylyshyn P, Organ LW (1978) Computerized graphic display of physiological data collected during human stereotactic surgery. Appl Neurophysiol 41: 183–187
6. Thompson CT, Hardy T, Bertrand G (1977) A system for anatomical and functional mapping of the human thalamus. Comp Biomed Res 10: 9–24

Acta Neurochirurgica, Suppl. 39, 13–14 (1987)

3-D Reconstruction of Cerebral Angiography in Stereotactic Neurosurgery

C. Giorgi*, U. Cerchiari[2], G. Broggi, and A. Passerini[1]

Division of Neurosurgery, Istituto Neurologico, Milano, Italy, [2] Department of Physics, Istituto Naz. dei Tumori, Milano, Italy, [1] Division of Radiology, Istituto Neurologico, Milano, Italy

Summary

A method is described, that enables the surgeon to appreciate the three-dimensional distribution of cerebral vessels within the stereotactic space.

Keywords: Computer graphics; cerebral angiography; stereotaxy.

Introduction

Knowing the precise spatial distribution of cerebral vessels within the stereotactic frame coordinates goes beyond the need of obtaining safe, avascular trajectories: vessels can be important anatomical landmarks[1, 3] both for cortical and for central structures. Diseases affecting the vessels can be approached with the aid of stereotactic technique, if precisely localized. Contrast angiography is still the best-means for imaging cerebral vessels, both for superior spatial and density resolution, and for the unique ability to distinguish between vascular "phases". After reviewing the work of other authors on the subject[2, 4], and trying different approaches, we propose a method, based upon stereoscopic perception, which seems to be suitable for clinical applications.

Description of the Method

The system is based upon the Philips digital angiograph (DVI 2), integrated with a Brown-Roberts-Wells stereotactic frame**. The frame's "z" axis is aligned with the center of rotation of the DVI "C" arm; the presence of marks that will show on all angiograms, allows calculation of the spatial relationship between the two (Fig. 1). The patient lies on the DVI table, the head positioned within the stereotactic frame in the usual way. A series (8 to 12) of angiographic pairs is shot on an arc of 120°, and arterial, capillary and venous phases are triggered by ECG. One extra plain image is taken at each angle to allow for subtraction of bone. The series of images, recorded on 8″ floppy discs is transferred to a Digital micro PDP 11/73, connected to a pair of graphic controllers ECS TESAK mod. EGP 414*** and two high resolution monitors.

Knowing the spatial relationship between the frame and the angiograph, the program allows for the projection of "phantom" surgical instruments on the

Fig. 1. The stereotactic frame is aligned with the axis of rotation of the angiograph; a temporary device is mounted on the probe holder: it allows the identification of the frame's position in the series of images

---

* Cesare Giorgi, M.D., Division of Neurosurgery, Istituto Neurologico, 11, via Celoria, I-20133 Milano, Italy.
** Radionics Inc. Burlington, Mass. USA.

*** Ecs Tesak spa-Sesto Fiorentino (FI), Italy.

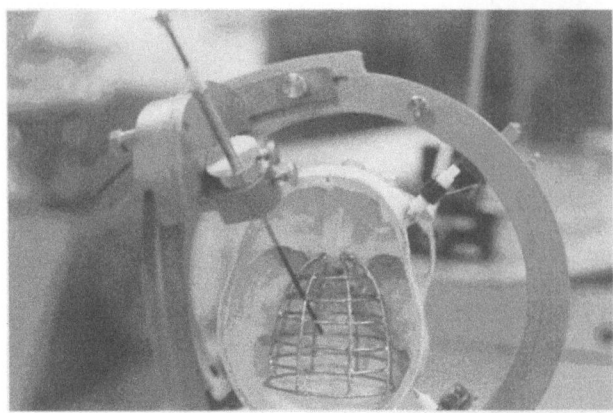

Fig. 2. A dry skull mounted on the frame. A metal cage is positioned inside: it simulates a vascular structure, transversed by the surgical probe

Fig. 3. Stereoscopic pair of the cage, obtained from the setup shown on Fig. 2. The "phantom" surgical probe (indicated by white arrows) can be positioned within the cage, perceived as being tridimensional; the angular values of the probe are automatically given

stereoscopic angiographic pairs. Vascular images can be "rotated", by rapidly switching between successive frames, shown on the graphic monitors. Alternatively, the probe can be freely moved within the images: its position is automatically converted into frame coordinates. A test trial is shown, obtained with a metal cage positioned in a dry skull, fixed in the stereotactic frame (Figs. 2 and 3).

References

1. Steiner L, Leksell L, Forster DMC, Greitz T, Backlund EO (1974) Stereotactic radiosurgery in intracranial arterovenous malformations. Acta Neurochir (Wien) [Suppl] 21: 195–209
2. Suetens P, Oosterlink A, Haegemans A, Gybels J (1982) Three dimensional reconstruction of the blood vessels of the brain. Proceedings ISMIII (IEEE Computer Society) Berlin
3. Szikla G, Bouvier G, Hori Y (1975) Localization of brain sulci and convolutions by arteriography: a stereotactic anatomoradiological study. Brain Res 95: 497–502
4. Vignaud J, Rabischong P, Yver JP, Pardo P, Thural C (1979) Multidirectional reconstruction of angiograms by stereogrammetry and computer. Neuroradiology 18: 1–7

Acta Neurochirurgica, Suppl. 39, 15–17 (1987)

Angiographic Localizer Ring for the BRW Stereotactic System

D. Vandermeulen*,[1], **P. Suetens**[2], **, **J. Gybels**[1], and **A. Oosterlinck**[2]

[1] Department of Neurology and Neurosurgery, U.Z. Gasthuisberg, Leuven, Belgium, [2] Department of Electrical Engineering, E.S.A.T., University of Leuven, Heverlee, Belgium

Summary

An accessory locating device to the existing BRW stereotactic system is presented. It can be used as a reference device to locate angiographic data with respect to the BRW stereotactic system. Hence, the projection of target points onto angiograms, visible on CT scans, are easily calculated, as well as the stereotactic coordinates of a set of points (*e.g.*, AVM) indicated on at least two angiograms. As a final result integrated images of cerebral blood vessels and an outline of tumor lesions can be generated using more sophisticated computer equipment.

Keywords: Angiography; CT scan; 3-D reconstruction.

Introduction

The increasing number of diagnostic medical images (CT, MRI, PET, SPECT, ultrasound and digital radiography) has called for the development of procedures to obtain spatially integrated images. More specifically, in the field of stereotactic neurosurgery, the integration of CT data and cerebral angiographies is of tremendous help during both stereotactic biopsies and cerebral irradiations. In the former case the knowledge of the exact position of the cerebral blood vessels guarantees a safer electrode trajectory. In the latter case arteriovenous malformations visible on angiograms can be located in the stereotactic frame and eventually irradiated. One possible way to get a 3 D integrated image is to obtain the spatial position of the blood vessels with respect to a frame fixed to the patient's head. The CT coordinates of the tumor have then to be superimposed on the 3 D image of the blood vessels. Our method has been described in full detail in[1].

Method

The solution adopted is based on the following principle: during the whole procedure (CT examination, angiography, operation) the patient wears a base ring fixed to his head. On this base or reference ring different accessory rings can be mounted. In the BRW stereotactic apparatus, the set of accessory rings consists of a CT-localizer ring, a NMR-localizer ring, an arc-system ring with the electrode. The CT-localizer ring is used for the transformation of CT coordinates (*e.g.*, tumor outline) to BRW-head ring coordinates.

We have built a similar accessory ring (Fig. 1), attachable onto the BRW-base ring, enabling the transformation between plane radiographic coordinates and BRW-head ring or stereotactic coordinates. This prototype reference ring contains 20 little metal markers (reference points, 10 for A-P, 10 for lateral angiographies), with known BRW-head ring coordinates, visible on an arbitrary angiogram.

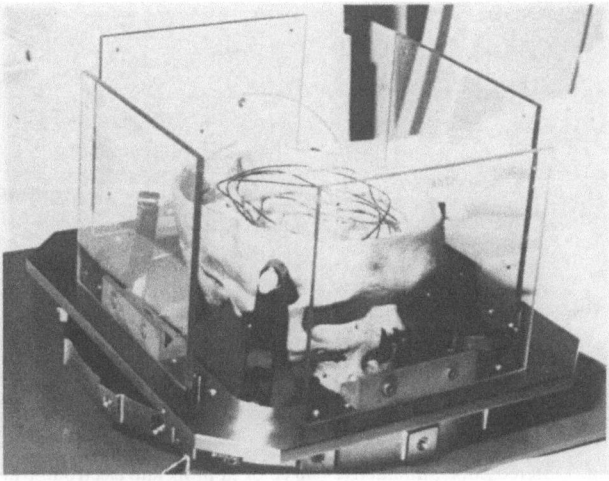

Fig. 1. Angiographic localizer ring

* D. Vandermeulen, Department of Electrical Engineering, E.S.A.T./MI2, University of Leuven, de Croylaan 52 B, B-3030 Heverlee, Belgium.
** P. Suetens is associated with the National Fund for Scientific Research (N.F.W.O.).

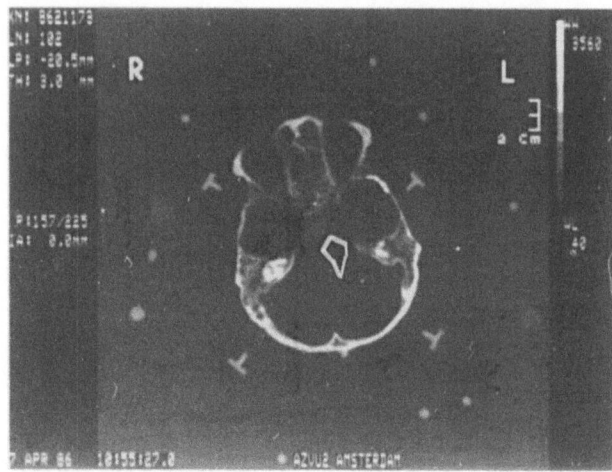

Fig. 2. CT scan and indicated target points

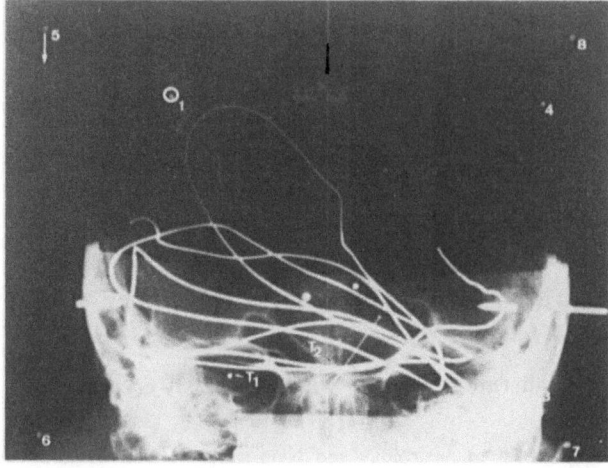

Fig. 3. Radiographic image of control points

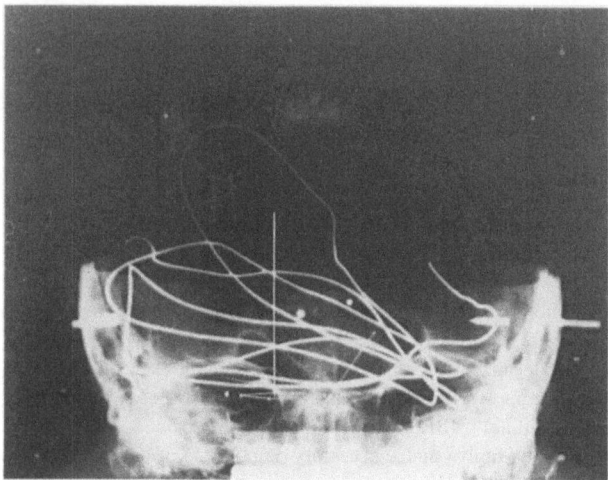

Fig. 4. CT target points transferred onto angiogram

(Fig. 2) are calculated by means of the BRW standard software package. Next, the angiogram is made with the angiographic localizer ring attached to the BRW base ring. The angiographic coordinates (X, Y) of at least 6 reference points are then read by means of a digitizer or by superimposing a transparent millimeter paper onto the angiogram (Fig. 3) and entered into the microcomputer, which then calculates the transformation matrix (transformation between BRW coordinates and radiographic coordinates). Finally, the computer asks for BRW coordinates of one or more CT points and prints the corresponding radiographic coordinates. These points can be plotted onto the angiogram or located on the millimeter paper overlaid onto the radiograph (Fig. 4).

The Reconstruction mode allows the surgeon to identify target points visible on standard A-P and lateral radiographs or angiograms. Hence, the BRW coordinates of any point can be calculated from the position of this point on a lateral and a frontal radiograph. Since there is no restriction on the relative position of both radiographs, one can as well use a stereoscopic pair. As a result, it is possible to interactively choose an appropriate electrode trajectory in between the blood vessels and calculate the BRW settings for the entry point (Fig. 5).

A major advantage of this approach is the fact that there is neither a restriction on the relative position of X-ray tube and sensor, nor is there any knowledge required of this position.

Basically, two different modes of operation are available: Projection mode and Reconstruction mode.

The Projection mode calculates the projective transformation of any object point with known BRW coordinates, onto an arbitrary radiograph or angiogram. First, the BRW coordinates of CT points

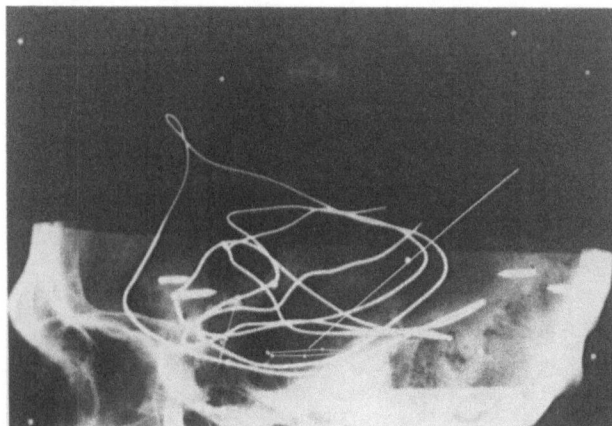

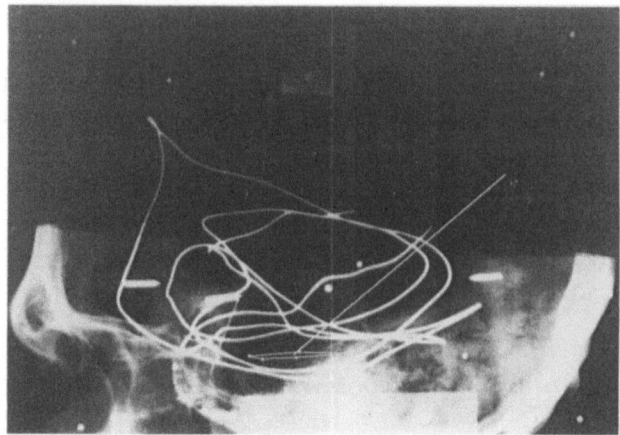

Fig. 5. Stereoscopic, interactive choice of appropriate electrode trajectory

Results

In a first phase, the procedure has been tested using mathematical simulations[2]. Using a set of eight control points, positioned at the vertices of a parallelepiped like frame, encompassing the skull, the positional accuracy (of a set of a few thousand points randomly spread over the entire stereotactic space) in the Projection mode was better than 0.5 mm. In the Reconstruction mode the positional accuracy was better than 0.5 mm when using an orthogonal pair of radiographs, whereas the accuracy was better than 5 mm when using a stereoscopic pair.

In a second phase a prototype angiographic localizer ring has been built. On top of the Talairach accessory ring four plexiglass plates each holding 5 metal reference points are mounted. Using the BRW phantombase, both projection and reconstruction procedures have been tested (for a set of 4 positions of the phantom base pointer). We have also tested the procedure with 2 metal target points placed into a phantom skull. The radiographic data have been entered using a Summagraphics digitizer tablet with a positional accuracy of 0.2 mm. Table 1 lists the errors both for the projection and reconstruction mode.

The systematic error listed in Table 1 is probably accounted for by a positional displacement error of the control points.

Table 1. *Projection and Backprojection Errors for a Set of Test Points, Using the BRW Phantom Base*

	Systematic error (mm)	Random error (mm)
Projection mode	1.1	0.15
Reconstruction mode		
orthogonal pair	0.92	0.27
stereoscopic pair	0.92	0.35

Discussion

Both simulation and prototype tests reveal sufficient operational accuracy (less than 1 mm) in the integration of angiographic data into the BRW stereotactic frame, using the angiographic localizer ring presented in this paper. Both the CT as well as the angiographic localization procedure are flexible in use. At the same time the system is versatile in its applications, such as functional stereotaxy, stereotactic biopsy and stereotactic irradiation. Finally, using more powerful computer hardware equipment, such as digitizer tablets, plotters, graphical display devices, interactive input devices, the procedure can be used to obtain graphically integrated images in a more or less automatic way.

Acknowledgments

The authors would like to thank the following persons and companies. The Radionics, Inc. company for the construction of the localizer ring. The Copharm, b.v., company, especially Mr. Th. Westerlage for their practical help and their having lent us a BRW apparatus. We are also grateful to the neurosurgical and neuroradiological team of the Academic Hospital in Amsterdam, especially Dr. Wolbers, for their kind cooperation during the prototype test phase.

This research is supported by the F.G.W.O (Belgium) under grant number 3.0009.85.

References

1. Suetens P, Baert AL, Gybels J, Haegemans A, Jansen P, Oosterlinck A, Wilms G (1984) An integrated 3D image of cerebral blood vessels and CT view of tumor. Frontiers in European radiology, vol 3, pp 81–100
2. Vandermeulen D, Suetens P, Gybels J, Oosterlinck A (1985) A new software package for the microcomputer based BRW stereotactic system: integrated stereoscopic views of CT data and angiograms. Medical Image Processing, Suctens P, Young IT (eds) Proc. SPIE 593, pp 103–114

Acta Neurochirurgica, Suppl. 39, 18–20 (1987)
© by Springer-Verlag 1987

Intraoperative CT Monitoring During Stereotactic Brain Surgery

S. Uematsu[*,1], **A. E. Rosenbaum**[2], **Y. S. Erozan**[3], **P. K. Gupta**[3], **H. Moses**[4], **H. J. Nauta**[1], **W. H. Rigsby**[2], **A. Wang**[1], **L. Weiderman**[1], and **A. J. Kumar**[2]

[1] Departments of Neurosurgery, [2] Radiology, [3] Pathology and [4] Neurology of the Johns Hopkins Medical Institutions, Baltimore, Maryland, U.S.A.

Summary

This paper reports our experience in performing the entire stereotactic surgical procedure with a CT scanner, which shows that target shifting can occur as the probe approaches the target. Therefore, when sampling of small targets or specific sites within larger targets is desired, CT confirmation of the probe's position ensures that the specific area seen on CT is biopsied.

Keywords: CT stereotaxy, brain biopsy; computed tomography.

Introduction

The recent advent of computerized tomography (CT) and magnetic resonance imaging (MRI) has enabled the detection of small brain lesions where the patient's clinical deficit is subtle and undetectable on clinical examination. In these cases, the neurosurgeon chooses the lowest-risk technique for approaching the lesion[2, 5].

Equipment and Method

So that the entire stereotaxic procedure can be performed at the gantry site (Siemens Somatom DR 3), the CT environment has been designed to meet standard operating-room requirements. Conventional sterile technique is strictly followed.

After placement of the CT-compatible Leksell frame and fiducial reference plates, intravenous contrast material is injected and the patient is scanned to localize the lesion, select the target, and compute the stereotaxic coordinates. Based on the coordinates, the probe entry site is marked on the scalp. The probe is advanced through a burr hole toward the target. When the probe has been advanced to 8 mm short of target, a CT slice is obtained. Corresponding probe advancement and CT scanning are made 4 mm above the target, at the target, and then 4 and 8 mm below the target, so that typically 5 biopsies are obtained (Figs. 1 A and B).

The specimens are examined by a pathologist at the scanner area, providing the surgeon with immediate information of the nature and adequacy of each specimen. The remaining specimens are transferred

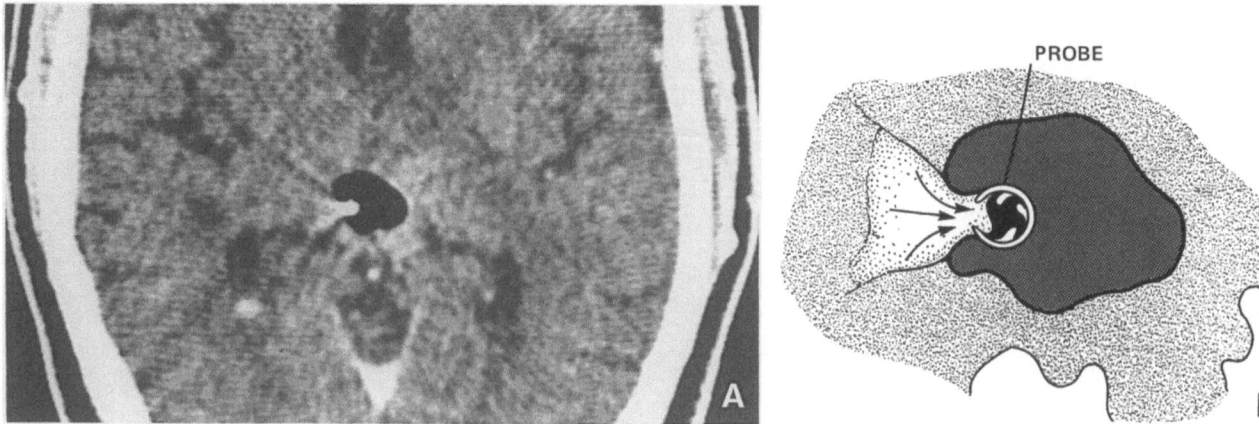

Fig. 1. (A) The probe has been passed very deeply into a mass that lies closely juxtaposed to the veins of Galen. A small amount of air ($1.0\,cm^3$) was injected through the probe and the probe tip became elegantly outlined. The artist's sketch (B) shows the probe in place and how the probe cuts tissue samples.

* S. Uematsu, M.D., Department of Neurosurgery, Johns Hopkins Medical Institutions, Baltimore, Maryland, U.S.A.

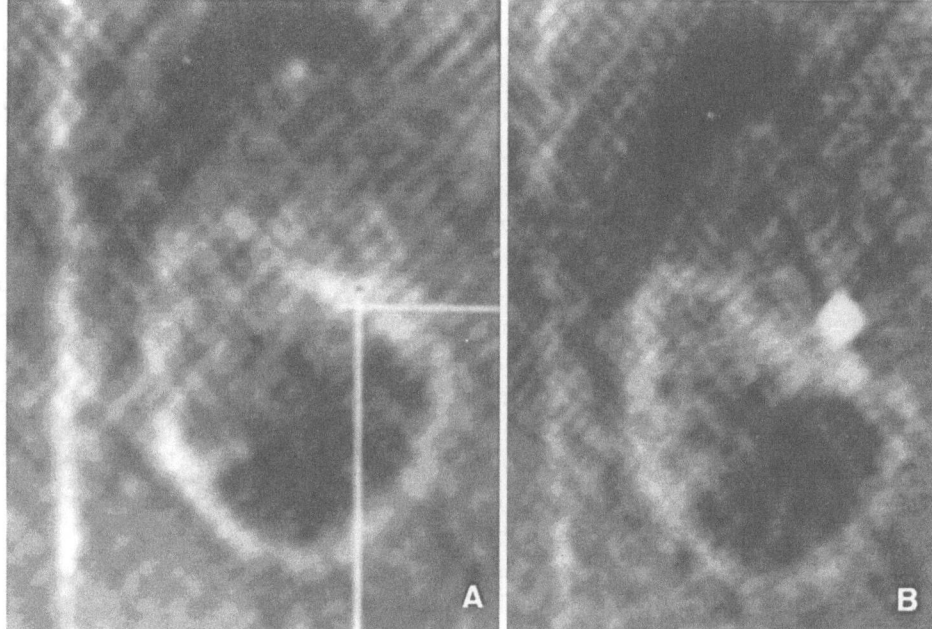

Fig. 2. (A) A target site was chosen in the enhanced ring of the lesion, as indicated by the crossing-point of the X and Z coordinate lines. CT taken after the placement of the biopsy probe disclosed the displaced wall of a neoplastic cyst. (B) Deformation of the cyst by the probe is apparent

to the pathology laboratory for further processing. The surgical closure and final CT scan follow. The final diagnosis is made after the examination of the entire material.

Results

Sixty-one stereotaxic procedures were carried out on 56 patients with intracranial abnormalities. The oldest patient was 77 years old and the youngest 7. The most common neurological sign prompting the procedure was mild hemiparesis. The second was focal seizures. The duration of the neurological symptoms in 50% of the cases was less than 1 month. Fifty-three cases were

referred for biopsy, 3 for -^{125}I seed implantation, another 3 for aspiration of a brain abscess, and 1 patient had placement ventricular catheter.

Surgical entry was most frequently made through conventional frontal or parietal burr holes. A case with a vermis lesion was approached through a suboccipital burr hole with the patient in the prone position under general anesthesia. Lesions were usually infiltrated in or lay adjacent to the sensory or motor cortex. The next most common sites were the deep basal ganglia adjacent to the third ventricle. In this series, 50% of the lesions were in the speech-dominant hemisphere. Cyto-

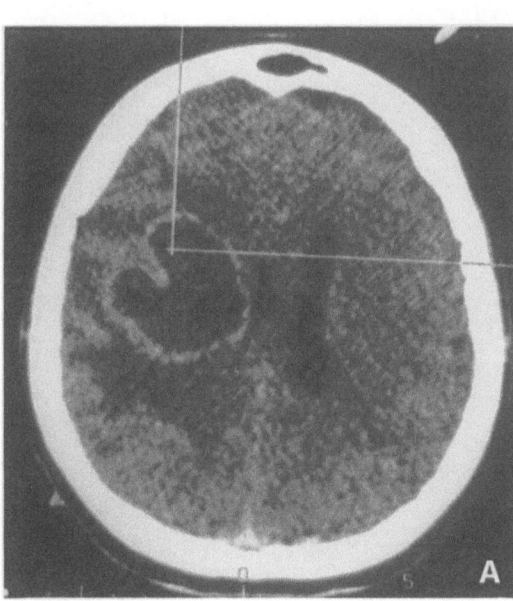
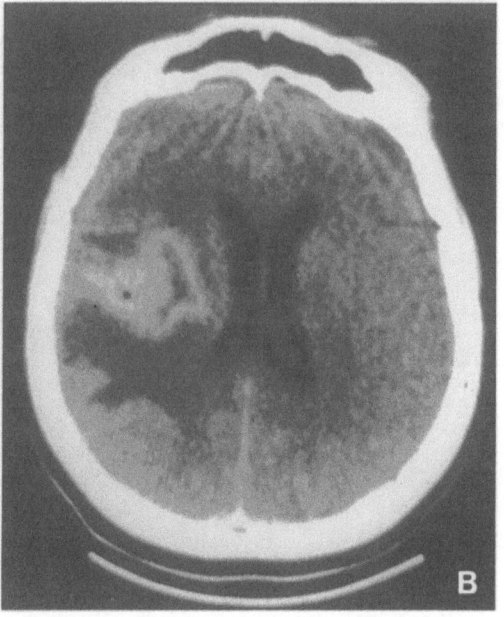

Fig. 3. (A) CT scan of a brain abscess before aspiration. The target site is indicated by the crossing-point of the horizontal (X) and vertical axes (Z). (B) CT shows the collapsed abscess cavity, with a hole created by the small amount of air trapped in the probe trajectory. (A and B) Note the proximity of the ventricular wall and the abscess wall

and histopathological examinations were performed in 50 cases, including 28 gliomas, 4 metastatic neoplasms, and 5 lymphomas[3]. Among the 28 cases of glioma, 20 cases were high-grade (grades III or IV) astrocytomas.

Discussion

The aim is to obtain multiple sampling of specific radiologically distinguishable tissues.

Intraoperative scanning at the level of the tip of the probe before application of an aspiration force increases both the precision and safety of the procedure. Intraoperative CT monitoring has demonstrated that the biopsy probe shifts the target. Being able to visualize the probe tip allows the surgeon to plan the direction of the side opening of the probe in reference to the tissue to be sampled (Figs. 1 A and B). Our intraoperative CT monitoring biopsies have also shown target shifting by the biopsy probe (Fig. 2).

During drainage of brain abscesses, intraoperative CT shows the progressive collapse of the abscess cavity and its relationship to the original structures. Without examination of these progressive changes, the probe might inadvertently penetrate more deeply and reinfect the adjacent brain or even extent into the ventricle, with catastrophic results (Figs. 3 A and B). Our patient group and the type of pathologic entities we found are similar to those reported by others[1]. However, the 5 cases of unsuspected lymphoma prior to histological examination are noteworthy. Two of these cases were patients with AIDS.

In the cases of brain tumor in our series there were no permanent neurological deficits or deaths. This level of morbidity and mortality matches that reported by others: Ostertag *et al.* [4] report 2.3% mortality and transient deterioration in 3% of 302 patients. In our hospital, the length of the patient stay is dramatically shorter than that for craniotomy. The average stay was 4.4 days, rather than 21.2 days. Although the stereotactic approach involved minimal surgical intervention, it is obviously not a cure for malignant brain tumors. However, neurosurgeons should employ the least invasive palliative technique that provides their patients with an opportunity for a resonable quality of life. CT-guided stereotaxis, with progressive refinement, satisfies this goal.

References

1. Apuzzo MLJ, Sabshin JK (1983) Computed tomographic guidance stereotaxis in the management of intracranial mass lesions. Neurosurg 12 (3): 376–380
2. Brown RA (1979) A computerized tomography-computer graphics approach to stereotaxic localization. J Neurosurg 50: 715–720
3. Erozan YS, Gupta PK, Rosenbaum AE, Kumar AAJ, Uematsu S (1985) Cytopathology of brain lesions sampled using computed tomography-guided stereotaxic needle aspirations. Acta Cytol 29: 929
4. Ostertag CB, Mennell HD, Kiessling M (1980) Stereotactic biopsy of brain tumors. Surg Neurol 14: 275–283
5. Rosenbaum AE, Lunsford LD, Perry JHK (1980) Computerized tomography-guided stereotaxis: A new approach. J Appl Neurophysiol 43 (3–5): 172–173

Acta Neurochirurgica, Suppl. 39, 21–24 (1987)

Magnetic Resonance Planned Thalamotomy Followed by X-Ray/CT-guided Thalamotomy

S. Uematsu*[,1], A. E. Rosenbaum[2], M. R. Delong[1], Ch. M. Citrin[3], W. R. Jankel[1], A. J. Kumar[2], J. C. McArthur[3], H. J. Nauta[1], J. Sherman[3], and H. Narabayashi[4]

Departments of [1] Neurosurgery, [2] Radiology, and [3] Neurology of the Johns Hopkins Medical Institutions, Baltimore, Maryland, U.S.A., [3] Magnetic Resonance of Washington, Chevy Chase, Maryland, U.S.A., and the [4] Department of Neurology of Juntendo University, Tokyo, Japan

Summary

Magnetic resonance imaging (MRI) is currently the optimal neuroradiologic technique for visualizing the anterior and posterior commisure for defining the AC-PC line. CT is the optimal technique for electrode and probe guidance during stereotactic thalamotomy. Various possibilities of transferring or overlying MRI and CT are outlined which in some future might result in more refined methods of CT-MRI guidance for stereotactic surgery.

Keywords: Stereotaxy; thalamotomy; magnetic resonance imaging.

Introduction

To date, target determination for stereotactic treatment of tremor has been carried out radiologically, with ventriculography. However, none of the target structures can be visualized with this technique. The critical target nucleus, the Vim, is in fact situated immediately adjacent to the medial rim of the internal capsule and the outer edge of the thalamus[1].

MRI provides excellent delineation of the grey and white matter, the internal capsule, basal ganglia, and thalamus[2]. Moreover, it provides high-contrast midsagittal sections showing the commissures and the adjacent structures that can be used for calculation of the target. But ferromagnetic instruments may endanger the patient and personnel during MRI scanning, so surgery cannot be performed at the scanner site.

The technique described in this paper combines the advantages of both CT or MRI for determining the target noninvasively (on MRI) and allowing performance of the thalamotomy (under CT guidance), with final confirmation of the metallic probe in the vicinity of the target structure obtained before the therapeutic lesion is made.

Method

The first step is precise delineation of the anterior-posterior commissure line (AC-PC line) by a midsagittal MRI scan. The AC-PC line is drawn on the MRI film. The line is then extended anteriorly to the frontal and posteriorly to the occipital skin surface. The distance from the frontal skin intersection to the upper edge of the frontal sinuses is measured on the MRI image. The distance from the occipital skin intersecting point to the inion is also measured on MRI image (Fig. 1 A). These measurements are then transferred to the scalp as follows (Fig. 1 B).

Under fluoroscopic observation, the scalp directly over the inion and superior margin of the frontal sinus are infiltrated with a 1% local anesthetic (xylocaine). A through-and-through wire suture is placed in deep enough to anchor the wire directly into the galea over each of these bony landmarks. The sutures are placed perpendicular to the sagittal plane of the head. The suturing needle should go deep enough at the area of the bony landmark to minimize displacement of the wire marking by the scalp. Where the projected AC-PC line intersects the frontal and occipital scalp are marked on the scalp at the distances measured from the landmarks on the MRI film (Fig. 1 B). The frontal and occipital AC-PC markings are connected by a wire suture placed on the patient's scalp and wrapped around the head.

The patient's head is then placed in the Leksell stereotactic frame. The frame is prepared with a vertical wire stretched between the zero coordinates of the two horizontal arms of the frame. The vertical wire on the frame and the wire for the projected AC-PC line on the scalp are aligned under fluoroscopic guidance until they are perfectly superimposed by adjusting the patient's head in the frame. The head is fixed in this position in the frame by four pins drilled into the skull, two in the front and two occipitally (Fig. 1 C).

The CT gantry is positioned at 0 degrees (*i.e.*, without gantry tilt). Precise orientation of the scanning plane and AC-PC line is

---

* S. Uematsu, M.D., Department of Neurosurgery, Johns Hopkins Medical Institutions, Baltimore, Maryland, U.S.A.

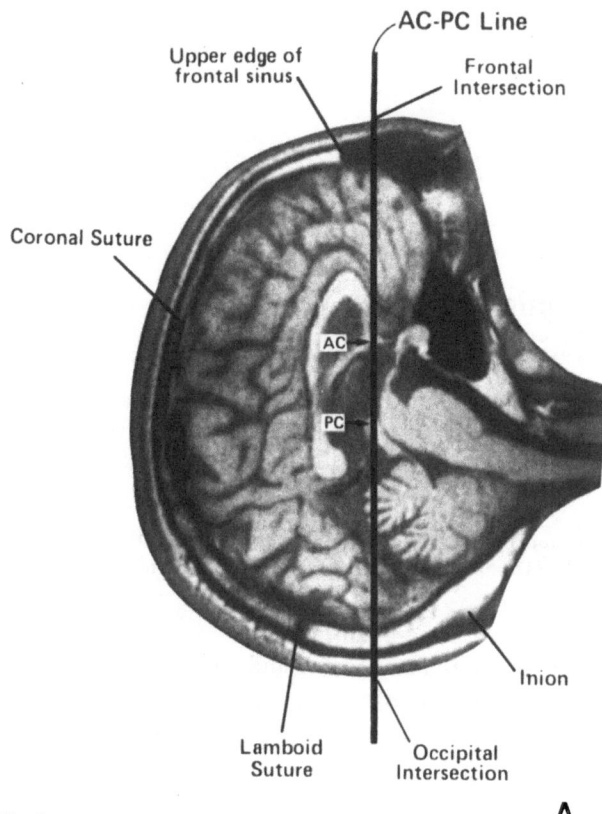

Fig. 1a   **A**

Fig. 1. A) Artist's depiction of a midsagittal MR scan of the case described here. The distance from the frontal intersection of the AC-PC line to the upper edge of the frontal sinus was 13.0 mm, and the distance from the occipital intersection of the line to the inion was 34 mm. B) The AC-PC line is externalized by wire wrapped around the head. C) The head is fixed onto the Leksell frame

confirmed when the localizing light from the gantry is directly superimposed on the AC-PC line defined on the patient's scalp, and a topography is taken at this level (Fig. 2). Serial nonoverlapping axial scans 2 mm thick are made over the line. The scans are reviewed on the screen and an optimal film is selected. Fig. 3 shows a film of an image made 2 mm above the AC-PC line and in which a theoretical target was selected. First, the CT film and the previously obtained MRI film are compared by superimposing the CT film on the MRI film, in order to identify the internal capsule, particularly its medial rim. Superimposition of a transparent brain map over the CT film assists in defining the appropriate thalamic nucleus, Vim.

After the target site is selected on the CT screen, the cursor is superimposed over the target. The fiducial sites are targeted and assigned to integrate the scanner, frame, and plane of CT sections. The CT coordinates (X, Y, Z) are then displayed on the screen by the built-in computer. Finally, with the use of the calculated X, Y, and Z coordinates, the surgeon is able to proceed with the functional stereotactic procedure (Fig. 3).

Case Report

A patient with multiple sclerosis underwent unilateral stereotactic thalamotomy for severe tremor of the arms in November 1985. A

radiofrequency probe (Radionics, 1.1 mm in diameter, 2 mm exposure) was advanced into the target. A CT scan confirmed the precise placement of the probe (Fig. 4). Responses to electrical stimulations are shown in Table 1.

Serial scans were taken between the stimulations and the heat application to reassess whether change in the position of the probe had supervened.

By one day after the procedure, the intention tremor in the right arm had improved so much that the patient was able to drink from a cup. Clinical follow-up six months after surgery showed that the reduction of the tremor in the right arm had been maintained. MRI six months after the surgery showed the thalamotomy lesion in the location of Vim (Fig. 5).

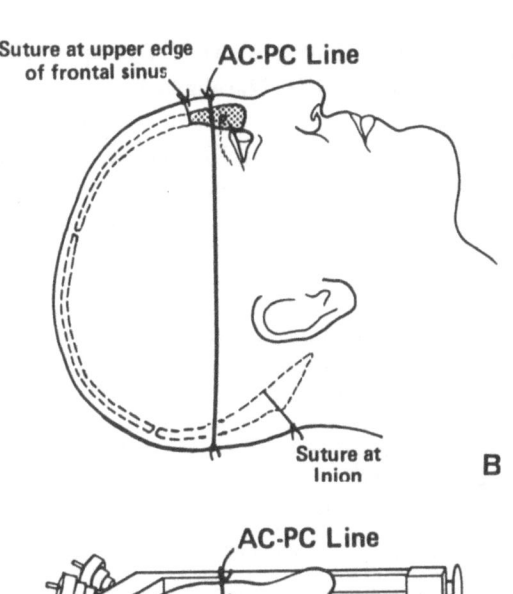

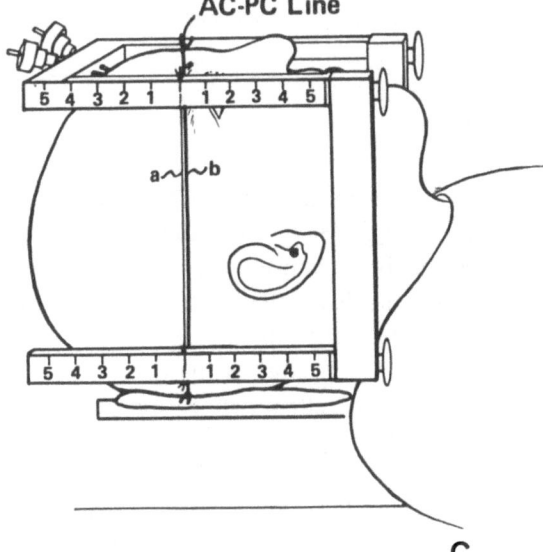

Figs. 1 b and c

Discussion

Magnetic resonance imaging is currently the optimal neuroradiologic technique for safely and readily visualizing the anterior and posterior commissures for defining the AC-PC line. CT is the optimal technique for electrode and probe guidance during the stereotac-

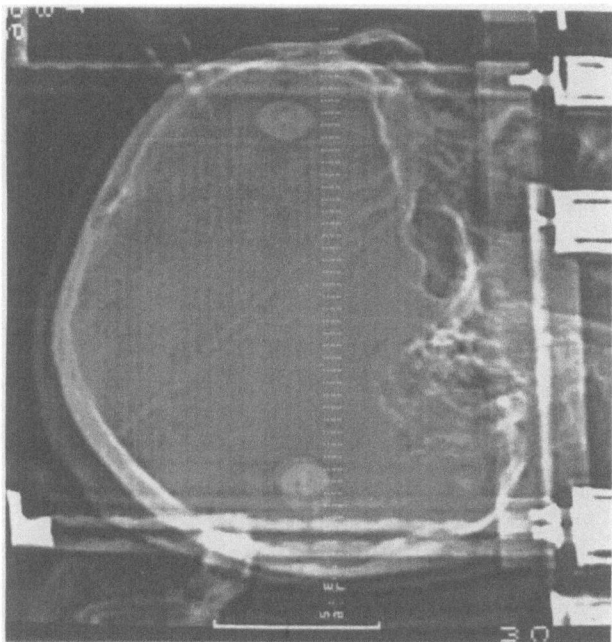

Fig. 2. A computed radiograph showing the AC-PC line encoded into the CT system. Each scan consists of a 2 mm slice, made along the AC-PC line

tic thalamotomy. Transferring or overlaying MRI and CT information currently is a tedious, variably accurate 3-step process. The MRI mapped AC-PC line for defining the thalamus becomes simply transferred to CT and CT-guided stereotactic program by using the localizing light laser or computed radiograph to align the plane of the patient so that the entire metallic band around the scalp is visualized on a single thin slice. The electrophysiological assistant then verifies the position of the probe along the final pathway traversed by the probe and/or electrode[3].

Two more refined methods of CT-MRI guidance for stereotactic surgery are likely to become available. The first, an indirect method, would require the MRI-defined AC-PC plane to be externalized so that the skin can be accurately marked, as in CT. Then the various MRI slices would be matched with CT slices to confirm target location. A CT stereotactic software program would again define the stereotactic coordinates. The second, a direct method, would require the AC-PC plane and level to be defined on MRI (with or without the stereotactic frame *in situ*), after which the MRI scanning data would be readable by the CT scanner. Then the stereotactic coordinates would be mapped from high-contrast resolution MRI and stereotactic surgery would be performabled by CT, which does not have the hazards of ferromagnetic instrument usage peculiar to magnetic resonance.

Fig. 3. An axial scan 2 mm above the theoretical AC-PC line. The frame coordinates X, Y, and Z are shown. The target site, the Vim, 2 mm above the AC-PC line, 16 mm from the midline, and 7 mm anterior to the posterior commissure, is shown as the crossing point of the horizontal line (Z) and vertical line (X)

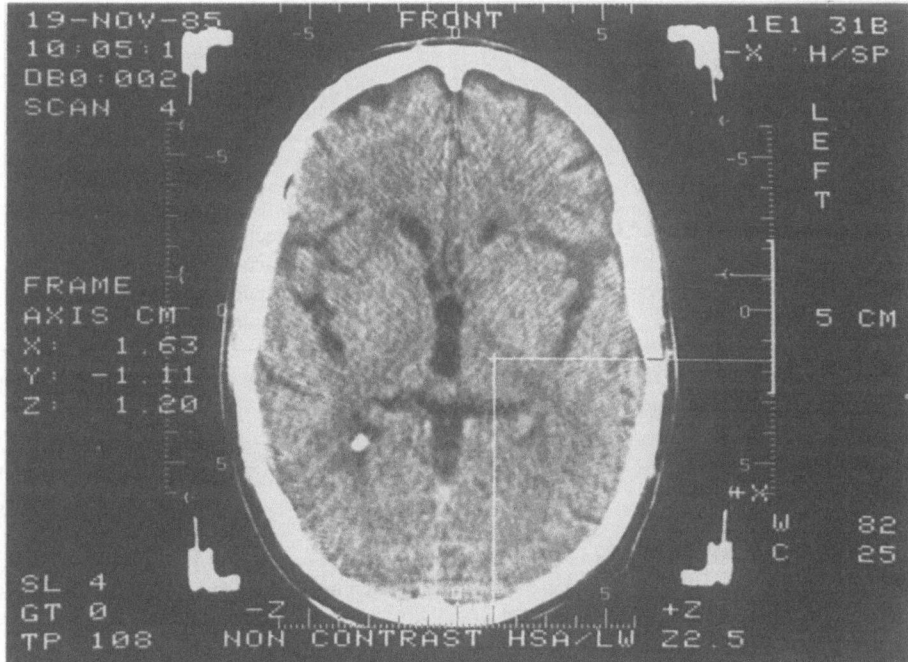

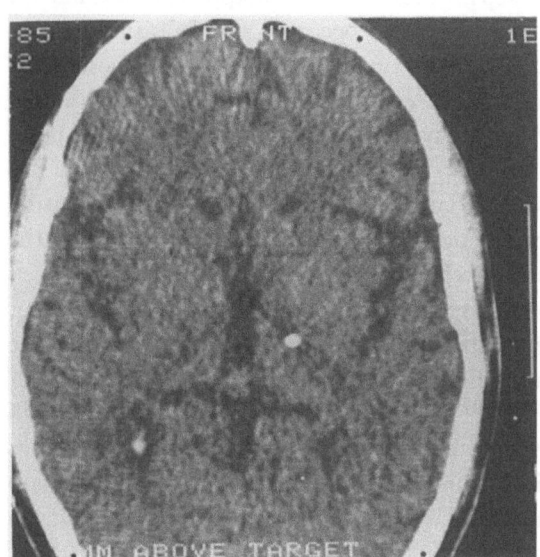

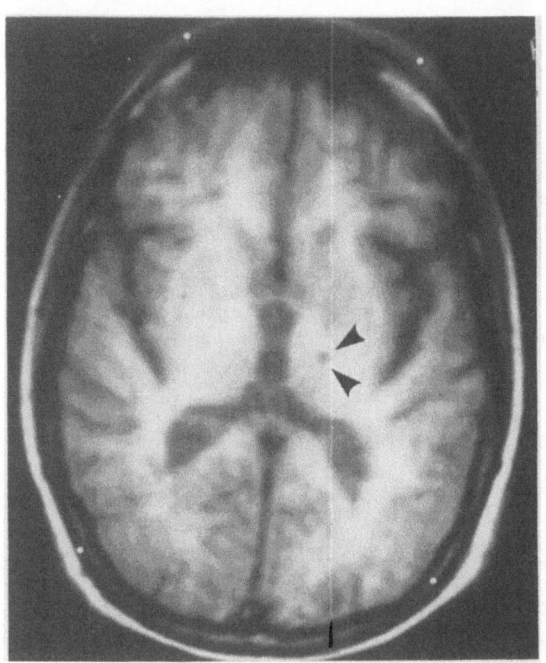

Fig. 4. One of the serial intraoperative CT scans is shown confirming probe tip location during the electrophysiological study and before radiofrequency therapeutic lesion-making

Fig. 5. A MRI image taken 6 months after the thalamotomy. The rounded lesion in the region of Vim is seem and matched well with the probe position seen in the earlier CT scan

Table 1. *Probe Location and Response to Electric Stimulation*

Electrode tip location	Stimulation		Response	Location	RF Lesion	
	Frequency	Volt			Temperature	Time (seconds)
2 mm above target	2 c/sec	2	—			
		4	—			
		5	twitch	face		
	50	2	tingle	hand		
Target	50	2	tingle	face, hand		
5 mm above	50	2	twitch	face		
			tingle	hand	70 °C	15
						15
						30
						60
2.5 mm above target	50	2	contraction hand			
	50	2	contraction hand			
	50	2	contraction hand			
	50	2	contraction hand *		70 °C	100

* During the making of this lesion, with the patient's right arm elevated, the tremor appeared to cease completely. No sensory or motor impairment was noted on neurological examination.

References

1. Albe-Fessard D, Arfel G, Guiot G (1963) Activités électriques caractéristiques de quelques structures cérébrales chez l'homme. Ann Chir 17: 1185–1214

2. Leksell L, Leksell D, Schwebel J (1985) Stereotaxis and nuclear magnetic resonance. Technical note. J Neurol Neurosurg Psychiatr 48: 14–18

3. Ohye C (1982) Depth microelectrode studies. In: Schaltenbrand G, Walker AE (eds) Stereotaxy of the human brain. Thieme-Stratton Inc, New York, pp 372–386

Acta Neurochirurgica, Suppl. 39, 25–27 (1987)

MRI-directed Stereotactic Biopsy of Cerebral Lesions

R. Bradford[1], **D. G. T. Thomas***, [1], and **G. M. Bydder**[2]

[1] Gough-Cooper Department of Neurological Surgery, Institute of Neurology, London, U.K., [2] Department of Radiology, University of London, Royal Postgraduate Medical School, Hammersmith Hospital, London, U.K.

Summary

Using a modified CT-directed stereotactic system we have performed MRI-directed biopsies in five patients whose intracerebral lesions were not clearly shown by CT. Tissue sampling from targets defined on the MR image enabled a histological diagnosis to be made in four of the five cases.

Keywords: Stereotaxy; brain biopsy; MR imaging.

Introduction

Stereotactic biopsy of cerebral lesions under computed tomographic (CT) control is now well established and has become a routine neurosurgical procedure[4, 8—10]. Recently it has become apparent that magnetic resonance imaging (MRI) may reveal cerebral lesions which are not clearly visualized by CT[3]. In order to obtain tissue samples from such cases, using stereotactic technique it has been necessary to use a stereotactic system modified for MRI compatibility.

Materials and Methods

The stereotactic frame used for biopsy under MRI control was a prototype developed by Trent Wells (Southgate, CA) and distributed by Radionics (Burlington, MA). This frame was a modification of the Brown-Roberts-Wells (BRW) CT-directed stereotactic system[2]. For MRI compatibility the aluminum alloy base ring of the BRW frame was replaced by one constructed of nonmetallic paxolin and carbonfiber. The ferromagnetic stainless-steel skull fixation pins were replaced with titanium. The localizing frame for MRI consisted of twelve carbon-fiber tubes containing either copper or gadolinium salts. The MR imaging system used with the stereotactic frame has been described previously[11, 13]. Single slice contiguous IR 1500/500/44 and SE 1500/80 sections of 8 mm thickness were imaged. Target and fiducial coordinates were obtained directly from the MR visual display unit. Calculations were then made with this data using BRW software on an Epson HX-20 computer. Using the BRW CT-directed system targets defined on CT can be biopsied with an accuracy of less than 2 mm[1]. We compared the accuracy of the modified MRI system with the CT using the latter as the standard. A perspex phantom with 6 target points contained within it was attached to the modified base ring and then imaged with MR. The target coordinates in stereotactic space were then calculated using target and fiducial data obtained from the MR image. The phantom was then scanned with CT and the coordinates of the same 6 targets recalculated using CT data. The mean differences between MR- and CT-calculated coordinates were 1.0 ± 0.9 mm (SD) in the X axis, 3.75 ± 1.2 mm in the Y axis and 5.0 ± 2.2 mm in the Z axis.

Patients

Five patients with a progressive neurological syndrome underwent MRI-directed stereotactic biopsy during the period January 1985 to April 1986 (Table 1). These patients represent a subset (8%) of a total of 63 CT-directed biopsies performed during this period. Three of the patients had brain stem lesions clearly seen with MRI but not adequately delineated by CT scanning. Two patients had lesions revealed by CT but which were difficult to interpret. One patient's CT (case 4), despite a progressive neurological syndrome, had remained unchanged since a cerebral infarct 4 years previously. In case 5 a previous open biopsy of a cerebral lesion and subsequent radiotherapy made CT scans difficult to interpret. MR imaging of these patients revealed their lesions in greater detail and gave more information with regard to target sites for biopsy.

The base ring was attached to the skull under local anesthesia together with intravenous midazolam. Biopsies were carried out under general anesthetic through burr holes using the BRW arc system and a side cutting blunt nosed cannula 1 mm in diameter with a 10 mm port. Brain stem biopsies were carried out via a transfrontal approach. Histological examination was performed by immediate smear and paraffin sections of multiple biopsies from one or more targets.

* D. G. T. Thomas, M.D., Gough-Cooper Department of Neurological Surgery, Institute of Neurology, Queen Square, London WC1N 3 BG, U.K.

Table 1. *MRI-Directed Stereotactic Biopsy*

Case	Age, sex	Symptoms	CT findings	MRI findings	Pathological diagnosis
1	59, male	left pontine syndrome	reported as normal	malignant lesions in L. pons	low grade astrocytoma
2	24, female	left facial weakness ataxia	isodense mass in pons	pontine tumour clearly shown	low grade astrocytoma
3	38, male	right VI nerve palsy diplopia	irregular high density in pons ? artifact	large pontine lesion	hematoma
4	68, female	right hemiparesis*	old left frontal lesion	left posterior frontal mass	oligodendroglioma
5	38, male	epilepsy**	scar from biopsy subthalamic calcification	diffuse right hemisphere abnormality	mild reactive changes only

   * Previous cerebral infarct. Epilepsy.
  ** Previous failed open biopsy. External beam radiotherapy.

Results

Pathological material was obtained from all five cases submitted to MRI-directed biopsy. In four cases a definite histological diagnosis was made which enabled appropriate therapy to be instituted. In the remaining case the nature of the pathological process could not be identified. There was no surgical morbidity associated with the procedure.

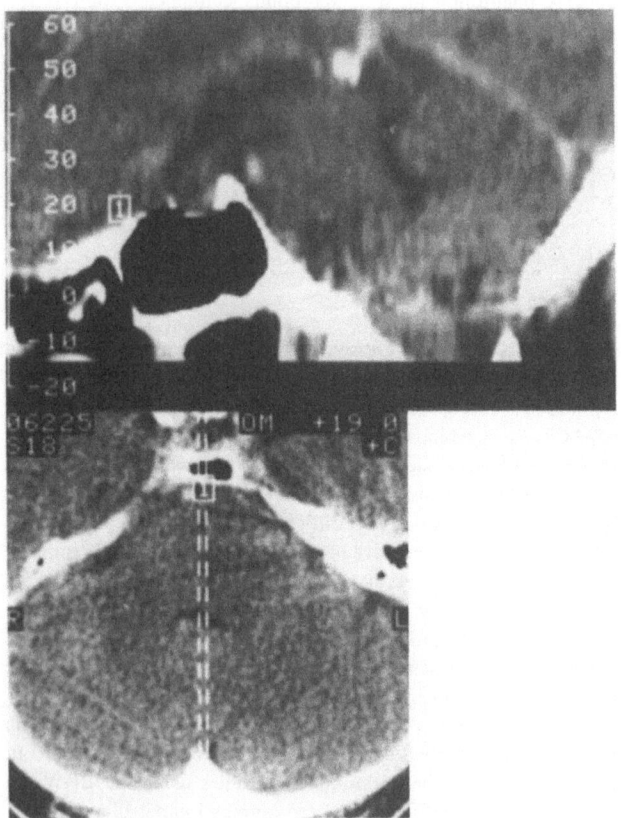

Fig. 1. CT scan of the posterior fossa with a sagittal reconstruction reported as normal

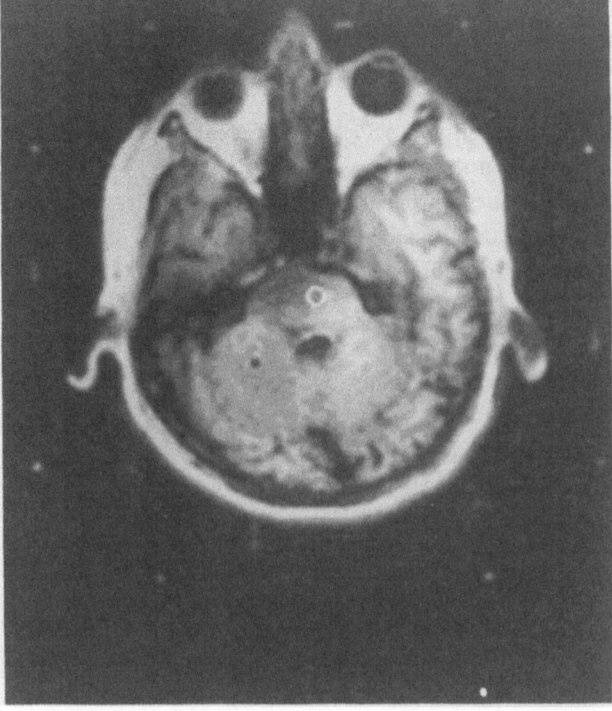

Fig. 2. MR image showing an intrinsic lesion of long T 1 in the left side of the pons. A biopsy target site is shown

Illustrative Case Report (case 1)

A 59-year-old man had complained of intermittent discharge from the left ear for ten years. Two months prior to admission to hospital he developed deafness in the left ear, progressive numbness of the left side of the face and an unsteady gait. On examination he had a rotatory nystagmus on lateral gaze and a depressed corneal reflex on the left. Sensation was diminished in the distribution of the left trigeminal nerve. A CT scan (Fig. 1) was reported as normal. Subsequent MR imaging revealed an intrinsic lesion of long T 1 in the left side of the pons suggestive of an infiltrating glioma (Fig. 2). MRI-directed biopsy of the pontine lesion was performed via a right frontal burr hole. Histological examination of the biopsy samples confirmed a diagnosis of low grade astrocytoma infiltrating the pons.

Discussion

The majority of cerebral lesions revealed by CT scanning or MR imaging require a tissue diagnosis before appropriate therapy can be commenced. Brain biopsy with CT stereotactic techniques have reduced mortality to less than 1% and acheived a diagnostic accuracy rate as high as 96%[6]. Recently MRI has been integrated with modified stereotactic systems to biopsy cerebral lesions[5, 7, 11]. Previous reports [7], however, have used MRI-directed biopsy in cases where the lesion is adequately seen on the CT scan. We have reserved MRI-directed biopsy for a small subset of patients who require stereotactic biopsy of lesions which are not shown at all or are not clearly delineated on CT scanning. In these cases MR imaging has revealed the lesion or given more precise information with regard to selecting target sites for biopsy. At present there is little information regarding the accuracy of MRI for stereotactic surgery. Two studies[7, 12] have been performed using a modification of the Leksell system. Wyper *et al.*[12] using an agar filled phantom, into which aluminum pellets were inserted stereotactically to act as targets, found MRI produced an accuracy of better than 2 mm. Lunsford[7] performed intraoperative CT in 3 cases immediately following MRI-directed biopsy and claimed an accuracy for the Leksell system of 1 mm. Using the modified BRW system and defined targets within a phantom we have found some discrepancy in the calculated coordinates derived from either CT or MRI. If CT is used as the standard then the MRI system would appear to be less accurate. This was particularly so for the Z axis and is most likely due to the 8 mm slice width used by our MR imaging system. We feel the accuracy is adequate for obtaining tissue samples from cerebral mass lesions but further studies

on the linearity of MRI are needed before its use in functional neurosurgery is considered.

In our limited series it has been technically feasible to perform MRI-directed stereotactic biopsy in cases where CT controlled biopsy would not be possible and in the majority of cases to obtain a pathological diagnosis. Future modifications of the current localizing frame will allow stereotactic biopsy of targets defined in the coronal and saggital planes which may improve the accuracy in the Z axis.

Acknowledgment

We thank Trent Wells and Eric Cosman for supplying the prototype MR localizer.

References

1. Brown RA (1979) A computerized tomography-computer graphics approach to stereotaxic localization. J Neurosurg 50: 715–720
2. Brown RA, Roberts TS, Osborn AG (1980) Stereotaxic frame and computer software for CT-directed neurosurgical localization. Invest Radiol 15: 308–312
3. Bydder GM, Steiner RE, Young IR *et al* (1982) Clinical NMR imaging of the brain: 140 cases. AJR 139: 215–236
4. Leksell L, Jernberg B (1980) Stereotaxis and tomography. A technical note. Acta Neurochir (Wien) 52: 1–7
5. Leksell L, Leksell D, Schwebel J (1985) Stereotaxis and nuclear magnetic resonance. J Neurol Neurosurg Psychiatry 48: 14–18
6. Lunsford LD, Martinez AJ (1984) Stereotactic exploration of the brain in the era of computed tomography. Surg Neurol 22: 222–230
7. Lunsford LD, Martinez AJ, Latchaw RE (1986) Stereotaxic surgery with magnetic resonance- and computerized tomography compatible system. J Neurosurg 64: 872–878
8. Mundinger F, Birg W, Klar W (1978) Computer-assisted stereotactic brain operations by means including computerized axial tomography. Appl Neurophysiol 41: 169–182
9. Ostertag CB, Mennel HD, Kiessling M (1980) Stereotactic biopsy of brain tumours. Surg Neurol 14: 275–283
10. Thomas DGT, Anderson RE, du Boulay GH (1984) CT-guided stereotactic neurosurgery: experience in 24 cases with a new stereotactic system. J Neurol Neurosurg Psychiatry 47: 9–16
11. Thomas DGT, Davis CH, Ingram S *et al* (1986) Stereotaxic biopsy of the brain under MR imaging control. AJNR 7: 161–163
12. Wyper DJ, Turner JW, Patterson J, Condon BR (in press) Accuracy of stereotaxic localization using MRI and CT. J Neurol Neurosurg Psychiatry
13. Young IR, Burl M, Clarke GJ *et al* (1981) Magnetic resonance properties of hydrogen: imaging the posterior fossa. AJR 137: 895–901

Acta Neurochirurgica, Suppl. 39, 28–33 (1987)

The Spatial and Morphological Assessment of Cerebral Neuroectodermal Tumors Through Stereotactic Biopsy

M. Scerrati*, **G.F. Rossi,** and **R. Roselli**

Istituto di Neurochirurgia, Università Cattolica, Roma, Italy

Summary

Stereotactic biopsies were carried out in an attempt to define the grading of 99 neuroectodermal tumors with respect to their modalities of growth and their degree of malignancy. Valuable data has been obtained in all explored tumors. The reliability of the information provided by stereotactic biopsy depends on the careful planning of the procedure by exploring different parts of the tumor as well as of the surrounding brain tissue.

Keywords: Neuroectodermal tumors; stereotactic biopsy; classification.

Introduction

The knowledge of the evolutive stage of brain neuroectodermal tumors can be of crucial importance both for the planning of therapy and for establishing the prognosis. Such a knowledge is based mainly on the correct assessment of: 1. the modalities of growth of the tumor, as reflected by its spatial arrangement and by its relation with the surrounding brain tissue[1-3, 10-13] and, 2. its degree of malignancy and of progression, as indicated by the morphological appearance of the tumoral cells and by the uniformity of such an appearance in the different parts of the tumoral mass[8, 9, 11, 13-15].

Material and Methods

Out of a series of 135 stereotactic biopsies of brain lesions, 99 turned out to be neuroectodermal tumors. The Talairach stereotactic apparatus, upgraded for polar and for orthogonal approaches in our laboratory[12], was used in most patients; the Leksell stereotactic instrument was utilized in some cases of pediatric age. The spatial

---

* Massimo Scerrati, M.D., Istituto di Neurochirurgia, Università Cattolica S. Cuore, Largo A. Gemelli, 8, I-00168 Roma, Italy.

location of the tumor was basically defined on CT scans[5]. In all patients stereotactic cerebral angiography, and in some cases stereotactic ventriculography, were performed[5]. The biopsies were carried out along up to three tracks. Multiple specimens were collected stepwise (5–10 mm intervals) from the core of the tumor as defined neuroradiologically and from its periphery and the surrounding brain tissue. The specimens were fixed in 10% formalin, embedded in paraffin, cut at 4 micron and stained with standard or, if necessary, special methods (10–12). Smear preparations were also used in certain cases for rapid intraoperative diagnosis[4-6].

Results

The histological diagnoses (WHO classification)[16-17] of the neuroectodermal tumors were: pilocytic astrocytoma in 3 cases, 1 subependymal giant cells astrocytoma, 28 fibrillary astrocytomas, 15 oligodendrogliomas, 17 anaplastic astrocytomas, 1 anaplastic oligodendroglioma, 29 glioblastomas, 2 ependymomas, 1 choroid plexus carcinoma, 1 pineocytoma, and 1 ganglioglioma.

a) The modalities of growth are related to the proliferative and the infiltrative properties of the tumor[7, 9, 13-15]. 1. Growth by proliferation without invasion of the surrounding brain tissue; borders well defined, sharply demarcated and easily distinguishable (*benign tumors, WHO grade I, 8 cases*) (Fig. 1). 2. Growth by proliferation and infiltration of the surrounding brain (BAT, brain adjacent tumor); borders poorly demarcated and hardly recognizable, because of the rather differentiated morphological appearance of the neoplastic cells (*semibenign tumors, WHO grade II, 43 cases*) (Fig. 2). 3. Growth by proliferation and infiltration of the surrounding brain (BAT, brain adjacent tumor); borders poorly demarcated but recognizable because of the anaplastic cytologic and structural characters of the tumoral tissue (*malignant tumors, grade III–2V, 48 cases*) (Fig. 3 C).

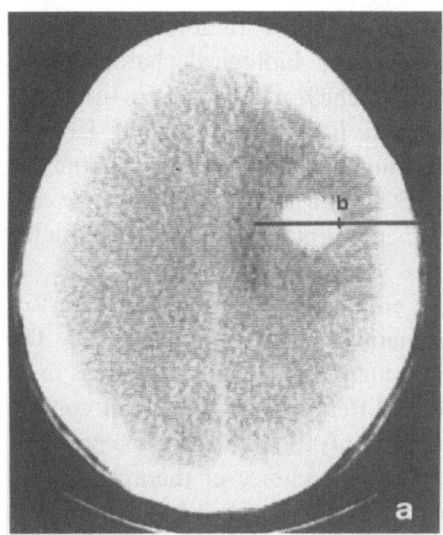

Fig. 1. CT scan of a right deep frontal pilocytic astrocytoma and biopsy track (a); histology of the specimen taken from the boundary of the tumor (b; × HE, X310)

b) The degree of malignancy and progression is related to the level of anaplasia and to the qualitative instability of the tumor[7-9, 13-15]. Malignancy can involve the whole tumoral mass or parts of it, *i.e.* the irreversible transition of the tumoral cells towards more heterogeneous stages[7, 8, 15]. The phenomenon of progression is very frequent among neuroectodermal tumors and its recognition is crucial for a correct definition of malignancy[8, 9, 13-15]. Its evidence in our series (Fig. 3) was proved in the following tumor groups: 1. *semibenign astrocytomas, grade II (28 cases)*: foci of increased cellular density and slight nuclear abnormality (suggesting a higher degree of malignancy) in 10 cases (36%); 2. *anaplastic astrocytomas, grade III (17 cases)*: more differentiated areas (fibrillary, protoplasmatic, gemistocytic astrocytoma, or oligondendroglioma) in 10 cases (59%); 3. *glioblastomas, grade IV (29 cases)*: more differentiated areas (astrocytoma or oligondendroglioma) in 20 cases (69%). In the remaining malignant tumors (7 anaplastic astrocytomas, 9 glioblastomas, 1 anaplastic oligodendroglioma and 1 choroid plexus carcinoma) anaplasia diffusely and steadily involved the whole tumor.

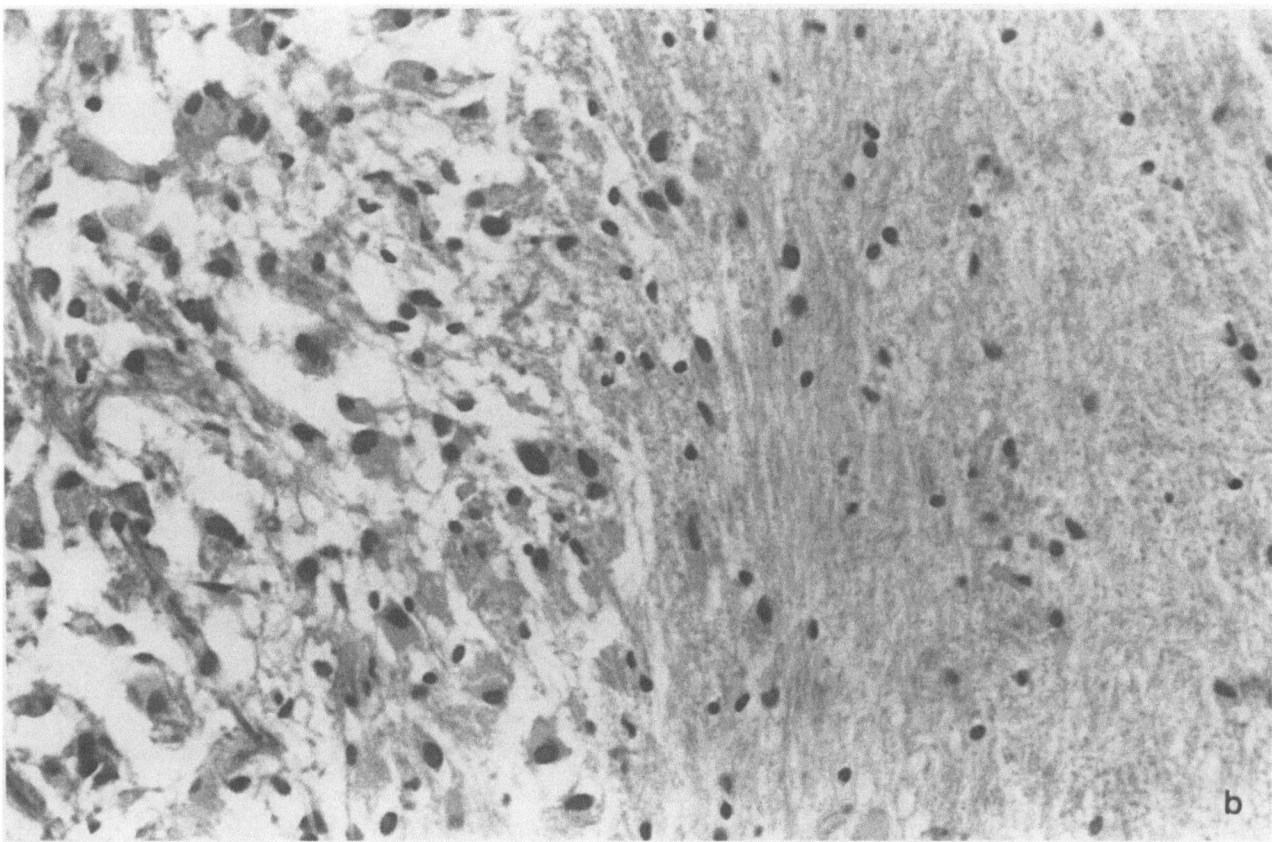

Fig. 1 b

Discussion

Growth characteristics and the degree of malignancy have to be known for the definition of the evolutive stage of neuroectodermal tumors. We found that

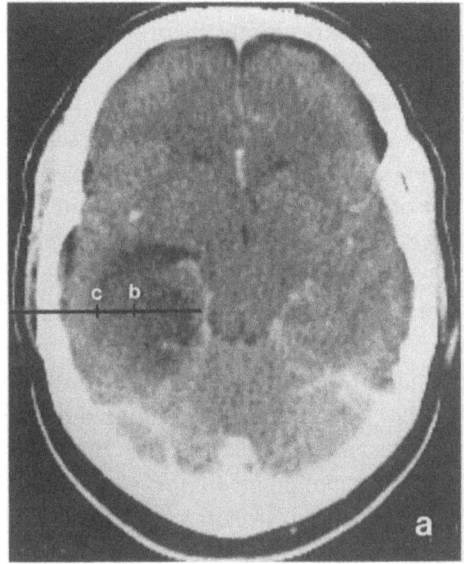

Fig. 2. CT scan of a left medio-posterior temporal fibrillary astro-cytoma and biopsy track (a); histology of the specimens taken from the core (b; HE, × 310) and from the boundary (c; HE, × 310) of the tumor

stereotactic biopsy can yield valuable information on both in all the explored tumors.

As far as the definition of the modalities of growth is concerned, the main problem was found in the group of semibenign tumors (WHO grade II), particularly in astrocytomas. Although these tumors have rather benign morphological and biological characteristics, they have infiltrative behavior[9, 13, 14, 16, 17]. In certain cases their borders are hardly recognizable. Isolated cells are not only found scattered far away from the tumoral core but it is also difficult to identify these cells as neoplastic elements due to their differentiated character[1–3].

Our experience confirms that malignant transformation is evident mainly in astrocytomas[8, 9, 13–15]. The recognition of more anaplastic elements or areas in an otherwise relatively differentiated tumor can confuse the grading and is certainly important for the prognostic evaluation and for the choice of therapeutic procedures. This is particularly true when foci of anaplasia are found in semibenign astrocytomas.

The reliability of the information provided by stereotactic biopsy depends strictly on the careful planning of the procedure[1–6, 10, 11]. It is mandatory that samples be taken from different parts of the tumor as well as from the surrounding brain. The planning of

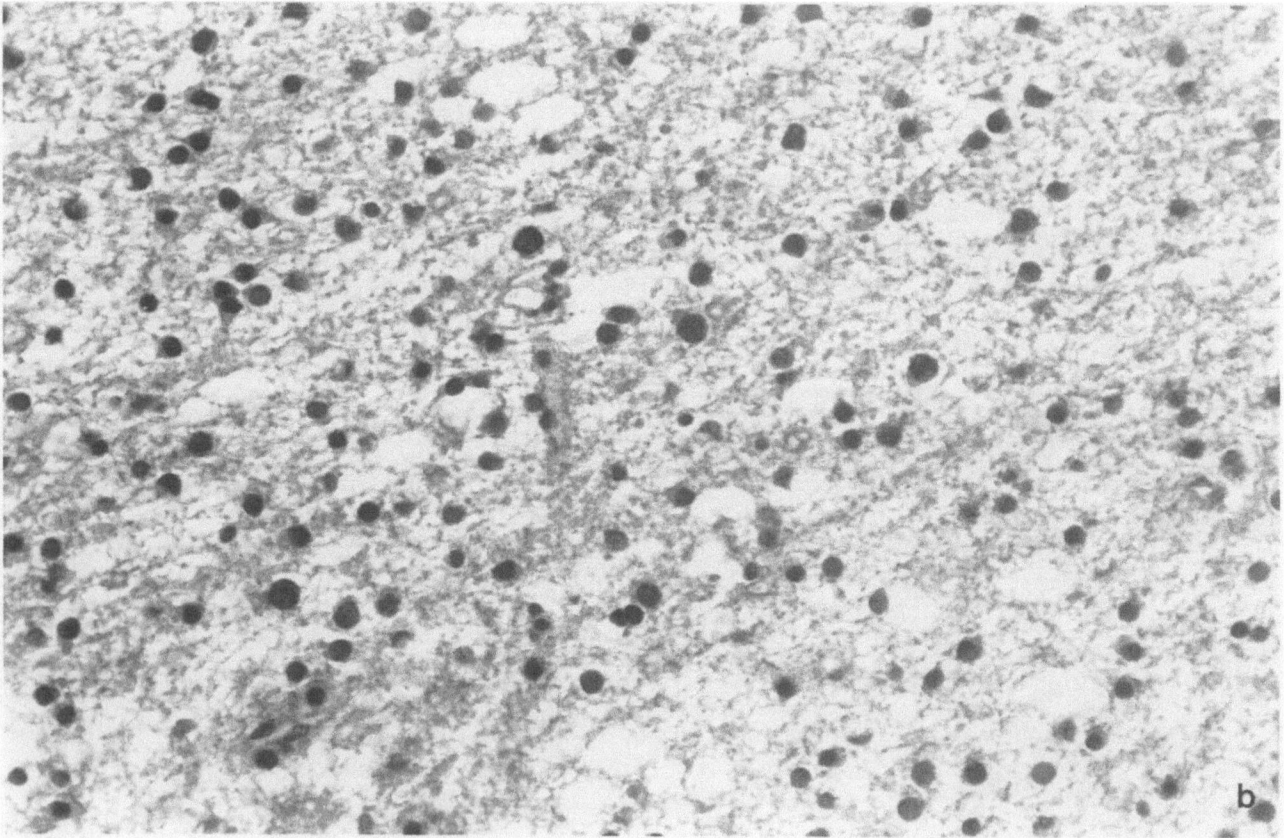

Fig. 2b

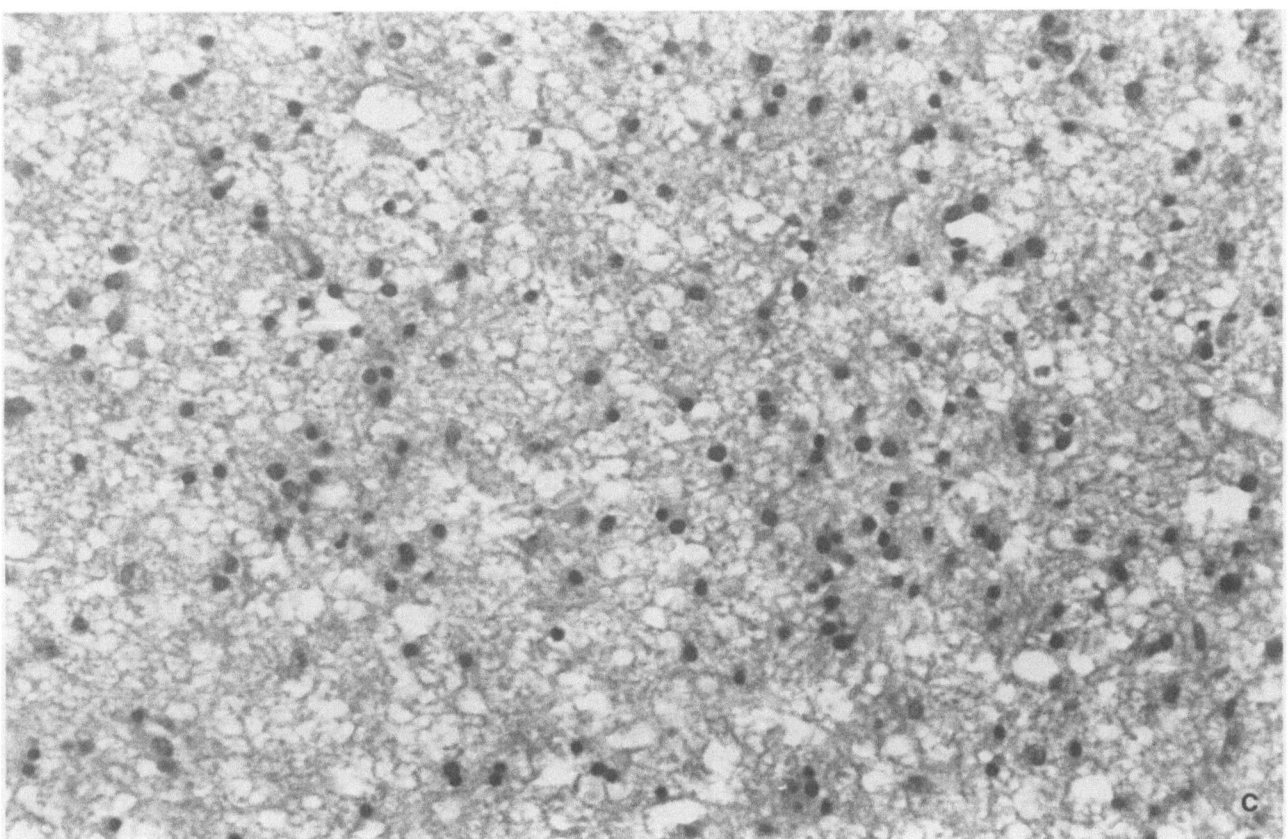

Fig. 2c

sample collection is largely based on the findings of the neuroradiological studies, which are performed in stereotactic conditions[1-3, 5, 6, 10, 11], integrated with clinical history and neurological examination.

Acknowledgements

The autors are grateful to Prof. Dr. Giorgio Macchi (Neurological Institute of Catholic University, Roma) for his kind assistance in the neuropathological examinations.

This research is partially supported by a grant of Ministry of Education.

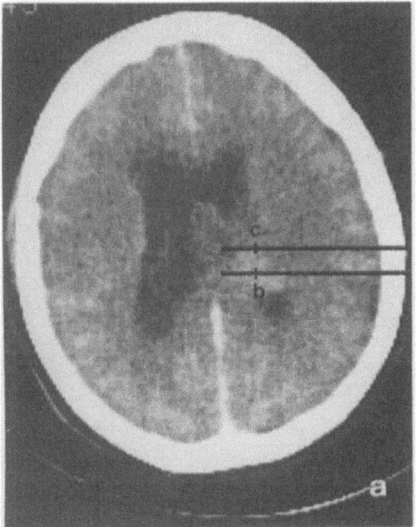

Fig. 3. CT scan of a deep seated astrocytoma of the right hemisphere and biopsy tracks (a); histology of the specimens taken from the anaplastic core (b; HE, ×125) and from its boundary with a more differentiated area of the tumor (c; HE, ×125)

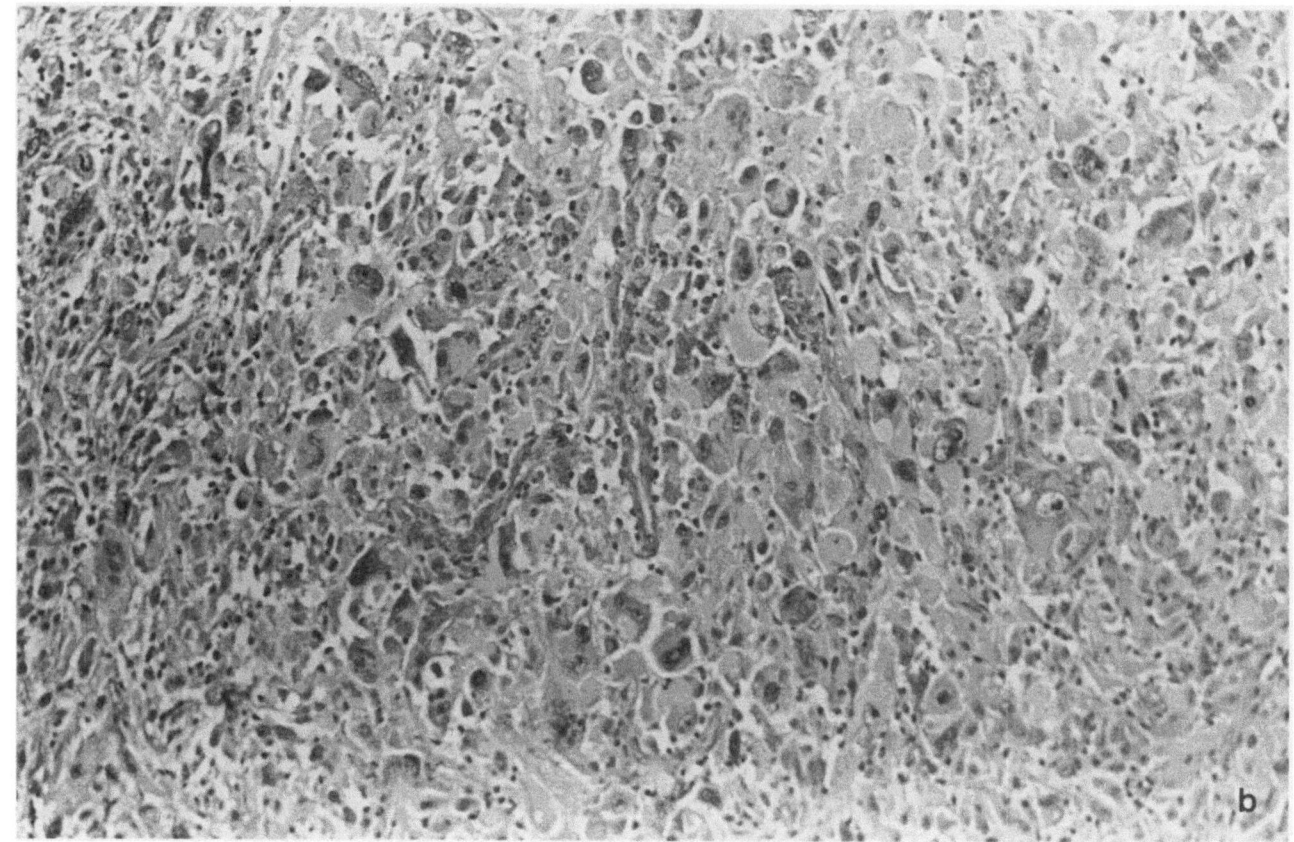

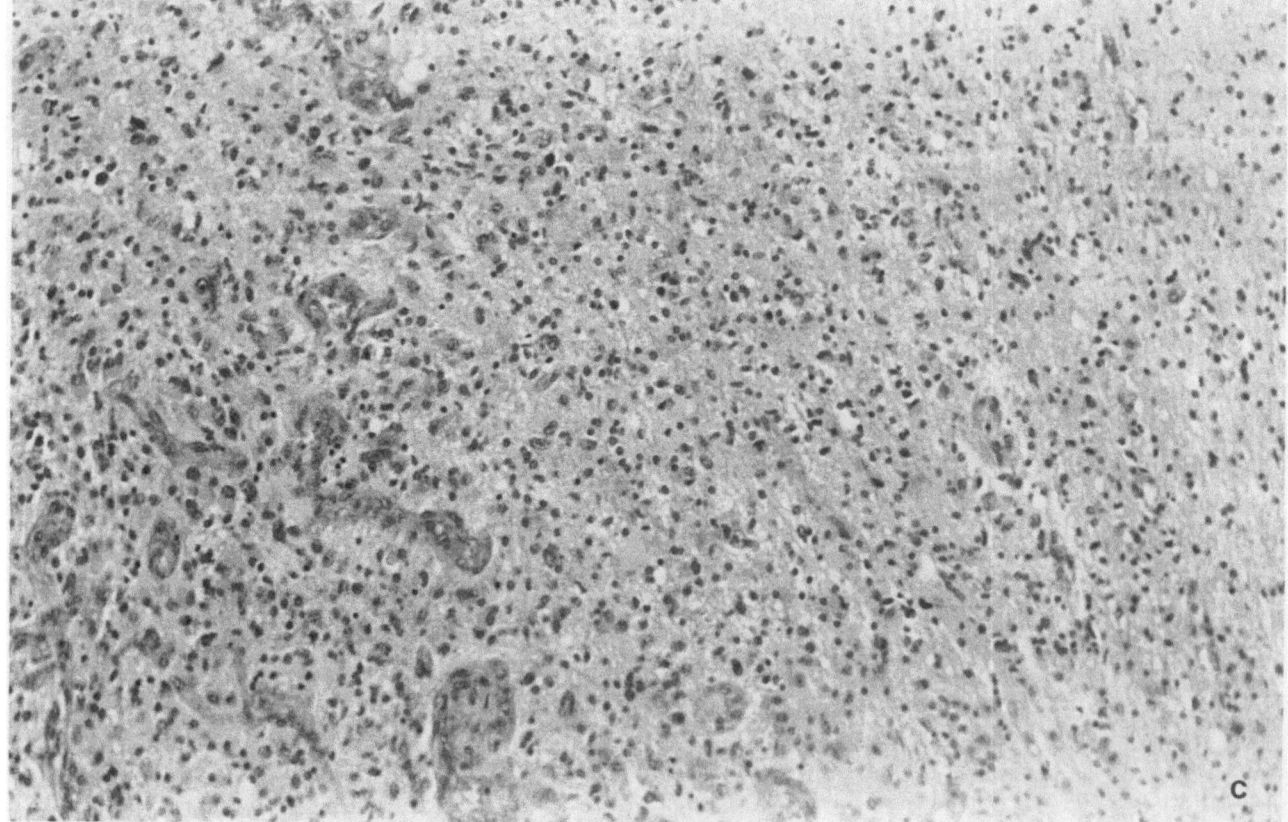

Figs. 3b and c

References

1. Daumas-Duport C, Vedrenne C, Szikla G (1979) Contribution of stereotactic biopsies to the 3-D localization of brain tumors, gliomas in particular. In: Szikla G (ed) Stereotactic cerebral irradation. Elsevier, North Holland, Amsterdam New York Oxford, pp 33–41
2. Daumas-Duport C, Szikla C (1981) Délimitation et configuration spatiale des gliomes cérébraux. Neurochirurgie 27: 273–284
3. Daumas-Duport C, Monsainegeon V, N'Guyen JP, Missir O, Szikla G (1984) Some correlations between histological and CT aspects of cerebral gliomas contributing to the choice of significant trajectories for stereotactic biopsies. Acta Neurochir (Wien) [Suppl] 33: 185–194
4. Kleihues P, Volk B, Anagnostopoulos J, Kiessling M (1984) Morphologic evaluation of stereotactic brain tumor biopsies. Acta Neurochir (Wien) [Suppl] 33: 171–181
5. Moschini M, Scerrati M, Colosimo C, Fileni A, Luna R, Roselli R, Rossi GF (1984) Integrazione tra neuroradiologia e neurochirurgia stereotassica nella diagnosi e nella definizione spaziale degli espansi endocranici. In: De Dominici R *et al* (eds) Radiologia-Firenze 1984, Monduzzi, Bologna, pp 221–224
6. Ostertag CB, Mennel HD, Kiessling M (1980) Stereotactic biopsy of brain tumors. Surg Neurol 14: 275–283
7. Prodi G (1977) La biologia dei tumori. Ambrosiana, Milano
8. Rubistein LJ, Herman MM, van den Berg SR (1984) Differentiation and anaplasia in central neuroepithelial tumors. In: Rosenblum ML *et al* (eds) Brain tumor biology. Progress in experimental tumor research, vol 27. Karger, Basel New York, pp 32–48
9. Russel DS, Rubistein LJ (1971) Pathology of tumors of the nervous system. Arnold, London
10. Scerrati M, Pizzolato GP (1981) The role of stereotactic biopsy in the 3-dimensional localization of cerebral tumors. Acta Neurochir (Wien) 57: 305–306
11. Scerrati M, Rossi GF (1984) The reliability of stereotactic biopsy. Acta Neurochir (Wien) [Suppl] 33: 201–205
12. Scerrati M, Fiorentino A, Fiorentino M, Pola P (1984) Stereotactic device for polar approaches in orthogonal systems. J Neurosurg 61: 1146–1147
13. Scherer JH (1940) The form of growth in gliomas and their practical significance. Brain 63: 1–34
14. Schiffer D, Fabiani A (1975) I tumori cerebrali. Pensiero Scientifico, Roma
15. Walker MD (1983) Oncology of the nervous system. Martinus Nijhoff, Hingham
16. Zülch KJ (1979) Types histologiques des tumeurs du système nerveux central. Classification histologique internationale des tumeurs. OMS, Genève
17. Zülch KJ (1980) Principles of the new World Health Organization (WHO) classification of brain tumors. Neuroradiology 19: 59–66

Acta Neurochirurgica, Suppl. 39, 34–37 (1987)
© by Springer-Verlag 1987

Treatment of Cystic Astrocytomas with Intracavitary Phosphorus 32

T. W. Hood[*, 1], **B. Shapiro**[2], and **J. A. Taren**[1]

[1] Section of Neurosurgery, Department of Surgery and [2] Division of Nuclear Medicine, Department of Internal Medicine, University of Michigan Hospitals, Ann Arbor, Michigan, U.S.A.

Summary

Cyst formation by astrocytomas can cause progressive neurological deficit and can necessitate multiple surgical procedures. Before the advent of computed tomography (CT) preoperative diagnosis of cystic astrocytomas was difficult and stereotactic management of these lesions was limited. CT-guided stereotaxy provides a safe approach to all cystic astrocytomas including brain stem lesions. Based upon the experience of intracravitary radiation of craniopharyngioma cysts, the authors treated nine patients presenting with cystic astrocytomas utilizing colloidal chromium phosphorus 32 (^{32}P). Control of cyst formation was achieved in eight patients. Our preliminary data suggest that intracavitary ^{32}P may provide a significant adjunctive therapy in the management of cystic astrocytomas.

Keywords: Intracavitary irradiation; cystic astrocytoma; colloidal phosphorus.

Introduction

The management of extracerebellar cystic astrocytomas has received little attention in the neurosurgical literature. Primarily, case reports have detailed various treatment methods without direct comparison on different techniques[1, 4, 10, 16]. In 1985, Loftus *et al.* reviewed the general neurosurgical management of 25 cases of supratentorial cystic astrocytomas[8]. He reported a success rate of 50% with resection, fenestration, and radiotherapy, however, multiple procedures were required to accomplish a 70% success rate. Computed tomography (CT) has greatly improved the preoperative diagnosis of cystic lesions and has allowed increased stereotactic application in the treatment of cystic astrocytomas. Frequently, cystic astrocytomas are now diagnosed in regions not amenable to resection and treatment by aspiration requires multiple procedures with attendent risks. This preliminary report details our experience in the treatment of nine cystic astrocytomas with intracavitary colloidal chromium phosphorus 32 (^{32}P).

Clinical Materials and Methods

Nine patients, 3 male and 6 female, ages 12–66 years, presented with cystic astrocytomas at the University of Michigan from 1984 to 1986. Four lesions were located in the cerebral hemispheres, two in the diencephalon and three in the brain stem. Four lesions were grades I or II astrocytomas and the remaining five tumors were malignant (Fig. 1). Preoperatively all patients were evaluated with high resolution computed tomography on multiple occasions. Seven patients had previously received external radiotherapy. The surgical indication was progressive cyst enlargement and neurological deficit. Three patients had previously placed cyst catheters connected to subgaleal Rickham reservoirs which permitted direct radioisotope instillation. All other patients had delivery of their intracavitary radiation stereotactically. A Leksell stereotactic frame, Model CT (AB Eleckta Instruments, Decatur, Georgia) was used with either GE-8800 or 9800 CT/T units (General Electric Medical Systems, Milwaukee, Wisconsin). Volumetric determination of the cyst was initially obtained by an isotope dilution technique with 99Te sulfur colloid. Subsequently volume was measured by analysis of contiguous 3 mm CT images of the cyst.

Initially, a dosage of colloidal chromium ^{32}P calculated to deliver 20,000 rads to the cyst wall was used[15]. After the initial 2 patients required reinstillation a dosage of 40,000 rads was utilized. After withdrawing approximately 20% of the cyst volume for decompression, the radioactive suspension was barbotaged into the cyst. Procedures were performed under local anesthesia except in children under 15 years. Bremsstrahlung scans were obtained on the first postoperative day and at one week (Fig. 2). Follow-up CT scans were obtained at 3, 6, and 12 months postoperatively or with change of neurological status.

* Terry W. Hood, M.D., 2128 Taubman Health Care Center, Box 0338. University of Michigan Hospitals, 1500 E. Medical Center Drive, Ann Arbor, MI 48109, U.S.A.

Results

Progressive cyst reduction was seen in seven patients and cyst volume stabilization in one patient. In one case progressive cyst enlargement occurred despite intracavitary radiation. There was no operative morbidity or mortality. Postoperative Bremsstrahlung scans did not reveal extravasation or liver uptake of the radioactive isotope.

Of the four patients with grade I or II astrocytoma cyst, three had persistent cyst involution and one obtained cyst volume stabilization. Reversal of neu-

Discussion

The treatment of cystic tumors with intracavitary radiation has been reported by multiple authors for over 30 years[3, 7]. With the exception of case reports, the vast majority of the literature, however, documented the management of cystic craniopharyngiomas[14]. The use of intracavitary radiation for other tumors was limited since diagnosis and volume determination was limited by the imaging techniques of pneumoencephalography and angiography. Before the advent of CT the preoperative diagnosis of small cystic gliomas was

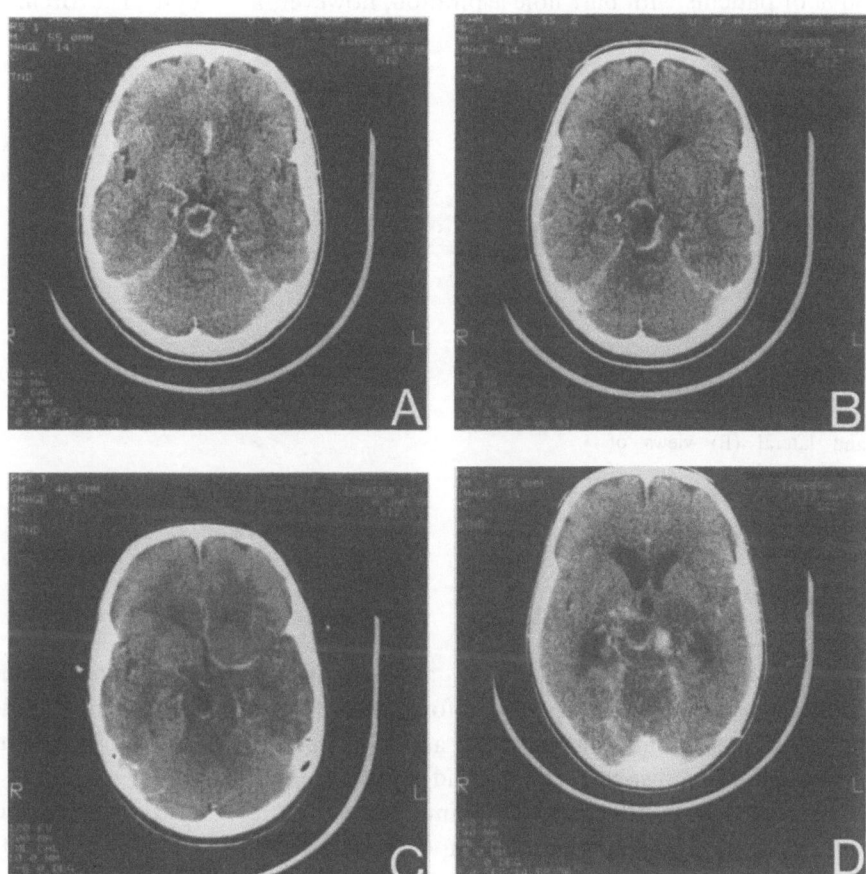

Fig. 1. Cystic grade III astrocytoma of the pontomesencephalon in a 13-year-old male. A) Patient presents with progressive left hemiparesis which reversed with stereotactic aspiration performed 3 weeks later. B) Recurrent left hemiparesis 3 weeks following decompression with interval cyst enlargement. Cyst aspirated and intracavitary ^{32}P administered with resolution of hemiparesis. C) Cyst size unchanged 2.5 months after radioisotope placement. D) Continued cyst involution 7 months after ^{32}P despite recurrent solid tumor

rological deficits were seen in three patients. Follow-up has ranged from 2 to 21 months with a mean of 10 months. Of the five patients with malignant glioma cysts, four obtained persistent cyst diminution following intracavitary [32]P. Improvement of neurological function was seen in four patients following cyst decompression. The one failure of the series was a large hemispheral grade IV astrocytoma which received two dosages of 20,000 rads. There were three deaths in this group secondary to progression of solid tumor. Follow-up has ranged from 3 to 9 months with a mean of 6 months.

difficult. With high resolution CT the diagnosis and volumetric analysis of cystic lesions can be determined accurately[9]. Additionally, cystic lesions in the thalamus and brain stem have been decompressed with minimal risk utilizing the CT-guided stereotactic technique[2, 6].

The frequency of cyst formation in astrocytomas has been documented by several reports. Gol reported a 25% incidence of cyst formation in low grade astrocytomas of the cerebral hemispheres[5]. In 1981 Mercuri *et al.* reported a 54% incidence of cyst formation in 41 low grade hemispheral astrocytomas in children[11]. Cyst formation in malignant astrocytomas has been reported to occur at an incidence of 10% or less[1, 13]. Therefore, the incidence of cyst formation in astrocytomas is not uncommon and may cause neurological deficit independent of solid tumor growth.

Surgical management of astrocytoma cyst formation has included fenestration, burr hole aspiration, cyst-peritoneal shunting, and intermittent drainage using a reservoir[8]. Ventricular fenestration risks dissemination of tumor particularly in malignant lesions. Cyst-peritoneal shunting may result in extracerebral metastases and is limited by the high protein content of tumor cyst fluid. Loftus *et al.* reported cyst control in 50% of patients with burr hole aspiration, however, a

control. With this higher dosage we have not observed any neurological deterioration nor evidence of reactive cerebral edema by CT. Additionally, we have witnessed no evidence of subsequent radiation necrosis, however, longer follow-up will be necessary to further evaluate this potential complication. Since uneven distribution of isotope may result in neurological deficit and lack of control of cyst volume, it is important that the isotope be distributed evenly in the cyst. Since our radiopharmaceutical is administered in a volume of 1 ml, aspirated cyst fluid is used to dilute the colloidal ^{32}P. Secondly, the diluted isotope is barbotaged into the cyst. The Bremsstrahlung scan has revealed no sug-

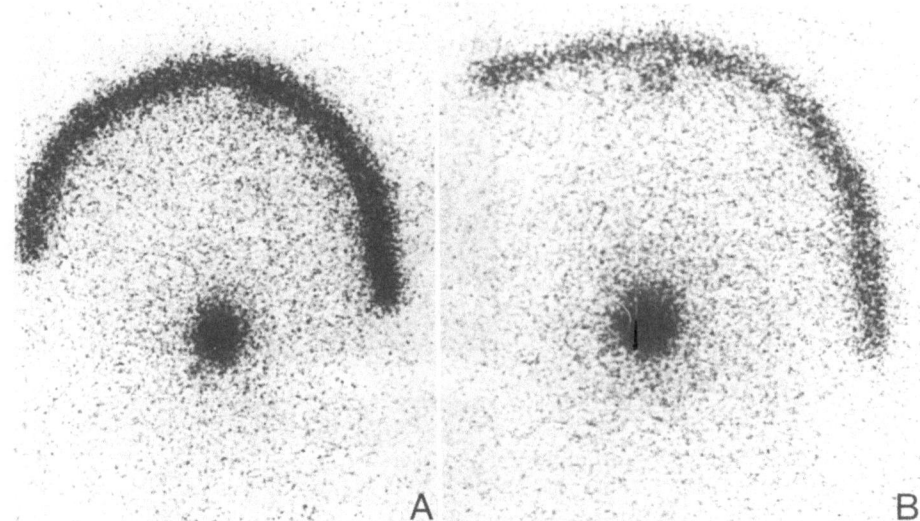

Fig. 2. Anterioposterior (A) and lateral (B) views of Bremsstrahlung scan of intracavitary ^{32}P in cystic grade I astrocytoma of the medulla in a 12-year-old female

70% cure rate required resection and oftentimes, multiple procedures. In our series four patients had recurrent cyst formation following aspiration alone. Although all hemispheral lesions had undergone primary resection, cyst formation continued to present a management problem. Since seven of our patients improved neurologically following cyst decompression, optimal management would include control of cyst volume by a single surgical procedure. Additional patients will be required to further determine the efficacy of intracavitary ^{32}P.

The first two patients of our series received an initial treatment of colloidal ^{32}P which delivered 20,000 rads to the cyst wall. This dosage was chosen from the experience obtained in the treatment of craniopharyngioma cysts. Both patients required additional radioisotope in the attempt to gain control of cyst volume. The remaining patients of our series received sufficient ^{32}P to deliver 40,000 rads to the cyst wall. All patients thusfar receiving the higher dosage have obtained cyst

gestion of uneven isotope distribution and thusfar no neurological deficits have occurred which have been previously reported[12].

In our series cyst control was obtained in 88% of patients without operative morbidity or mortality. Since our patient population is small and the follow-up is limited, these results must be considered preliminary. The only failure of the series was a hemispheral grade IV astrocytoma cyst. This therapeutic failure may be secondary to uneven isotope distribution, insufficient radioisotope, or excessively rapid fluid accumulation which would reduce the effective amount of radiation delivered to the cyst wall. All brain stem and diencephalic lesions have shown marked cyst reduction throughout follow-up or until death secondary to solid tumor growth. From this initial experience we would recommend intracavitary radiation utilizing colloidal chromium ^{32}P in the management of unresectable cystic astrocytomas.

References

1. Afra D, Norman D, Levin VA (1980) Cysts in malignant gliomas: Identification by computerized tomography. J Neurosurg 53: 821–825
2. Apuzzo MLJ, Sabshin JK (1983) Computed tomographic guidance stereotaxis in the management of intracranial mass lesions. Neurosurgery 12: 277–284
3. Backlund EO (1972) Stereotaktisk Behandling av Kraniofarygeom med Intracystiskt y-90 och Extern Co-60 Bestralning. Brödorna Layerström, Stockholm
4. Fox JL (1967) Intermittent drainage of intracranial cyst via the subcutaneous Ommaya reservoir: Technical note. J Neurosurg 27: 272–273
5. Gol A (1961) The relatively benign astrocytomas of the cerebrum: A clinical study of 194 verified cases. J Neurosurg 18: 501–506
6. Hood TW, Gebarski SS, McKeever PE, Venes JL (in press) Stereotaxic biopsy of intrinsic lesions of the brain stem. J Neurosurg
7. Leksell L, Liden K (1953) A therapeutic trial with radioactive isotopes in cystic brain tumors. In: Radioisotope techniques, Proc Isotope Technique Conf, Oxford, vol 1, HMSO, London, pp 76–78
8. Loftus CM, Copeland BR, Carmel PW (1985) Cystic supratentorial gliomas: Natural history and evaluation of modes of surgical therapy. Neurosurgery 17: 19–24
9. Lunsford LD, Levine G, Gumerman LW (1985) Comparison of computerized tomographic and radionuclide methods in determining intracranial cystic tumor volumes. J Neurosurg 63: 740–744
10. Mann KS, Yue CP, Ong GB (1983) Percutaneous sump drainage: A palliation for oft-recurring intracranial cystic lesions. Surg Neurol 19: 86–90
11. Mercuri S, Russo A, Palma L (1981) Hemispheric supratentorial astrocytomas in children: Long-term results in 29 cases. J Neurosurg 55: 170–173
12. Mundinger F (1970) The treatment of brain tumors with interstitially applied radioactive isotopes. In: Wang Y, Paoletti P (eds) Radionuclide applications in neurology and neurosurgery. Ch C Thomas, Springfield, Ill, pp 199–265
13. Poisson M, Phillippon J, van Effenterre R, Racadot J, Sichez JP (1977) Cerebral pseudocysts following chemotherapy of glioblastomas. Acta Neurochir (Wien) 39: 143–149
14. Schaub C, Askienazi S, Szikla G (1979 a) Endocavitary beta irradiation of glioma cysts with colloidal 186-rhenium. Abstr Symp Stereotactic Irradiations, Paris, pV30
15. Taasan V, Shapiro B, Taren JA, Beierwaltes WH, McKeever P, Wahl RL, Carey JE, Petry N, Mallette S (1985) ^{32}P therapy of cystic grade IV astrocytomas: Technique and preliminary application. J Nucl Med 26: 1335–1338
16. Wald SL, Fogelson H, McLaurin RL (1982) Cystic thalamic gliomas. Child's Brain 9: 381–393

Acta Neurochirurgica, Suppl. 39, 38–40 (1987)

New Technique for Three-Dimensional Linear Accelerator Radiosurgery

F. Colombo*[,1], A. Benedetti[1], A. Zanardo[2], F. Pozza[3], R. Avanzo[4], G. Chierego[4], and C. Marchetti[4]

[1] Department of Neurosurgery, City Hospital, Vicenza, Italy, [2] Institute of Applied Mechanics, University of Padova, Italy, [3] Department of Radiotherapy, City Hospital, Vicenza, Italy, [4] Service of Medical Physics, City Hospital, Vicenza, Italy

Summary

A new method for external stereotactic focal irradiation of three-dimensional irregular target volumes is proposed. In this method the target is irradiated by a linear accelerator set in various angular positions around the isocenter. During the irradiation the target is translated in a direction perpendicular to the beam. By controlling the velocity of the translation it is possible to modify the configuration of therapeutic isodoses so as to make them follow the borders of the target.

Keywords: Focal irradiation; linear accelerator.

Introduction

The rationale of stereotactic irradiation procedures is to deliver a high radiation dose inside the target volume with a steep dose gradient coinciding with target borders. Isotope implantation techniques[9, 11] as well as radiosurgical procedures[2–7] are particularly well-suited for the treatment of spherical or spheroidal target volumes attributable to the fixed three-dimensional configuration of resulting isodose surfaces. Irregularly shaped target volumes are more difficult to treat: one must resort to either implanting multiple isotope seeds or carry out multiple irradiation fields. In both cases, the calculation of resulting isodoses needs the use of a computer with high logical and numerical capacities. Moreover, the dose absorbed by the target shows a lack of homogeneity.

The aim of the proposed method is to provide the possibility of modifying the shape of therapeutic isodoses in a way that they follow as closely as possible the target shape defined by CT and verified by stereotactic biopsy.

Material and Methods

Since 1982 we have treated 80 patients with our standard technique of linear accelerator radiosurgery which is essentially based on multiple isocentric arc irradiations. The technique and the early clinical results have been published elsewhere[4–6]. The results were usually satisfying, they could have been better if we had had the possibility of modifying the shape of isodoses. In the proposed new technique, a single irradiation arc is substituted by a limited number of angular positions of the source (for example 12 positions, 15 degrees apart). While the source is radiating, the target is translated, always parallel to itself in a direction perpendicular to the therapeutic beam. The radiation beam is scanning the whole section of the target. The instantaneous velocity of the translation is modified according to a predetermined motion law. In this way it is possible to control the radiation dose absorbed by a single spatial element of the target.

With the described technique the relative movements of the radiation source and of the target are reciprocal to those employed in the "first generation" CT machines, with the exploring beam scanning the whole section of the target with rectilinear movements in various angular positions. Moreover, the proposed irradiation law by which the absorbed dose is build up, is very similar to the procedure of "back projection" employed for the reconstruction of CT images[8].

The problem of the determination of the best irradiation law with respect to target shape and dose gradient can be analytically solved only in cases of regular plane figures. Complex figures, generally found in clinical practice, do not find a solution in analytical terms.

We have empirically found a solution that seems more effective and easy to employ. It consists in modulating the irradiation by translating the stereotactic frame at an instantaneous velocity inversely proportional to the instantaneous length of the chord AA (Fig. 1), representing the thickness of the target section in the plane of irradiation, measured in the direction of the beam. If n be the number of angular positions in which the source is successively set, the nth part of the total dose is required for each translation.

---

* Federico Colombo, M.D., Centro Regionale Veneto di Neurochirurgia Stereotassica, Divisione di Neurochirurgia, Ospedale Civile, viale Rodolfi, I-36100 Vicenza, Italy.

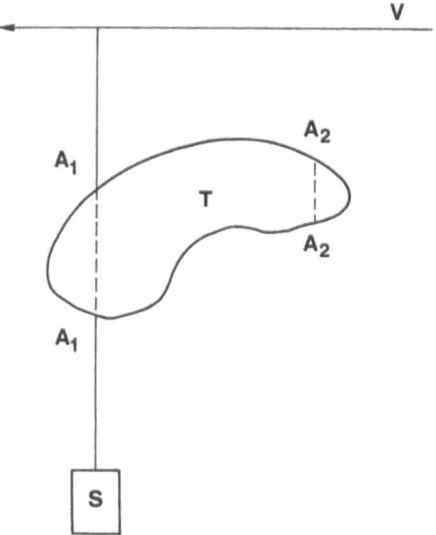

Fig. 1. Schematic view of target irradiation law. The radiation beam intersects the target giving rise to chord A_1–A_1. The stereotactic head frame moves at a velocity inversely proportional to the length A_1–A_1. At point 2 the frame shall move at a velocity inversely proportional to A_2–A_2. The nth part of the dose to be delivered will be given for each translation, being n the number of angular positions in which the source is set around the isocenter. T target section; S radiation source; V direction of head frame translation

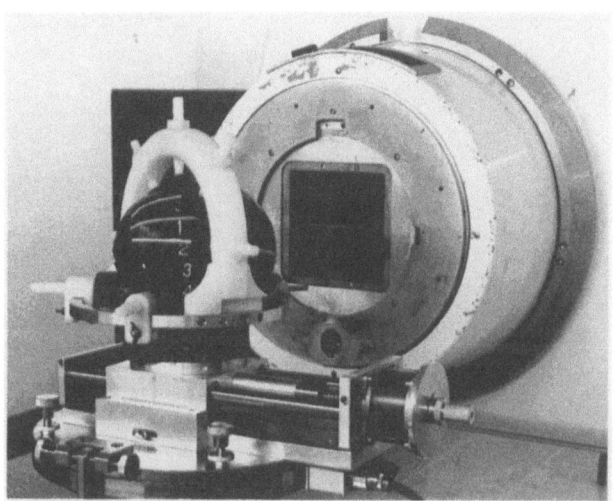

Fig. 2. Stereotactic head frame with Anderson Rando antropomorphic phantom. Computer driven step motor moves the head frame on a slide

Results

The proposed solution has been evaluated at the computer for figures that could be handled analytically, using a graphical system for more complex and irregular figures. We have considered the dose absorbed by a spatial element of the target to be directly proportional to the time of exposure. The time of exposure has been calculated as directly proportional to the instantaneous length of the chord so as to have, for each translation, the irradiation of a spatial element directly proportional to the chord. This situation can be easily resolved graphically. This situation, however, is verified only for the irradiation with high-energy radiation beams, which show little attenuation while passing intracranial tissue.

Once these preliminary steps were completed, we have built a working machine for radiodosimetric evaluation (Figs. 2 and 3). The essential part of the machine is a stereotactic head frame. This frame is connected with a slide moved by a step motor. The movement of the frame is controlled by a computer. The frame can also be rotated around a vertical axis passing through the linear accelerator's isocenter. Angular positions around this axis are measured by a goniometer.

For the irradiation procedure we have employed our Varian Clinac 4 MV linear accelerator. The beam cross-section has been reduced (with the standard built-in collimator) to a field measuring 2×5 mm at the isocenter.

Parameters for determining translation motion law have been calculated with the aid of a IBM model XT computer with 512 K byte memory. The profile of the target has been memorized by a drawing board. Special software has been developed for calculating the intersections between a series of parallel lines and the target profile. After a first series of data has been collected, the target profile is rotated a preselected degree and subsequent series of intersections are determined.

The complete series of data is stored on a floppy disk that can be read by the computer controlling the movements of the stereotactic frame. Radiodosimetric tests have been performed either by film dosimetry (Kodak Xomat-V for therapy verification) and by ionization chambers. Polyethylene and anthropomorphic phantomes (1) have been utilized.

The results confirm the possibility of obtaining irregularly shaped isodoses closely tailored to the borders of complex target volumes. By a combination of a series of irradiations in different angular approaches it is possible to obtain three-dimensional irregular isodose surfaces, which are adapted to the shape of the target volume. One of our results in concave targets is shown in Fig. 3.

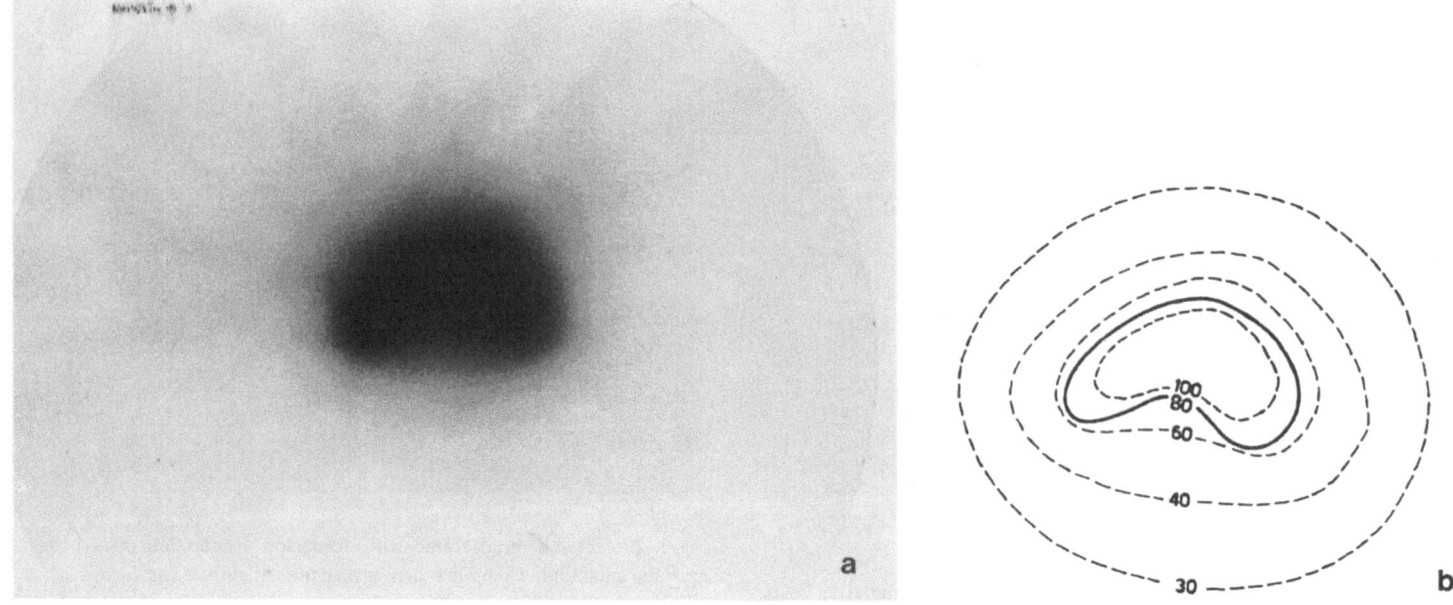

Fig. 3. Irradiation of a concave target in polyethilene head phantom. a) radiodosimetric film b) resulting isodoses read at the photodensitometer. Irradiation with translation on 12 angular positions, 15° degrees apart, around the isocenter. Linear accelerator set at a 60 rads/minute rate. Beam cross-section 2 mm at the isocenter

Conclusions

The substitution of arc irradiation with a combination of fixed angular positions and linear translations allows the modification of the shape of isodoses. This solution has been tested and found to be effective. We are now studying a computer driven movable stereotactic head frame for clinical use. This apparatus must be equipped with a complete series of safety devices which are not necessary for the experimental apparatus.

References

1. Alderson SW, Lanzl LH, Rollins M, Spira J (1962) An instrumented phantom system for analog computation of treatment plans. Am J Roentgenol 87: 185–190
2. Arndt J, Backlund EO, Larsson B, Leksell L, Noren G, Rosander K, Rahn T, Sarby B, Steiner L, Wennerstrand J (1979) Stereotactic irradiations of intracranial structures. Physical and dosimetric considerations. In: Szikla G (ed) Stereotactic cerebral irradiations. Elsevier, Amsterdam, pp 81–92
3. Betti O, Derechinski V (1983) Irradiation stéreótaxique multifaisceaux. Neurochir 29: 295–298
4. Colombo F, Casentini L, Benedetti A, Pozza F, Peserico L (1984) Biopsia e radioterapia stereotassica dei gliomi cerebrali. Minerva Medica 75: 73–77
5. Colombo F, Benedetti A, Pozza F, Avanzo RC, Chierego G, Marchetti C, Zanardo A (1985) Stereotactic external irradiation by linear accelerator. Neurosurg 16: 154–160
6. Colombo F, Benedetti A, Pozza F, Avanzo RC, Chierego G, Marchetti C (1985) Stereotactic radiosurgery utilizing a linear accelerator. Appl Neurophysiol 48: 133–145
7. Heifetz MD, Wexler M, Thompson R (1984) Single beam radiotherapy knife. J Neurosurg 60: 814–818
8. Katz M (1978) Principles and technique of image reconstruction with CT. In: Weisberg LA, Nice C, Katz M (eds) Cerebral computer tomography—a text atlas. Saunders, Philadelphia, pp 10–27
9. Kelly PJ, Bruce AK, Goerss S (1984) Computer simulation for the stereotactic placements of interstitial radionuclide sources into computer tomography defined tumor volumes. Neurosurg 14: 442–446
10. Leksell L (1971) Stereotaxy and radiosurgery: an operative system. Ch C Thomas, Springfield, Ill
11. Mundinger F (1979) Rationale and methods for interstitial Ir 192 brachycurietherapy and Ir 192 and I 125 protracted long term irradiations. In: Szikla G (ed) Stereotactic cerebral irradiations. Elsevier, Amsterdam, pp 101–117

Acta Neurochirurgica, Suppl. 39, 41–44 (1987)

Surgical Treatment of Hypertensive Intracerebral Haematoma by CT-guided Stereotactic Surgery

K. Amano*, H. Kawamura, T. Tanikawa, H. Kawabatake, M. Notoni, H. Iseki, T. Shiwaku, T. Nagao, Y. Iwata, T. Taira, Y. Umezawa, T. Shimizu, and **K. Kitamura**

Department of Neurosurgery, Neurological Institute, Tokyo Women's Medical College, Tokyo, Japan

Summary

Ninety consecutive cases of hypertensive intracerebral haematoma were treated with CT guided stereotactic evacuation. The patients were composed of 61 males and 29 females, ranging from 42 to 87 years old. The location of haematoma was either in the putamen (59 cases) or in the thalamus (31 cases). The average volume of the evacuated haematoma was 21.4 ml in the putaminal haematoma and 14.0 ml in the thalamic haematoma. Postoperative follow-up study in 46 patients showed good recovery of neurological deficits both in putaminal and thalamic group. Criteria of surgical indication of CT-guided stereotactic evacuation of intracerebral haematoma were advocated based on the author's clinical experience.

Keywords: Cerebral haematoma; stereotaxis; CT scan.

Introduction

CT-guided stereotactic evacuation of intracerebral haematoma (ICH) was first reported in 1978[1]. This method of treatment for ICH has advantage surpassing other types of surgical or medical treatment in its accuracy and lesser invasiveness. The authors made an apparatus, called the Iseki frame, for CT-guided stereotactic surgery[2]. This apparatus, which is made mainly of acrylic resin and carbon fiber and partly of aluminum alloy, povides intraoperative CT monitoring for stereotactic procedure in the CT room without noise on the CT image.

* Keiichi Amano, M.D., Associate Professor, Department of Neurosurgery, Neurological Institute, Tokyo Women's Medical College, 8 Kawada-cho, Shinjuku-ku, Tokyo, Japan.

Materials and Methods

Ninety patients with ICH were treated with the Iseki frame and Archimedes type of haematoma drill from January 1982 to May 1986. The patients were composed of 61 males and 29 females aged between 42 and 87 years (Tables 1–3). Fifty-nine patients had haematoma in the putamen (Table 4) and 31 patients in the thalamus (Table 5). The haematoma extended into the lateral ventricles in 21 patients, but the rest of the group (69 patients) had no ventricular haemorrhage (Tables 6 and 7). CT-guided stereotactic haematoma evacuation was performed under the following criteria of indication; 1. Haematoma either in the putamen or in the thalamus 2. Presence of local mass effect due to haematoma on CT regardless the time interval from the onset of haemorrhage 3. Presence of focal neurological deficits and/or altered level of consciousness. Chronic haematoma, if it meets the criteria 2, is considered to be indicated, whereas acute small haematoma, if it does not meet the criteria 3, is considered to be not indicated for CT-guided stereotactic evacuation.

Results

The average volume of aspirated haematoma was 18.9 ml in the overall group (90 cases), 18.9 ml in male patients (61 cases), 18.8 ml in female patients (29 cases), 21.4 ml in putaminal haematoma (59 cases), 14.0 ml in thalamic haematoma (31 cases), 19.5 ml in patients without ventricular haemorrhage (69 cases) and 16.6 ml in patients with ventricular haemorrhage (21 cases) (Tables 1–7). These results show the general tendency that haemorrhage in the putamen is more extensive than that in the thalamus. Because of the difference in viscosity between haematoma and ventricular fluid, the aspirated volume of the true haematoma is less in cases with ventricular rupture than those without ventricular rupture. The haematoma aspirated within the first 3 days from the onset comprises one third of the whole group. One fifth of the whole group was operated later than 11th day from the onset (Table 8). Seventy-three percent of the whole group showed the aspirated haematoma volume between 10 and 40 ml, and this size of haematoma on CT is considered to be the adequate standard of indication of CT-guided stereotactic evacuation in regard to the size of haematoma (Table 9). Postoperative follow-up study (ranging from 3 months to 19 months) was done in 46 patients. In regard to the improvement of altered level of consciousness, the percentage of the patients with the least impaired consciousness in the group increased from 68% (preoperative) to 93% (postoperative) in 28 patients with putaminal haematoma, and also increased in 18 patients of thalamic haematoma from 33% (preoperative) to 78% (postoperative). In regard to the improvement of motor function of the upper limb, the percentage of the patients with least impaired motor function in the group increased from 11% (preoperative) to 50% (postoperative) in 28 patients with putaminal haematoma and also increased from 17% (preoperative) to 56% (postoperative) in 18 patients of thalamic haematoma. In regard to the improvement of motor function of the lower limb, percentage of the patients with least impaired motor function in the group increased from 7% (preoperative) to 39% (postoperative) in 28 patients of putaminal haematoma and also increased from 0% (preoperative) to 27% (postoperative) in 18 patients of thalamic haematoma (Figs. 1 and 2).

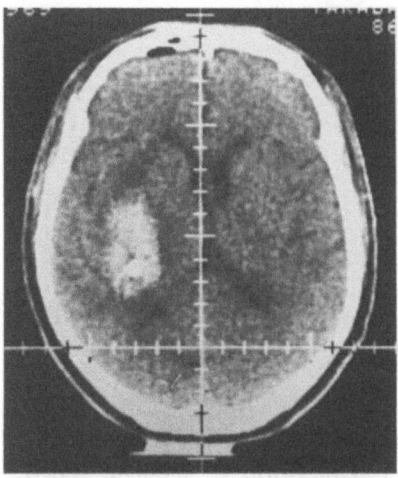
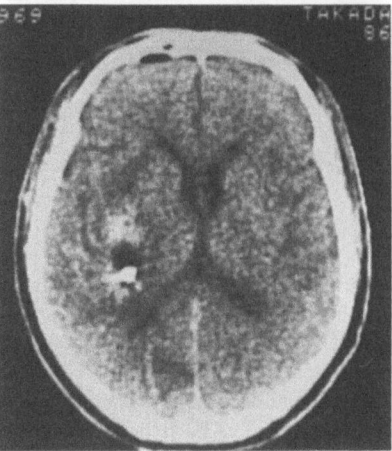

Fig. 2. Upper: Intraoperative CT before evacuation of ICH. Lower: Intraoperative CT immediately after evacuation of ICH. The haematoma drill is still in place within the brain

Discussion

In the authors opinion, CT-guided stereotactic evacuation of ICH surpasses the haematoma removal by craniotomy in that it does not require a general anesthesia, long operation time and corticotomy. This rather new method of stereotactic treatment is not only preferable to craniotomy in terms of accuracy of haematoma removal, but is also superior to medical conservative treatment in which one has to wait many weeks or months until the haematoma is spontaneously absorbed in the brain tissue. During that long period of time, the deeper structures of the brain surrounding the haematoma are under the influence of local mass effect and long standing tissue oedema which inevitably lower the function of the surrounding tissue, resulting in long-term rehabilitation with poor recovery of the patients. By taking out the haematoma at an early stage of the disease process with this less invasive surgical method, the time required for functional recovery is much shortened than by medical, conservative therapy.

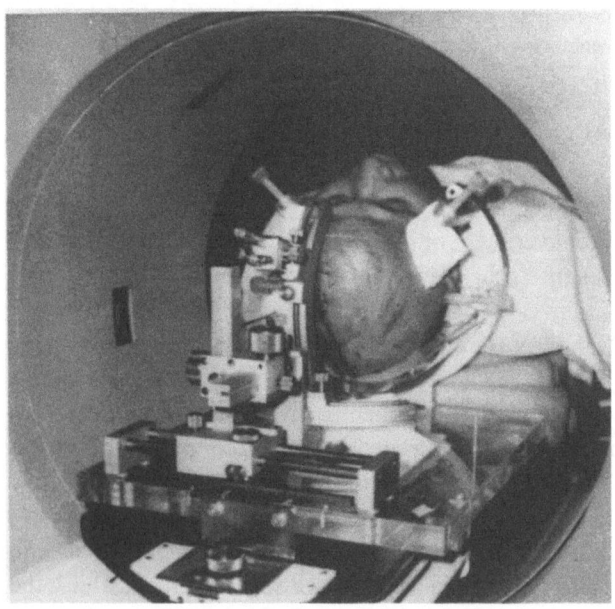

Fig. 1. Haematoma patient fixed to Iseki frame in CT gantry (Toshiba 60 A CT)

Disappearance of the local mass effect is obvious on CT immediately after the stereotactic evacuation of haematoma and the patients show surprisingly rapid recovery of neurological function. Currently, the authors are convinced that the medium-sized haematoma (10–40 ml) with local mass effect on CT should be treated by CT-guided stereotactic evacuation a large-sized haematoma (more than 40 ml) should be removed by decompressive craniotomy because of impending herniation in most of those cases, and small haematoma without neurological deficits should be subject to medical, conservative treatment. The Iseki frame for CT-guided stereotactic surgery has an advantage in that it does not produce noise on the CT image. CT-guided stereotactic evacuation of ICH is performed in the CT gantry under continuous CT image control in regard to the whole process of haematoma evacuation; how much is already removed or how much is left. It is unnecessary, therefore, to take a postoperative CT to assure that the haematoma is adequately removed. Most of the other types of CT-guided stereotactic frame commercially available, are unable to take intra-operative CT because of noise on the CT image resulting from the frame itself.

Table 1. *90 Cases of Hypertensive Intracerebral Haematoma Treated by CT-Guided Steretactic Surgery*

	Age (years)	Timing of operation (days)	Aspirated volume (ml)
Ave	59	7	18.9
SD	11	7	10.2
Max	87	60	48.0
Min	42	1	1.5

Table 2. *61 Male Patients with Hypertensive Intracerebral Haematoma*

	Age (years)	Timing of operation (days)	Aspirated volume (ml)
Ave	59	7	18.9
SD	11	8	10.4
Max	82	60	48.0
Min	42	1	1.5

Table 3. *29 Female Patients with Hypertensive Intracerebral Haematoma*

	Age (years)	Timing of operation (days)	Aspirated volume (ml)
Ave	60	6	18.8
SD	11	5	10.1
Max	87	23	44.5
Min	44	1	2.5

Table 4. *59 Patients with Putaminal Haematoma*

	Age (years)	Timing of operation (days)	Aspirated volume (ml)
Ave	58	7	21.4
SD	12	8	9.8
Max	87	60	48.0
Min	42	1	2.0

Table 5. *31 Patients with Thalamic Haematoma*

	Age (years)	Timing of operation (days)	Aspirated volume (ml)
Ave	60	6	14.0
SD	9	5	9.5
Max	81	23	44.0
Min	47	1	1.5

Table 6. *69 Haematoma Patients without Intraventricular Haemorrhage*

	Age (years)	Timing of operation (days)	Aspirated volume (ml)
Ave	59	8	19.5
SD	11	8	10.2
Max	87	60	48.0
Min	42	1	1.5

Table 7. *21 Haematoma Patients with Intraventricular Haemorrhage*

	Age (years)	Timing of operation (days)	Aspirated volume (ml)
Ave	60	5	16.6
SD	10	4	10.4
Max	79	21	44.0
Min	47	1	2.5

Table 8. *Timing of Surgery After Onset of Haemorrhage in 90 Cases*

Timing of operation	Number of cases	%
Less than 24 hours	4	4
At the 2nd day	11	13
At the 3rd day	14	16
At the 4th day	7	8
At the 5th day	14	16
At the 6th day	6	7
At the 7th day	4	4
At the 8th day	3	3
At the 9th day	3	3
At the 10th day	4	4
Later than 11th day	20	22
Total	90	100

Table 9. *Aspirated Volume of Haematoma in 90 Cases*

Aspirated volume (ml)	Number of cases	%
Less than 10	19	21
$10 \leq\ < 20$	27	30
$20 \leq\ < 30$	31	34
$30 \leq\ < 40$	8	9
More than 40	5	6
Total	90	100

References

1. Backlund E, Holst H (1978) Controlled subtotal evacuation of intracerebral haematomas by stereotactic technique. Neurol 9: 99–101
2. Iseki H, Amano K, Kawamura H, *et al* (1985) A new apparatus for CT guided stereotactic surgery. Appl Neurophysiol, 48: 50–60

Acta Neurochirurgica, Suppl. 39, 45–48 (1987)

Stereotactic Evacuation and Local Administration in Intracerebral Haematomas. A Comparative Study

L. Yagüe, G. Garcia-March, C. Paniagua, M. J. Sánchez-Ledesma, P. Diaz, D. Ludeña, A. Maillo, and J. Broseta*

Department of Neurosurgery, Hospital Universitario, Salamanca, Spain

Summary

Based on current controversies on optimal treatment for spontaneous intracerebral haematomas, chronic experiments to investigate the validity of open surgery, stereotactic evacuation and local urokinase administration in these lesions were performed in 52 dogs. Under general anesthesia diverse volumes of autologous blood were intracerebrally injected to produce the haematoma. A catheter was introduced and chronically implanted in the contralateral ventricle for intracranial pressure monitoring. The animals were divided in two groups of 26 dogs each, according to haematoma location in subcortical or basal ganglia structures. The natural history was studied in both groups. Different types of treatment consisting in surgery, stereotactic evacuation, urokinase injection within the clot and both latter techniques combined were carried out 24 or 72 hours following haematoma production. Clinical status, systemic arterial pressure, intracranial pressure and CT scanning were used for result evaluation. Brain specimens were submitted for pathological examination. Our results indicate that stereotactic evacuation performed during the first 24 hours after haematoma occurrence was the most effective and innocuous procedure for basal ganglia lesions. Local urokinase plus stereotactic aspiration showed a high efficacy in controlling delayed basal ganglia and subcortical blood collections. Other therapeutic approaches behaved almost as the natural history.

Keywords: Intracerebral haematoma; urokinase; stereotactic surgery; intracranial pressure.

Introduction

Optimal treatment for patients presenting with spontaneous intracerebral haematoma remains undetermined and controversial. Conservative conduct[4] as well as open surgery[8] are certainly indicated in several cases. The main difficulties arise when dealing with deep and axial intracerebral lesions. The use of stereotactic evacuation of this type of haematoma was introduced several years ago[1-3, 5, 7]. However conclusive criteria for patient selection and surgical timing as well as the relevance of degradation stage of blood and of location of lesions have not been definetively established. Recently, urokinase stereotactically administered within the haematoma was associated to stereotactic evacuation in an attempt to facilitate liquefaction of clots and thus aspiration[9].

Our group designed an experimental model of intracerebral haematoma and open surgery, stereotactic evacuation and local urokinase infusion were applied. The results obtained by using these techniques were analyzed and compared.

Experimental Material and Methods

Fifty-two adult mongrel dogs weighting 12–28 kg were used. Anesthesia was induced by intramuscular injection of pentobarbital (40 mg/kg body weight) and after endotracheal intubation maintained with pentobarbital endovenous infusion (30 mg/kg body weight) and atropine (0.05 mg/kg body weight). The right femoral artery was cannulated for arterial blood pressure monitoring and the left one isolated for blood extraction. A 0.6 mm diameter silicone catheter was placed in the right lateral ventricle and fixed to the burr hole for intracranial pressure chronic recording. Through a left 4 mm burr hole a 0.6 diameter needle was inserted into the brain 3 mm anterior to the midpoint of the nasion-inion line, 15 mm lateral to midline and 13 mm deep from the dura for subcortical locations, and at the midpoint of the nasion-inion line, 16 mm lateral to midline and 23 mm deep from the dura for basal ganglia lesions[10]. A mean of 3.5 ml and 2.6 ml of autologous blood were respectively injected during 30 minutes in small fractions with 10 minutes interval. The total volume of blood injected was sufficient to produce a sustained and permanent increase of the intracranial pressure.

The possible beneficial effect of open surgery, stereotactic evacuation, urokinase infusion and both latter techniques combined was tested in different groups of 4 dogs, having subcortical or basal

---

* J. Broseta, M.D., Departamento de Neurocirugía, Hospital Clínico Universitario, E-37007 Salamanca, Spain.

ganglia haematomas, and were compared with a control group of untreated animals, 10 in each location, to observe the natural history of the process.

Neurological condition and intracranial pressure were checked daily. Clinical status was scored as 0, normal; 1, stuporous; 2, superficial coma; 3, deep coma; and, 4. dead. CT scannings were performed following haematoma production and the last treatment. Brain specimens were submitted for gross examination and histological studies.

Results

Mortality Rate: Natural history showed a 70% mortality rate for subcortical blood collections and a 90% mortality for the basal ganglia ones after a 10-day period. In subcortical haematomas early open surgery carried out during the first 24 hours since haematoma occurrence did not show any benefit on mortality percentages; 72 hours delayed open surgery, early and 72 hours delayed stereotactic evacuation or solely local urokinase presented a 20% increasing in life expectancy; local urokinase infusion followed 24 hours later by stereotactic evacuation offered the best outcomes with a 45% reduction on mortality.

Neither open surgery performed 24 or 72 hours following haematoma production, nor 72 hours delayed stereotactic evacuation, nor local urokinase infusion reduced mortality rate in basal ganglia haematomas, whereas early stereotactic evacuation improved significantly life prognosis, not observing mortality in this group. Urokinase administration followed by stereotactic aspiration yielded an intermediate mortality percentage.

Neurological Condition: Alteration of consciousness, focal neurological deficits, handicaps and changes in animal habits were evaluated after haematoma production and daily during postoperative follow-up. Fig. 1 illustrates the correlation between pre- and postoperative global clinical scores, confirming that in subcortical haematomas association of local urokinase administration followed by stereotactic aspiration, and in basal ganglia lesions early stereotactic evacuation also offered the best outcomes.

Intracranial Pressure Changes: Following haematoma production, the intracranial pressure tended to increase a mean of 2.3 mm Hg per day in subcortical haematomas and of 3 mm Hg per day in basal ganglia ones, excepting in the next day. This natural course was significantly modified only by early stereotactic evacuation, which was followed by a notable and permanent reduction of intracranial pressure in both haematoma locations (Fig. 2).

During haematoma production and after treatment changes on arterial blood pressure were not observed.

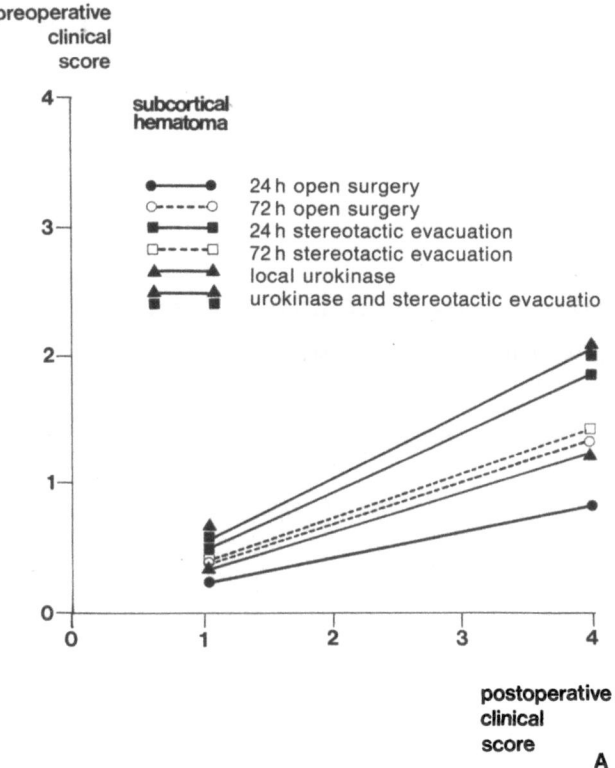

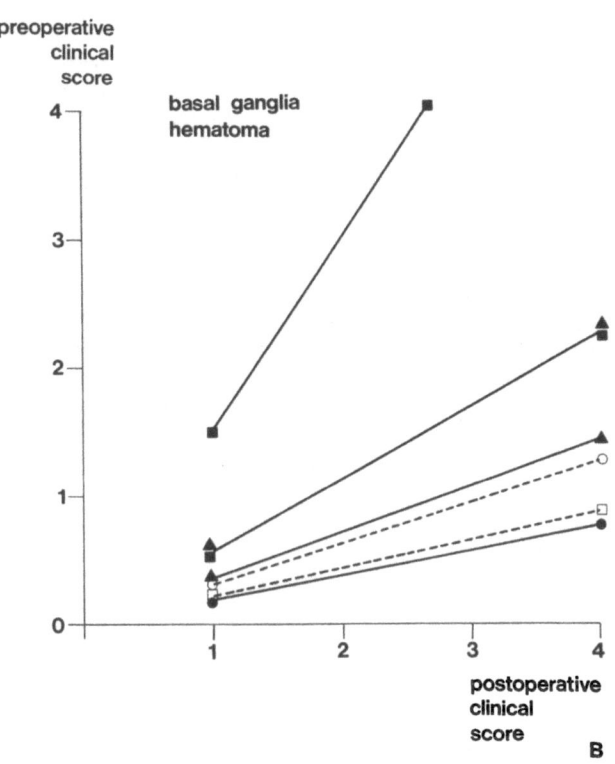

Fig. 1. Correlation between pre- and postoperative clinical status. A) In subcortical haematoma group urokinase administration followed after 24 hours by stereotactic evacuation is the most efficient technique. B) In basal ganglia haematoma group early stereotactic evacuation seems the best solution (legend as in Fig. 1A)

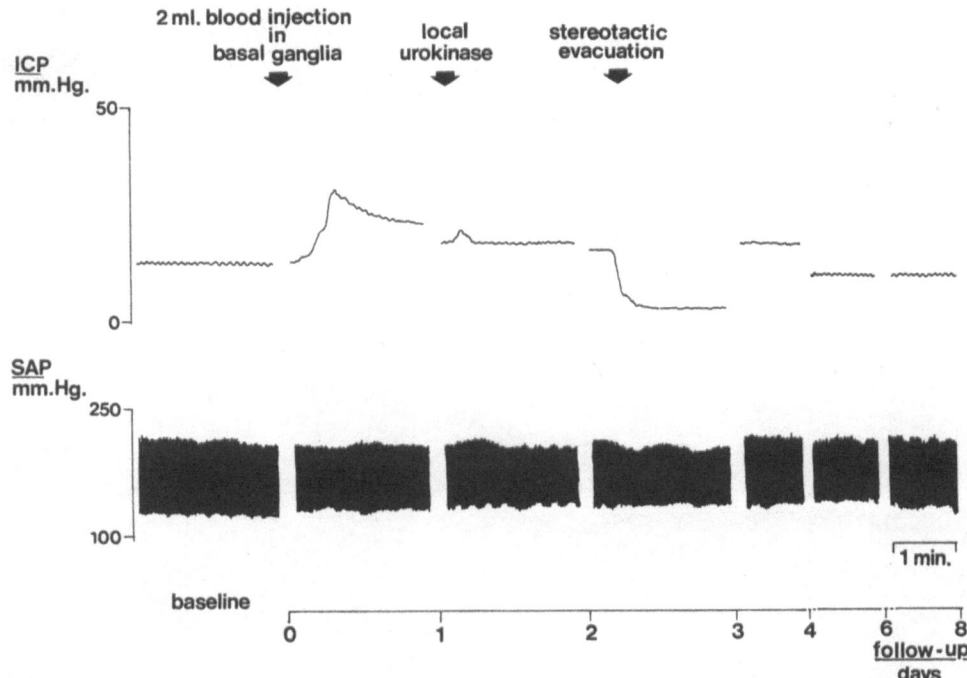

Fig. 2. Intracranial pressure and arterial blood pressure changes during haematoma production, urokinase infusion, stereotactic evacuation and follow-up

CT Scanning Observations: CT scans performed immediately after haematoma production revealed an area of moderate increased density in the site of the lesion, whose size was related with the volume of the injected blood. In subsequent controls done before treatment it was possible to determine the physical state of the blood collection. Postoperative CT scans disregarding which surgical procedure was used showed an almost total haematoma removal in most of the cases, the hyperdense image being replaced by a hypodense area. With our experimental facilities changes on perilesional edema could not be followed.

Pathological Findings: Gross examination of brain slices of untreated animals showed a delimitated organized blood collection in the target regions, though frequently extended to other neighboring structures, with a precise oedematous and congestive boundary. In most of the specimens from treated animals there were no important residual clots: a clean empty cavity was found in the animals with short survival and a glial scarring reaction tending to invade the previous lesions in survivors (Fig. 3).

Open surgery caused the greatest perilesional brain damage; cerebral herniation through the cranectomy was present. Less invasive techniques originate a much lesser secondary brain injury. This minor brain damage was also confirmed by histological studies, showing in these cases a reduction of glial proliferation, necrotic extension and cystic formation, edema and congestive reaction, hyaline degeneration and foam cells and leukocyte infiltration.

Discussion

Removal of intracerebral haematomas by open surgery can be highly hazardous, specially when dealing with patients in poor condition and/or presenting deep and axial lesions. This situation stimulated the search for less invasive solutions. Thus Backlund and von Holst[1] proposed stereotactic evacuation by using a new tool based on an archimedian screw to facilitate clot disgregation. Other series[2, 3, 5, 7] report a satisfactory average of successes when aspirating deep haematomas in late stages, but also a fair number of failures when evacuation attempts were done in early phases with the blood collection densely coagulated. In these last cases Matsumoto and Hondo administered urokinase within the clot every 6 to 8 hours to accelerate liquefaction of the clots previously to stereotactic aspiration.

In our experimental study an attempt was made to reproduce as close as possible the conditions found in humans. Subcortical and basal ganglia haematomas were produced and different types of treatment were applied at different times following haematoma production. In our experimental conditions open surgery produced an immediate decrease of intracranial pressure, which returned rapidly to preoperative levels, probably due to edema secondary to surgical trauma, thus confirming previous personal experience[3] and the results reported by Janny *et al.*[6]. Stereotactic evacuation proved to be efficient when it is performed in early stages, and administration of local urokinase can help when is followed by stereotactic aspiration.

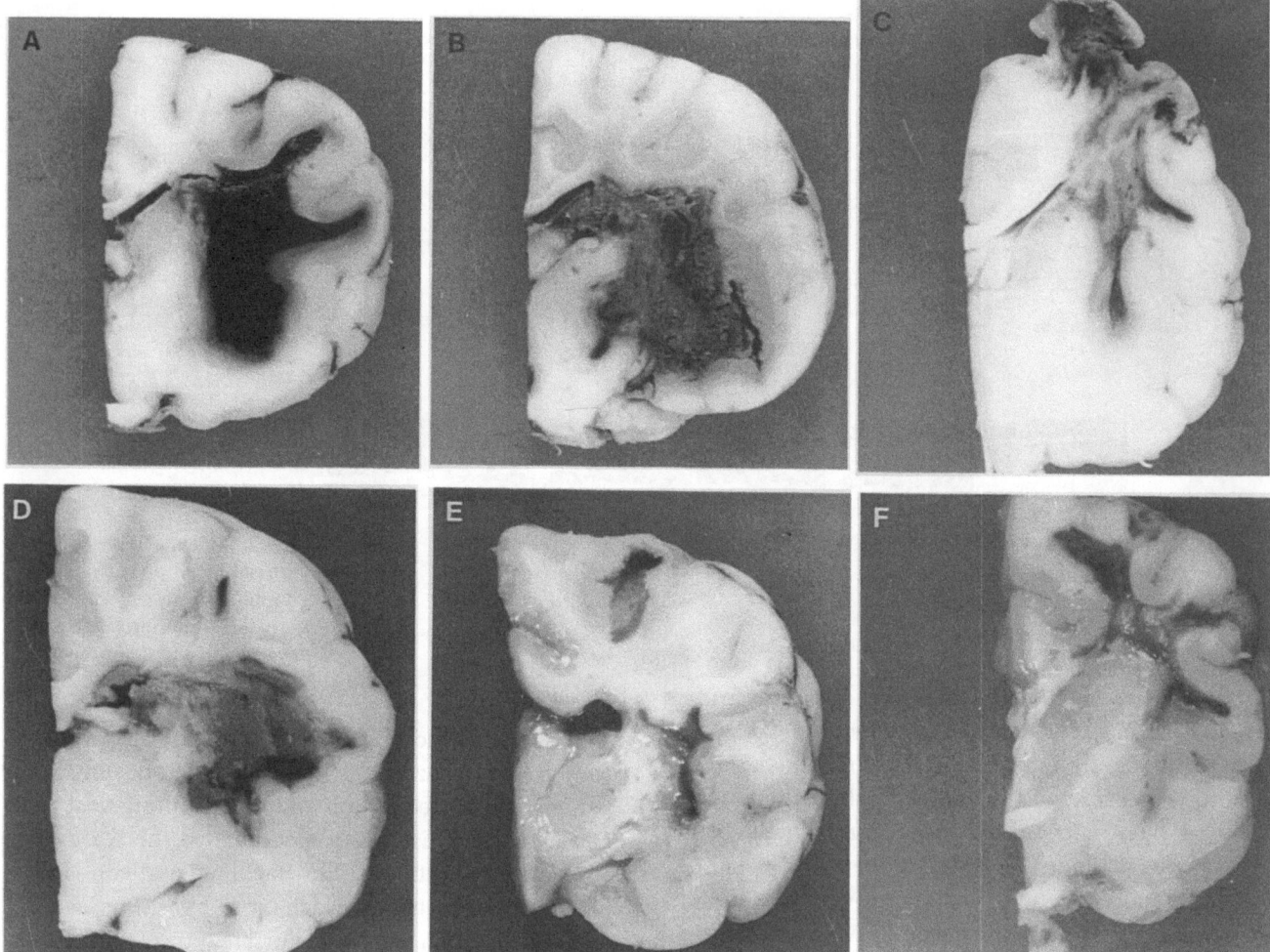

Fig. 3. Gross examination pathological findings. A) Subcortical haematoma. B) Basal ganglia haematoma. C) Open surgery (cerebral herniation through craniectomy deffect can be observed). D) Stereotactic evacuation (previous lesion has been invaded by glial reaction in survivors). E) Intracerebral urokinase administration within the clot. F) Urokinase administration followed 24 h. after by stereotactic aspiration.

References

1. Backlund EO, von Holst H (1978) Controlled subtotal evacuation of intracerebral hematomas by stereotactic technique. Surg Neurol 9: 99–101
2. Beatty RM, Zervas NT (1983) Stereotactic aspiration of a brain stem hematoma. Neurosurgery 13: 204–207
3. Broseta J, Gonzalez-Darder J, Barcia-Salorio JL (1982) Stereotactic evacuation of intracerebral hematomas. Appl Neurophysiol 45: 443–448
4. Duff TA, Ayeni S, Levin A, Javid M (1981) Nonsurgical management of spontaneous intracerebral hematoma. Neurosurgery 9: 387–393
5. Higgins AC, Nashold BS, Cosman E (1982) Stereotactic evacuation of primary intracerebral hematomas: New instrumentation. Appl Neurophysiol 45: 438–442
6. Janny P, Colmet G, Georget AM, Chazal J (1978) Intracranial pressure with intracerebral hemorrhages. Surg Neurol 10: 371–375
7. Kandel E, Peresedov V (1985) Stereotactic evacuation of spontaneous intracerebral hematomas. J Neurosurg 62: 206–213
8. Kaneko M, Tanaka K, Shimada T, Sato K, Chemura K (1983) Long-term evacuation of ultra-early operation for hypertensive intracerebral hemorrhage in 100 cases. J Neurosurg 58: 838–842
9. Matsumoto K, Hondo H (1984) CT-guided stereotactic evacuation of hypertensive intracerebral hematomas. J Neurosurg 61: 440–448
10. Suzuki J, Ebina T (1980) Sequential changes in tissue surrounding ICH. In: Pia HW, Langmaid C, Ziersky J (eds) Spontaneous intracerebral hemorrhage: advances in diagnosis and therapy. Springer, Berlin Heidelberg New York, pp 121–127

Section II

Movement Disorder and Spasticity

A. Movement Disorder

Acta Neurochirurgica, Suppl. 39, 51–53 (1987)

Stretch Reflexes and Parkinsonian Tremor

P. M. H. Rack*

Department of Physiology, The Medical School, Birmingham, U.K.

Summary

By coupling human limbs to a machine, which drives a single joint through sinusoidal flexion-extension movements, it is often possible to entrain the resting tremor of Parkinson's disease. Entrainment is most likely to occur when the imposed movement is large in amplitude and close in frequency to the spontaneous tremor.

Keywords: Reflexes; parkinsonism; tremor; EMG.

Introduction

Although patients with Parkinson's disease may have sluggish tendon jerks, many of them show enhanced reflex responses to slower and larger movements of their joints. These involve active contractions of muscles which have been stretched, and they can properly be described as stretch reflexes[4]. However they occur after a longer latency than the monosynaptic stretch reflexes, and are often described as "long loop" or "long latency" stretch reflexes.

Could these late and exaggerated contractions of the stretched muscles generate parkinsonian tremor by repeatedly re-exciting the motoneurone pool? This question has been debated for a long time, but in recent years the usual conclusion has been that peripheral mechanisms and reflexes play little[1] or no[6] part in the genesis of the tremor.

Particular objections to this peripheral loop hypothesis of parkinsonian tremor are the facts that a sudden joint displacement usually fails to reset the tremor[1], and that sinusoidal forces applied to the joint may fail to entrain the tremor[6]. However, a recent series of experiments[3] goes some way toward removing this objection.

When a limb was firmly fixed to a machine which drove a single joint through alternating flexion-extension movements, the resting tremor of Parkinson's disease could often be entrained by this imposed sinusoidal movement. When entrainment occurred, this implied that bursts of muscle activity responsible for the tremor were "tied to" afferent activity from the driven limb, and that this afferent activity was therefore playing a significant part in the tremor process. However, the subjects were not all the same; there were some in whom the tremor could be entrained readily by movements of a wide range of frequencies and amplitudes, whereas in others the tremor could only be entrained by large movements in a limited frequency range.

The experimental method is described in detail elsewhere[3]. Fig. 1 is a diagram of the mechanical arrangements for driving a wrist joint. Some records from the wrist flexor muscles are inset; these show bursts of EMG activity during the imposed movements. In interpreting the results, the major problem was to determine how far these bursts of EMG activity were locked to the driving movement.

In Fig. 2 some records are displayed in a way that shows the relationship between the bursts of EMG activity and the movement[3]. To make these records, the amplified signals were first rectified and stored on computer disc; the responses to successive cycles of movement were then arranged beneath each other to give a "ladder" pattern. The corresponding joint movement is shown below. It is easy to see that in Fig. 2 A the bursts of activity in successive cylces lie approximately beneath each other indicating that they occurred regularly, once in each cycle of movement. By contrast, the EMG bursts in Fig. 2 B ran on at a rate of

---

* P. M. H. Rack, Professor of Experimental Neurology, Department of Physiology, The Medical School, Birmingham 15, U.K.

their own which differed from the driven frequency; the result is a pattern of oblique bands across the figure. In this second case the muscle activity was clearly independant of the imposed movement.

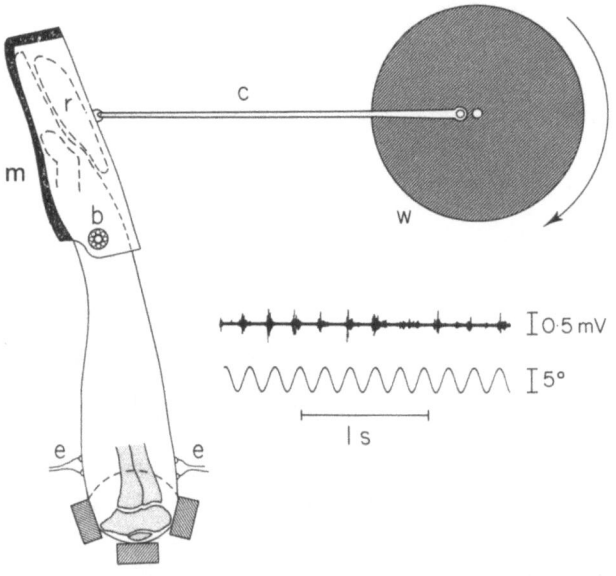

Fig. 1. The crank (c) converts rotary movement of a motor-driven wheel (w) into a reciprocating sinusoidal movement of the mould (m). The mould is mounted on bearigs (b) coaxial with the wrist, and the hand is firmly held within it by a pneumatic rubber bag (r). Surface electrodes (e) record EMGs. The section of record inset shows the EMG recorded from over the wrist flexor muscles during a 5.4 Hz movement

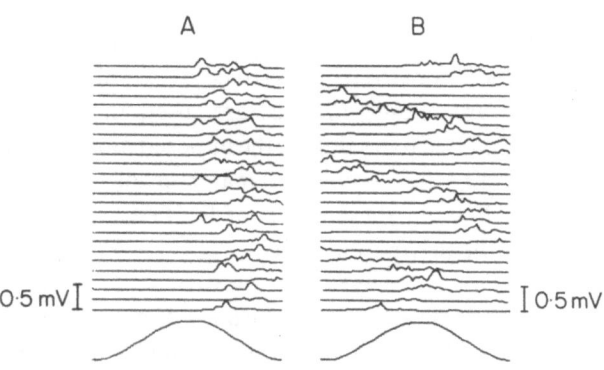

Fig. 2. Sections from two different experimental records. In each case the rectified EMGs from 26 successive cycles are displayed in sequence. The corresponding joint movements are shown below

Fig. 3 was prepared in the same way as Fig. 2. but it shows sections of record which are longer and more tightly packed; furthermore, the speed of the rotating wheel (Fig. 1) was progressively changed to show the effects of different frequencies of movement. In Fig. 3 A, a burst of EMG activity occurred once in each cycle of movement, and the burst remained locked to the movement even though the frequency was progressively increased from <3 Hz to >5 Hz. These EMG bursts remained locked to the movement even when the subject's attention was distracted and when he closed his eyes; there could be no doubt that the timing of this activity was determined by the imposed movement, and was (at least in part) a reflex response to movement of the limb.

By contrast, in Fig. 3 C, the bursts of EMG activity occurred at 3.8 Hz, and continued at that rate whatever the frequency of the imposed movement. Thus, the bursts of EMG activity formed oblique bands across the upper part of the figure where the frequency of movement was slower than 3.8 Hz, and across the lower part of the figure where the movement was faster than 3.8 Hz. When the frequency of the imposed movement was precisely 3.8 Hz, there were a few cycles of Fig. 3 C in which the bursts of EMG activity were aligned below each other, but this persisted for only the small number of cycles that one would expect to see as the frequency of movement was changed through the fixed frequency of the tremor.

Fig. 3 B illustrates an intermediate situation in which tremor activity was sometimes entrained by the imposed movement, but sometimes not. At frequencies of movement 3–5 Hz, the bursts of EMG activity were usually locked to the movement, as they had been in Fig. 3 A, and as one would expect of a reflex response; but at frequencies above 5 Hz, and below 3 Hz, the EMG bursts continued at their "own" rate, and the movement failed to entrain them.

The records shown in Fig. 3 have been chosen to illustrate the variety of responses that were seen in patients who all had clearly developed Parkinson's disease. There was, in fact, a continuous spectrum or responses. At the one end (Fig. 3 A) the activity was completely entrained, and could not be distinguished from an exaggerated stretch reflex response, whereas at the other end (Fig. 3 C) the movement failed to entrain the tremor. However, most of the subjects showed intermediate responses of the type seen in Fig. 3 B, in which the tremor was entrained only by movements that were large enough to be comparable with the usual tremor amplitude, and were fairly close to the tremor frequency. A small movement was used for Fig. 3 C, when the amplitude of movement was increased, some entrainment was also seen in this subject.

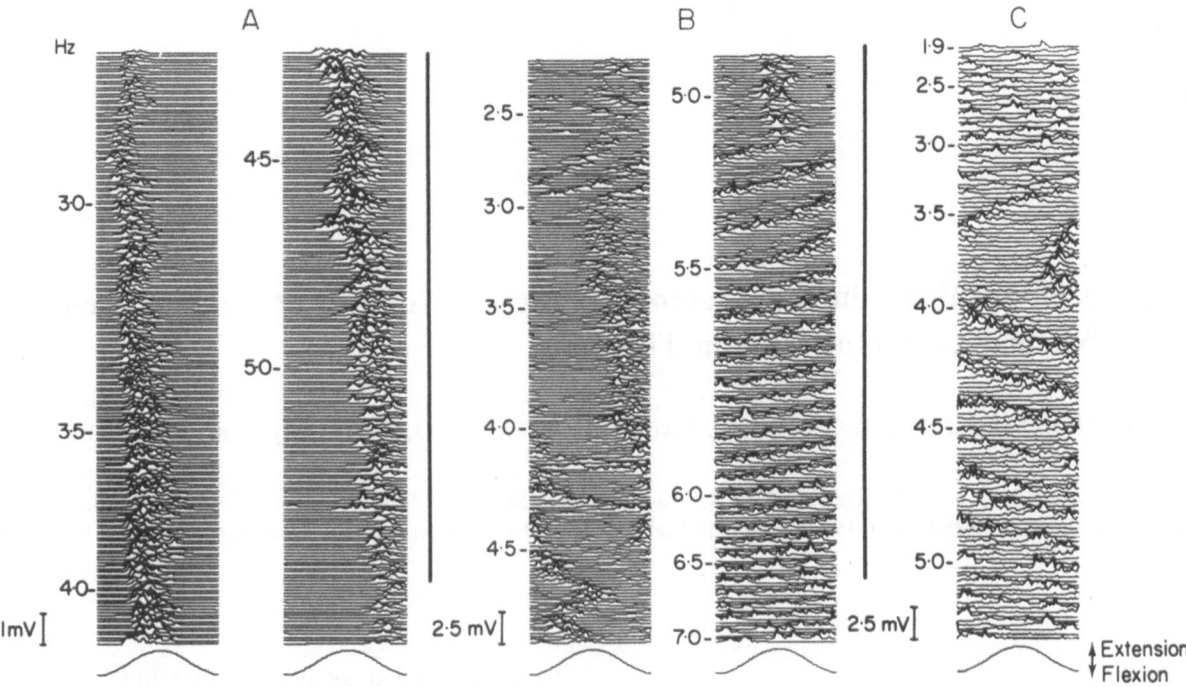

Fig. 3. EMG records from three different subjects during movements of progressively changing frequencies which are indicated beside the records. In A and B the movement was through ± 2.85°, but in C it was ± 0.5°. (In A and C the frequency was progressively increased, but in B it was reduced)

These results show quite clearly that reflex activity from moving muscles and joints does indeed play a part in parkinsonian tremor. However, the tremor seen in such records as Figs. 3 B and C could not all be explained in terms of reflexes which were local to that joint. Indeed, the wide variety of different responses suggests that a local reflex loop interacts with some other potentially oscillatory mechanism[1, 2, 5]. The nature of this other "oscillator" is discussed in detail elsewhere[3].

Acknowledgements

I am grateful to H. F. Ross who shared in the experimental work, and to S. R. Fellows for technical assistance.

References

1. Lee RG, Stein RB (1981) Resetting of tremor by mechanical perturbations: a comparison of essential tremor and parkinsonian tremor. Ann Neurol 10: 523–531
2. Llínas RR (1984) Rebound excitation as a physiological basis for tremor: a biophysical study of oscillatory properties of mammalian cantral neurones in vitro. In: Findley LJ, Capildeo R (eds) Movement disorders: tremor. Macmillan, London, pp 165–182
3. Rack PMH, Ross HF (1986) The role of reflexes in the resting tremor of Parkinson's disease. Brain 109: 115–141
4. Tatton WG, Lee RG (1975) Evidence for abnormal long-loop reflexes in rigid parkinsonian patients. Brain Res 100: 671–676
5. Teräväinen H, Evarts E, Calne D (1979) Effects of kinesthetic inputs on parkinsonian tremor. In: Poirier LJ, Sourkes TL, Bédard PJ (eds) Advances in neurology, vol 24. Raven, New York, pp 161–173
6. Walsh EG (1979) Beats produced between a rhythmic applied force and the resting tremor of parkinsonism. J Neurol Neurosurg Psychiatry 42: 89–94

Acta Neurochirurgica, Suppl. 39, 54–56 (1987)

The Role of Feedback in the Tremor Frequency Activity of Tremor Cells in the Ventral Nuclear Group of Human Thalamus

F. A. Lenz*[,1], S. Schnider[2], R. R. Tasker[1], R. Kwong[3], H. Kwan[4], J. O. Dostrovsky[4], and J. T. Murphy[5]

[1] Division of Neurosurgery, Toronto General Hospital, [2] Bell-Northern Research, Ottawa, [3] Department of Electrical Engineering, University of Toronto, [4] Department of Physiology and Institute of Biomedical Engineering University of Toronto, [5] Division of Neurology, Toronto General Hospital, Canada

Summary

Close loop system identification techniques have been used to identify the presence of feedback in the firing pattern of thalamic tremor cells recorded in parkinsonian tremor patients.

Keywords: Tremor; closed loop; thalamic; stereotactic.

Introduction

Previous studies have identified a population of thalamic tremor cells whose firing pattern is significantly correlated with EMG activity at tremor frequency during parkinsonian tremor[5]. While the demonstration of correlation establishes that the thalamic and EMG signals are linearly related, it does not allow us to draw any conclusions about the mechanism linking the two signals. The EMG signal might drive the thalamic signal by producing movement which is transduced and transmitted to thalamus so that the two signals are linked by a sensory process. Alternately, the thalamic signal might drive the EMG signal by transmission to the spinal motor nuclei so that the two are linked by a motor process. Thalamic tremor cells, which produce the thalamic signal, are likely to be involved in the generation of parkinsonian tremor since a lesion involving them arrests parkinsonian tremor. Hence the mechanism linking these cells with EMG activity during tremor may explain the mechanism of parkinsonian tremor.

In general, the mechanism of tremor and the properties of cells involved in the generation of tremor may be characterized as being either independent of or dependent upon sensory feedback[11]. In the first case, the cells generating tremor are assumed to act as autonomous pacemakers (central hypothesis). In the second case, the cells generating tremor are assumed to be involved in a sensory feedback system (feedback hypothesis). Any feedback control system can become unstable and oscillate if the feedback becomes too powerful, as in the case of clonus which is the oscillation produced by increased reflex feedback in patients with spasticity[11]. If tremor is produced by autonomous pacemakers then cells involved in the generation of tremor should be independent of feedback. If tremor is produced by transmission of afferent activity through an unstable feedback control system, then cells involved in the generation of tremor should show evidence of feedback. We have now applied linear systems analysis techniques to assess the significance of sensory feedback in the correlation between thalamic and EMG activity[10].

Methods and Results

Fig. 1 B shows digitized traces of the wrist flexor EMG and the spike train of a thalamic cell located in the neighborhood of cells with activity related to voluntary forearm movement[5]. Fast Fourier transforms of the EMG signal and an analog equivalent of the thalamic spike train[4] were performed and the results of this analysis used to compute autopower and coherence spectra. Fig. 1 C plots autopower spectra of the thalamic and EMG signals, which indicate the power contained in each signal as a function of frequency. Tremor frequency, determined from the EMG autopower spectrum, is indicated approximately by the symbol of a circle in the autopower

* F. A. Lenz, c/o Division of Neurosurgery, Toronto General Hospital, 200 Elizabeth Street, Toronto, Canada, M5G 2C4.

and coherence spectra. The probability of EMG and thalamic signals being correlated or linearly related was estimated from the coherence spectrum, which varies in value from 0 to 1. A coherence of greater than 0.45 indicates that the likelihood of the two processes being correlated at that frequency is statistically significant ($p < 0.05$,

Ref.[2]). In the example shown in Fig. 1, the coherence of the thalamic and EMG signals is 0.94 at tremor frequency (Fig. 1 D, left panel). This coherence value indicates a high probability that the system linking the two signals is linear and capable of description by a linear model.

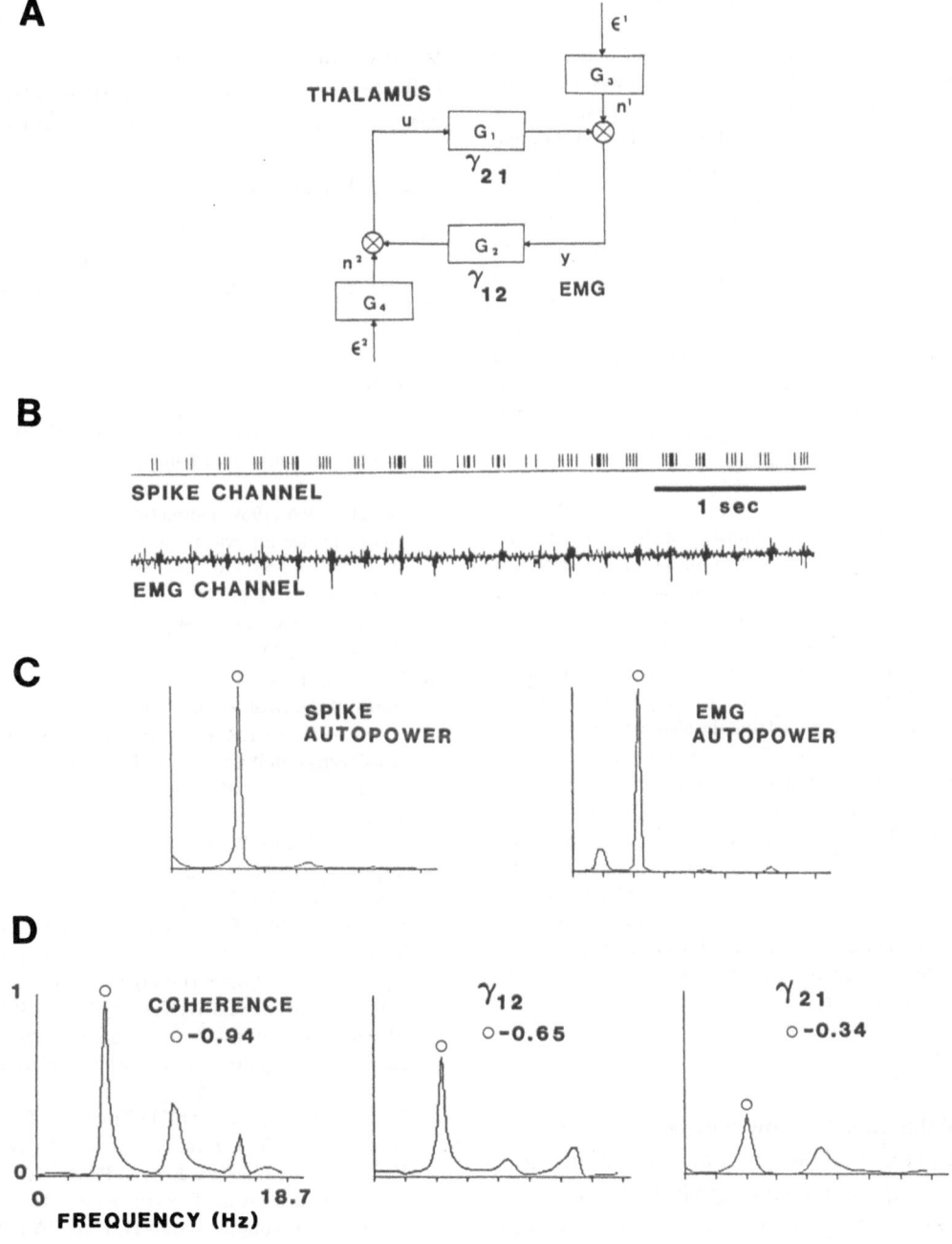

Fig. 1. Part A displays the linear model relating thalamic and EMG signals as described in the text. EMG activity and the activity of a tremor cell during tremor in a parkinsonian patient are shown in part B. Spike train and EMG autopower spectra are shown (part C) for the activity of the same thalamic tremor cell. Vertical scale of these spectra is in arbitrary units. The coherence spectrum (left) and two directed coherence spectra (right) for the same thalamic tremor cell are shown in part C (see text for details)

The proposed linear model of the relationship between thalamic unit and EMG activity is displayed in Fig. 1 A. The thalamic signal—u, and the EMG signal—y, are assumed to be the driving and output signals, respectively. G 1 is the linear transfer function which transforms the signal u into the signal y. In other words, G 1 is a motor process acting on the thalamic signal to produce the EMG signal. G 2 is the linear transfer function which transforms the signal y into the signal u. In other words, G 2 is a sensory process acting on the EMG signal to produce the thalamic signal. The noise signals, n 1 and n 2 represent signals which cannot be modeled by the transfer G 1 and G 2. The noise signals are modeled as outputs of transfer functions G 3 and G 4 acting upon white noise signals E 1 and E 2. This model is capable of describing the system assumed by both hypotheses of tremor as follows. If G 1 is zero the assumption that u is the driving signal is incorrect. If G 2 is zero the system is independent of feedback, a result consistent with the central hypothesis. If G 1 and G 2 are both nonzero then the system is dependent upon feedback, a result consistent with the feedback hypothesis.

The transfer functions G 1 and G 2 can be uniquely determined from the records of thalamic and EMG activity (Fig. 1 B), subject to precise conditions[1] which are satisfied in the present case[10]. With these theoretical conditions met, the transfer functions were then estimated by fitting an autoregressive model to the data. The fundamental assumption of this analysis is that the two signals u and y are not both driven by a third signal, not represented in the model. If such a third signal were present we would expect the two noise signals (n 1 and n 2) to be correlated, which was not the case as determined by evaluation of the residual covariance matrix for each cell studied[10].

In order to evaluate whether the transfer functions G 1 and G 2 were significantly different from zero, the transfer function matrices were normalized to produce the directed coherence functions $\gamma\,12$ and $\gamma\,21$[9]. $\gamma\,21$ can be thought of as the relative power in the EMG signal which is attributable to a motor process, acting upon the thalamic signal. $\gamma\,12$ can be thought of as the relative power in the EMG signal which is attributable to a sensory process acting upon the EMG signal. Simulation studies established that the directed coherence was significantly different from zero for $\gamma\,12$ or $\gamma\,21$ greater than 0.05[10]. In the example of Fig. 1, $\gamma\,12$ is 0.65 and $\gamma\,21$ is 0.20 at tremor frequency (note that the peak for $\gamma\,21$ occurs at a lower frequency than that of tremor—by 0.32 Hz). These results demonstrate that the sensory and motor processes are both nonzero, suggesting that the EMG and thalamic signals are linked by a feedback process. Similar analysis on eleven other cells recorded in two patients similarly found evidence of feedback processes linking thalamic and EMG activity. Despite demonstration of feedback in the case of these twelve cells, responses to somatosensory stimulation could not be found at the time of surgery for eight of the twelve cells.

Discussion

The results of this analysis show evidence of feedback in the system linking thalamic tremor cell activity and contralateral peripheral EMG activity, even for cells without obvious sensory input. Thus, tremor cells may be involved in an unstable feedback control system presumably characterized by exaggerated reflex activity. There is evidence that reflex activity occurring at longer latency than the tendon tap response (long latency reflexes, see Ref.[6]) is exaggerated in par-

kinsonian patients[8]. Furthermore, long latency reflex activity may be produced by transmission of afferent activity through a long reflex loop involving spinal sensory pathways, thalamus, motor cortex and motor neurons[3] as well as through spinal reflex pathways[7]. Although the present report supports the existence of a thalamic feedback circuit in the relationship between thalamic tremor cells and parkinsonian tremor, the significance of this feedback for the generation of tremor can only be established by further studies. If tremor is dependent upon feedback to thalamic tremor cells then both tremor cell activity and tremor should be reset by perturbing the tremorous limb peripherally.

Acknowledgements

Supported by the PSI Foundation, Toronto, Canada, the Parkinson's Foundation of Canada and MRC (Canada). F. A. Lenz was a MRC (Canada) Fellow and a Schering Scholar of the American College of Surgeons.

References

1. Anderson B, Gevers M (1982) Identifiability of linear stocastic systems operating under linear feedback. Automatica 18: 195–213
2. Benignius VA (1969) Estimation of the coherence spectrum and its confidence interval using the fast Fourier transform. IEEE Proc AU 17: 145–150
3. Cheney P, Fetz E (1984) Cortico-motoneuronal cells contribute to long-latency stretch reflexes in the rhesus monkey. J Physiol (Lond) 349: 249–272
4. French A, Holden A (1971) Alias-free sampling of neuronal spike trains. Kybernetik 8: 165–171
5. Lenz FA, Tasker RR, Kwan H, Schnider S, Kwong R, Murphy J (1985) Single unit analysis of the ventral tier of lateral thalamic nuclei in patients with parkinsonian tremor. Soc Neurosci Abstr 11: 1164
6. Lenz FA, Tatton W, Tasker RR (1983) The EMG response to displacement of different forelimb joints in the squirrel monkey. J Neurosci 3: 783–794
7. Mathews P (1984) Evidence from the use of vibration that the human long latency stretch reflex depends upon spindle secondary afferents. J Physiol (Lond) 348: 545–558
8. Tatton W, Bedingham W, Verrier M, Blair R (1984) Characteristic alterations in responses to imposed wrist displacements in parkinsonian rigidity and dystonia musculorum deformans. Can J Neurol Sci 11: 281–287
9. Saito Y, Harashima H (1981) Tracking of information within multichannel EEG records—Causal Analysis of EEG. In: Yamaguchi N, Fujisawa K (eds) Recent advances in EEG and EMG data processing. Elsevier, Amsterdam
10. Schnider S, Kwong R, Kwan H, Lenz FA (1986) Detection of feedback in the central nervous system of parkinsonian patients. Proceedings at the 25th IEEE Conference on Decision and Control. 1: 292–294
11. Stein R, Lee R (1981) Tremor and clonus. In: Brooks V (ed) Handbook of physiology: motor control, vol 1. American Physiological Society, Bethesda, pp 325–344

Acta Neurochirurgica, Suppl. 39, 57–60 (1987)
© by Springer-Verlag 1987

Computer-assisted Analysis of Functional Anatomy of Human Ventrolateral Thalamus

P. Birk*, [1], A. Struppler[1], H. G. Lipinski[1], C. Giorgi[2], U. Cerchiari[2], and H. Riescher[1]

[1] Neurologische Klinik, Technische Universität, München, Federal Republic of Germany, [2] Division of Neurosurgery, Istituto Neurologico, Milano, Italy

Summary

A three-dimensional map was created by a computer-assisted analysis of functional and somatotopic organization of the target area in the human ventrolateral thalamus.

Stimulation in the target area mostly elicited increased tone in skeletal muscles, with a concomitant decrease or stop of tremor.

Despite averaging of all responses, no clear somatotopic organization could be demonstrated for the tonifying stimulation effects. In addition, somatosensory-evoked potentials were recorded, indicating an afferent projection to the target area.

Keywords: Thalamus; 3-d anatomy; computer mapping.

Introduction

Stereotaxic treatment of movement disorders can be improved by computer-assisted analysis of functional anatomy of the human ventrolateral thalamus. A three-dimensional map reflecting averaged patient data may help to predict and to localize the correct site of the lesion with more precision and minimize side-effects. The map of the target area was created from stimulation and recording data, gathered during the intra-operative localization procedure. Functional and somatotopic organization was analysed.

Methods

The recording and stimulation sites of 196 patients, operated upon for parkinsonian tremor and segmental dystonia during the last decade, were digitized in reference to the stereotaxic system. 28 patients, operated during the last two years, were selected for the study. 20

patients suffered for parkinsonian tremor, 4 patients had torticollis spasmodicus, 4 had multiple sclerosis with a concomitant cerebellar tremor. Stimulation and recording was performed intraoperatively during the localization procedure. An electrode with a tip diameter and length of about 1 mm was used for stimulation. Constant voltage stimulus trains with a frequency of 50 Hz and 0.5 ms duration were applied. The amplitude ranged from about 1 to 5 V respectively.

The transition of neural noise was recorded during the penetration of the thalamus. A tungsten semimicroelectrode was used for that procedure. Thalamic evoked potentials following peripheral nerve stimulation were recorded by using a multielectrode. The multielectrode had 8 active areas (silver chloride) with a length of 1 mm and a diameter of 1.6 mm respectively. The active areas were separated by a distance of 1 mm and had an impedance of about 1–2 kOhm. 400 sweeps were averaged and filtered in a range of 0.5 Hz to 3 kHz. The target point was calculated by the use of frontal and sagittal positive ventriculograms of the 3rd ventricle. A reference ventriculogram and the Schaltenbrand-Wahren atlas were digitized by a video camera.

Results

The spontaneous neural noise profile was recorded along the probe trajectory in all 28 patients. The borderline of internal capsule and the thalamic reticular nucleus as well as the lower border of thalamus were identified by transition of the spontaneous neural noise activity. In 4 patients however the lower border could not be clearly identified, because the modulation of noise was not significant enough. Figs. 1 a and b demonstrate the spatial coordinates of noise modulation in reference to the 3rd ventricle. The upper border of the patients thalamus was correlated to that of the atlas thalamus. Fig. 2 demonstrates the result of the computerized transformation process. This procedure provided "interindividual" normalization of the pa-

---

* P. Birk, M.D., Neurologische Klinik, Technische Universität, D-8000 München, Federal Republic of Germany.

tients thalamus according to Schaltenbrand-Wahren atlas. Stimulation in the target area mostly elicits "tonization" of groups of muscles or single muscles with a consecutive decrease or stop of tremor. Often agonistic and antagonistic muscles are "tonified" simultaneously. In addition to the observed motor effects sensory effects were also described by the patients. Mostly these occurred simultaneously with the motor ones, but at different threshold amplitude. The stimulation effects were registrated by an operation protocol table using an appropriate key and then transferred to the data base of the computer (Figs. 3 and 4). Only the 75 stimulation effects elicited at the threshold amplitude were choosen for further analysis of func-

tional and somatotopic organization. 75% of the effects were to be motor, 25% sensory. No clear segregation of both groups was apparent, but the majority of the sensory effects occurred below the AC/PC plane, whereas motor effects were scattered over the whole thalamic and subthalamic area. Figs. 5 and 6 demonstrate sensory and motor effects plotted in the sagittal and frontal plane respectively.

No clear somatotopic distribution could be derived from the mean values of the spatial distribution of motor stimulation effects. They cluster in a narrow area of the thalamic and subthalamic border and there is no clear segregation neither in laterality nor depth nor anterior-posterior direction.

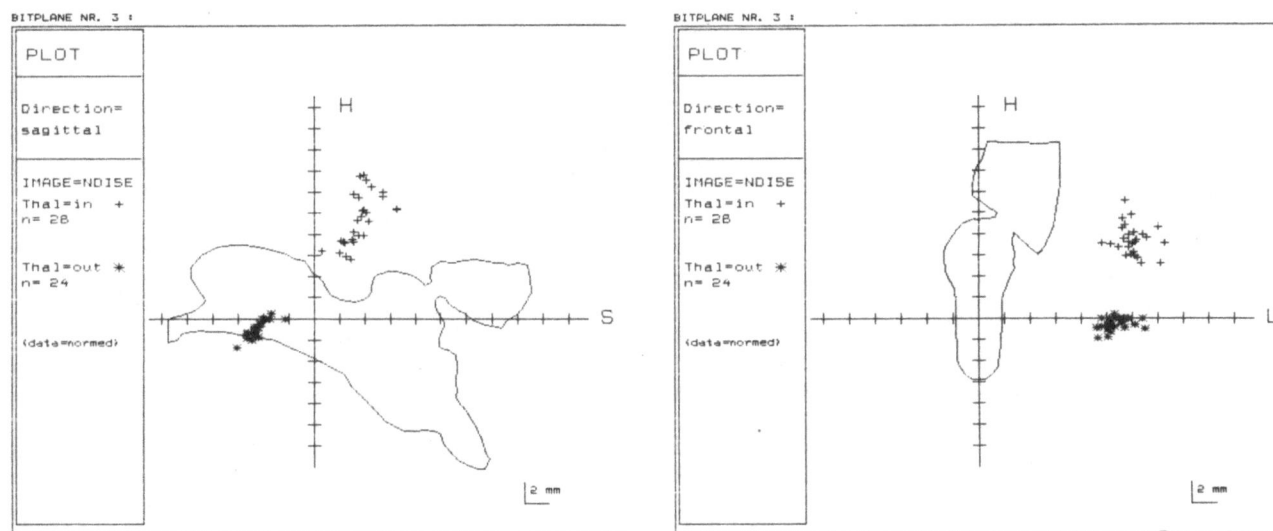

Figs. 1 a and b. Reference ventriculogram in the sagittal plane (a) and frontal plane (b). Crosses reflect the spatial coordinates of the upper thalamic border, stars of the lower one, according to transition in spontaneous neural noise. Data are displayed in reference to the Schaltenbrand-Wahren atlas. For further explanation see also text

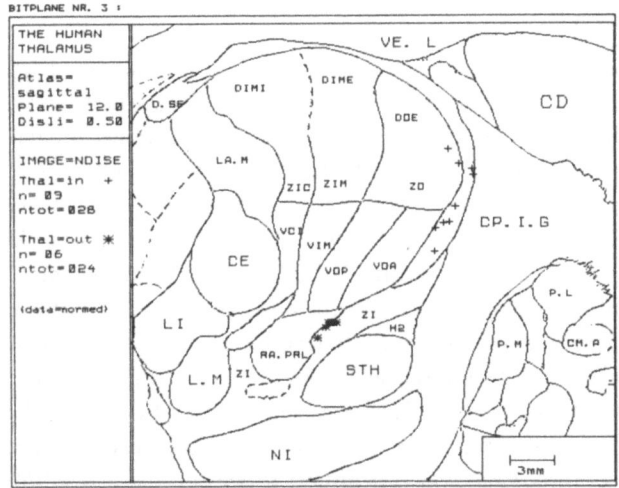

Figs. 7 a and b reflect the mean values plotted in reference to the corresponding sagittal and frontal planes. The afferent connections of the target area were examined by recording thalamic evoked potentials following mixed nerve stimulation. Fig. 8 represents evoked potentials recorded in a patient operated for parkinsonian tremor. The main early positivity in subthalamic area occurs at a latency of less than 20 ms. In addition the corresponding main negativity within the nucleus is registered with about the same latency. The phase of the potentials reverses at the border of the subthalamic to thalamic area.

Fig. 2. Spatial coordinates of upper border of thalamus (reticular thalamic nucleus) normalized in reference to the Schaltenbrand-Wahren atlas

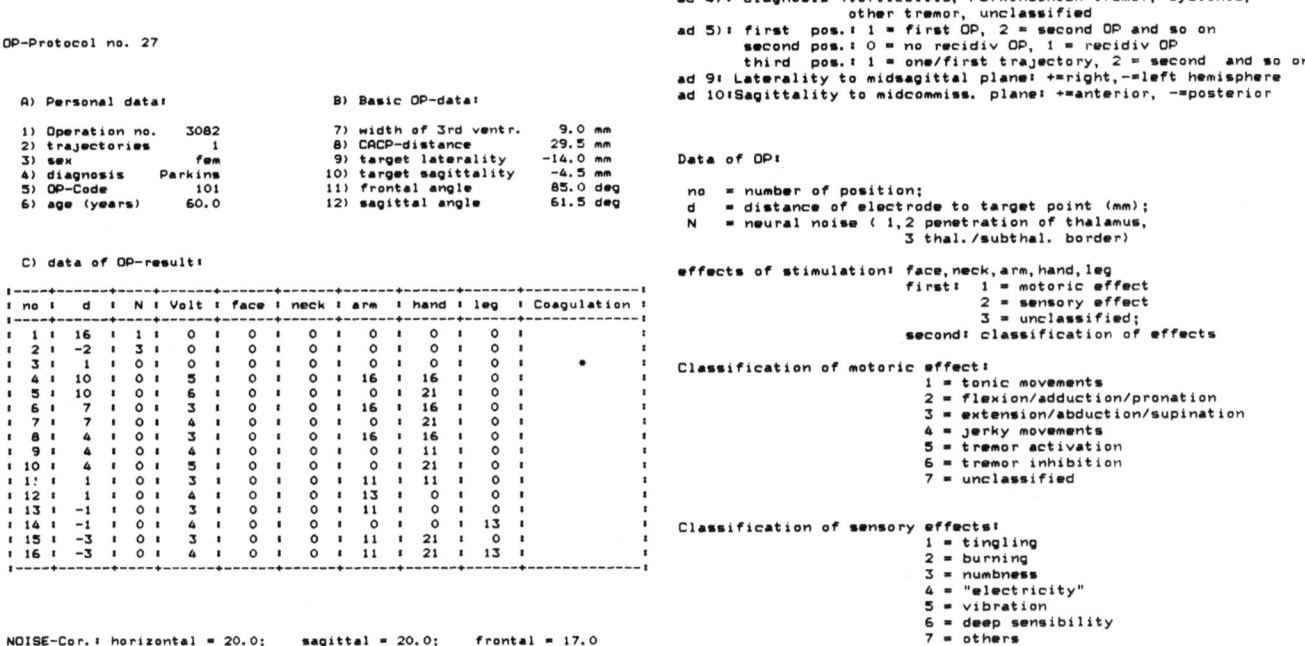

Fig. 3. Example of an operation protocol as created by the computer. For further explanation of the symbols see Fig. 4

Fig. 4. Key of the operation protocol, which is used for the transfer of data to the computer data base

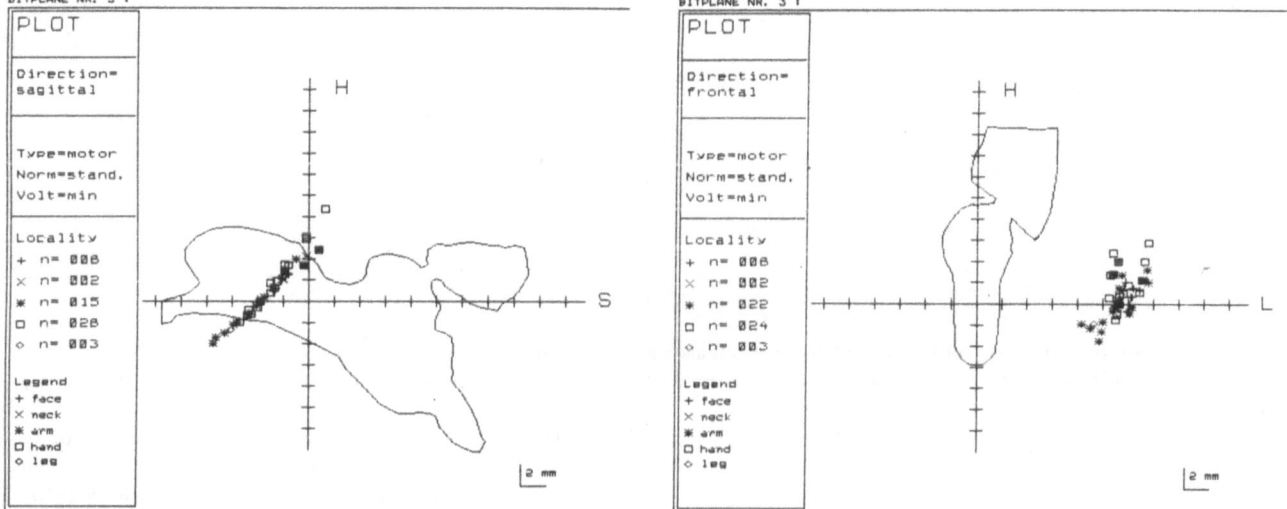

Figs. 5a and b. Sagittal (a) and frontal plane (b) of the reference ventriculogram. Localization of motor effects elicited at threshold amplitude, according to the site of propagation at the body. The symbols are explained in the lower part of the table at the left

Discussion

Stimulation in the target area mostly elicited motor effects. It resulted in "tonization" of skeletal muscles with a concomitant tremor inhibition. This suggests that the therapeutic lesions is set in a system facilitating muscle tone. In addition, the evoked potentials recorded in the target area following peripheral nerve stimulation, may reflect afferent input originating in the periphery[1, 2]. Individual stimulation in the target area, especially with selective electrodes, reflects somatotopic organization. "Interindividual" data pooling and normalization did not reproduce that observation. Further refinement is achieved by the use of selective electrodes and by correlating CT imaging and MRI with the stereotaxic system.

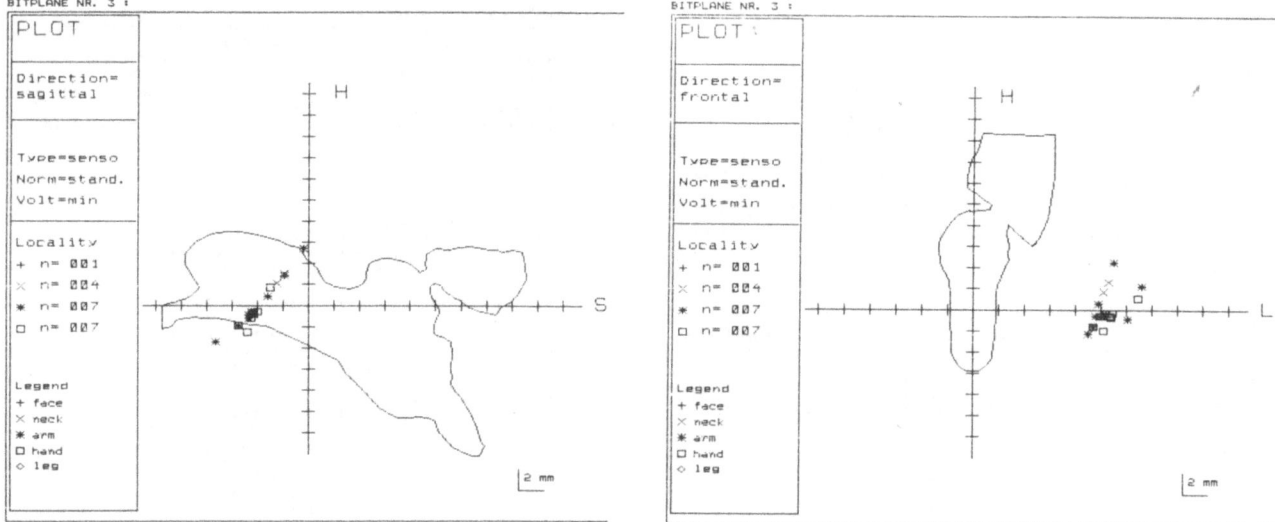

Figs. 6 a and b. Sensory effects elicited at threshold amplitude, sagittal plane (a) and frontal plane (b). For further explanation see also legend of Fig. 5

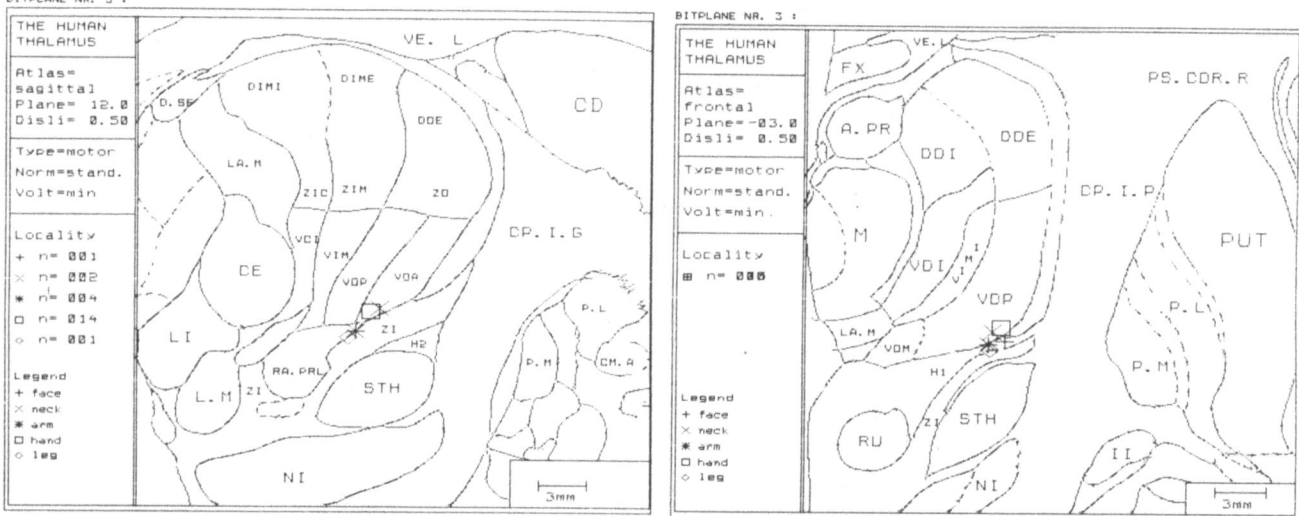

Figs. 7 a and b. Main values of spatial distribution of motor effects according to the propagation at the body. The data are displayed in reference to the Schaltenbrand-Wahren atlas. Sagittal atlas plane (a) and frontal atlas plane (b). The symbols are explained in the lower part of the table at the left

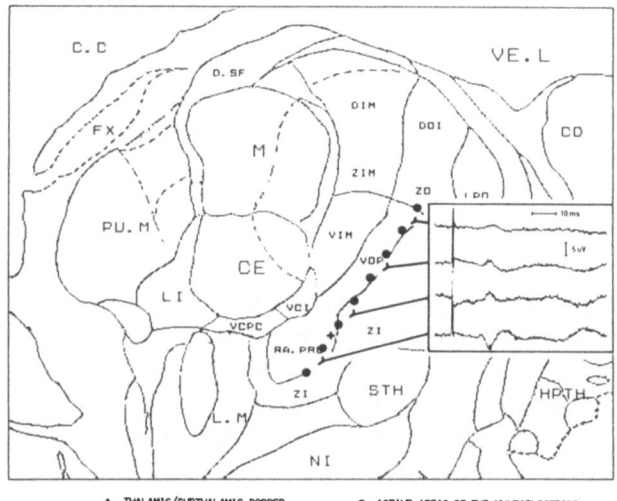

Fig. 8. Thalamic evoked potentials following stimulation of the contralateral median nerve, recorded by a multielectrode. Dots symbolize the active areas of the multielectrode. The cross symbolizes the subthalamic border, as it was identified by recording spontaneous neural noise. Spatial coordinates of the recording sites are displayed in reference to the Schaltenbrand-Wahren atlas

References

1. Birk P, Riescher H, Struppler A, Keidel M (1986) SEP and muscle responses related to thalamic and subthalamic structures in man. In: Struppler A, Weindl A (eds) Sensory-motor integration: implications for neurological diseases. Springer, Berlin Heidelberg New York Tokyo, in press
2. Struppler A, Birk P (1986) Functional anatomy of sensory-motor afferences of the thalamus. In: Samii M (ed) Surgery in and around the brainstem and the third ventricle. Springer, Berlin Heidelberg New York Tokyo, in press

Acta Neurochirurgica, Suppl. 39, 61–65 (1987)

Thalamotomy for Movement Disorders: a Critical Appraisal

E. Hitchcock*, G. A. Flint, and **N. J. Gutowski**

Department of Neurosurgery, University of Birmingham, U.K.

Summary

Symptomatic and functional assessments have been made on a number of patients with a variety of involuntary movement disorders. Difficulties of assessment and their relevance to outcome are discussed. Almost all groups showed a substantial symptomatic improvement but functional improvement was less pronounced.

Keywords: Involuntary movement; thalamotomy; functional assessment; follow up.

Introduction

In Parkinson's disease surgery results in an immediate dramatic improvement in tremor in between 80 to 90%. Lesser degrees of improvement are achieved for other features of the disease but the beneficial effect on tremor alone is usually taken to indicate an excellent surgical result. Unfortunately, natural progression of the disorder and persisting bradykinesia means that the overall gain in terms of function may not be as great. In assessing the overall benefits of stereotactic surgery it is necessary for us to evaluate the patients' general or functional ability as well as the symptomatic improvement. This is particularly necessary if comparisons are to be made between surgical and medical treatment.

To this end many workers, including members of this society, have devised assessment scales. These vary considerably in detail, ranging from a few general grades to lists of multiple features. Commonly, they are directed at particular symptoms of particular diseases. Unfortunately, the simplest schemes lack the necessary sensitivity to detect small but significant changes whilst the more detailed regimes are complex and time-consuming. A major drawback is that none is directly comparable with another.

The Department of Neurosurgery has, over the past several years, explored various assessment schemes. Most, although suitable for the individual condition to which they were applied, were judged unsuitable for other disorders related or otherwise. Two schemes have been adopted to evaluate the usefulness of thalamic surgery for movement disorders in general.

Method

The first scheme involved a simple Activities of Daily Living scale; the second was the more widely known Karnofsky scale. Both of these were combined with a simple grading of symptomatic improvement in terms of "cured", "improved" or "unchanged". The first scheme was applied to patients up to the first 12 months after thalamotomy. The second scheme covered the postoperative results and progress annually thereafter.

Of 79 patients who had thalamotomy for movement disorder between 1979 and 1984 55 were available for follow-up assessment over a four-year period. 44 of these patients had tremors, 19 being parkinsonian, 10 essential or other nonparkinsonian tremors and 30 patients had ataxic tremors due to demyelination. Seventeen patients had a variety of other involuntary movement disorders in the form of hemiballism, athetosis, chorea or dystonia, or combinations of these.

Results

All movement disorders (Figs. 1 and 2). Immediate symptomatic improvement was noted in 93% of patients. Despite some deterioration over the ensuing years 55% maintained their improved status at four years. Corresponding figures for disability scores on the Karnofsky scale were 58% immediate postoperative improvement, followed by a gradual decline to 33% by four years.

---

* Professor E. Hitchcock, Midland Centre for Neurosurgery and Neurology, Holly Lane, Smethwick, Warley, West Midlands, U.K.

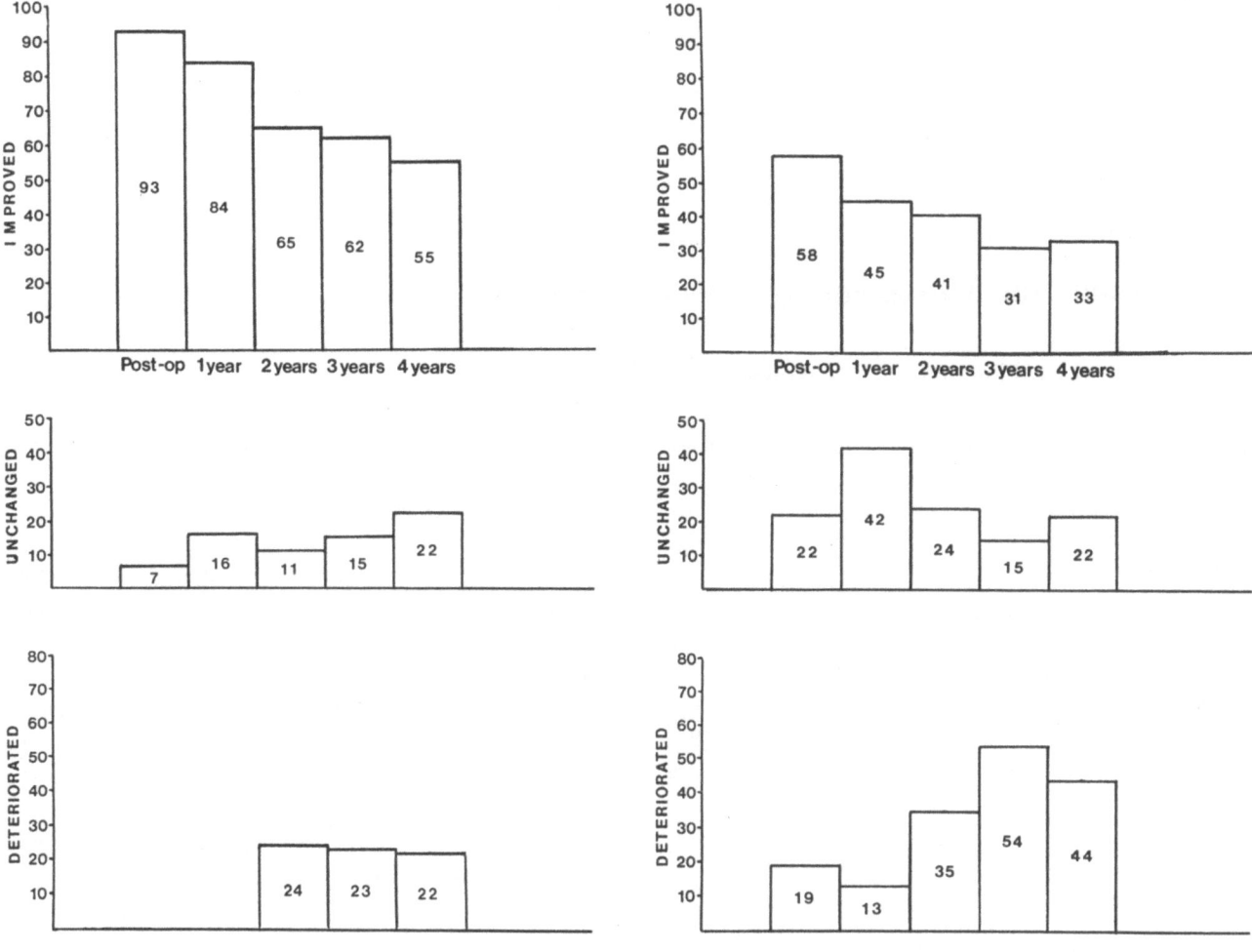

Fig. 1. Symptoms

Fig. 2. Disability

Parkinsonian and essential tremors (Fig. 3). This subgroup displayed immediate symptomatic improvement in 92% of cases, which was maintained at one year but fell to 67% by two years. An insufficient number of patients has been followed beyond that period for meaningful percentage figures to be given. 76% gained immediate functional improvement but over the next two years this figure fell to 47%.

Demyelination tremor (Fig. 4). An initial dramatic symptomatic improvement in 100% of cases declined rapidly to 50% by two years. Functional improvement was not nearly as marked being only 33% immediately and falling to 25% by two years.

Dystonia, athetosis and chorea (Fig. 5). These patients showed an 87% symptomatic improvement initially falling to 60% by two years. Again, functional improvement was less marked at 50% immediately, falling to 40% at two years.

Time course of improvement (Fig. 6). Application of the activities of Daily Living scale in the first 12 months postoperatively gave different absolute percentage figures from the above, reflecting the different sensitivity of the scales. A tendency for improvement in function to occur in the 12 months following lesion placement was noted in the dystonia—athetosis—chorea group.

Effect of age, disability and duration of symptoms. No clear correlation was noted between the duration of the patient's symptoms preoperatively and the likelihood of functional improvement postoperatively. There was a suggestion of a positive correlation between preoperative status and the likelihood of functional gain, the chances of improvement being better in the less disabled patients. Unfortunately, numbers were two small for the validity of this observation to be assured. There was, however, surprisingly, an appar-

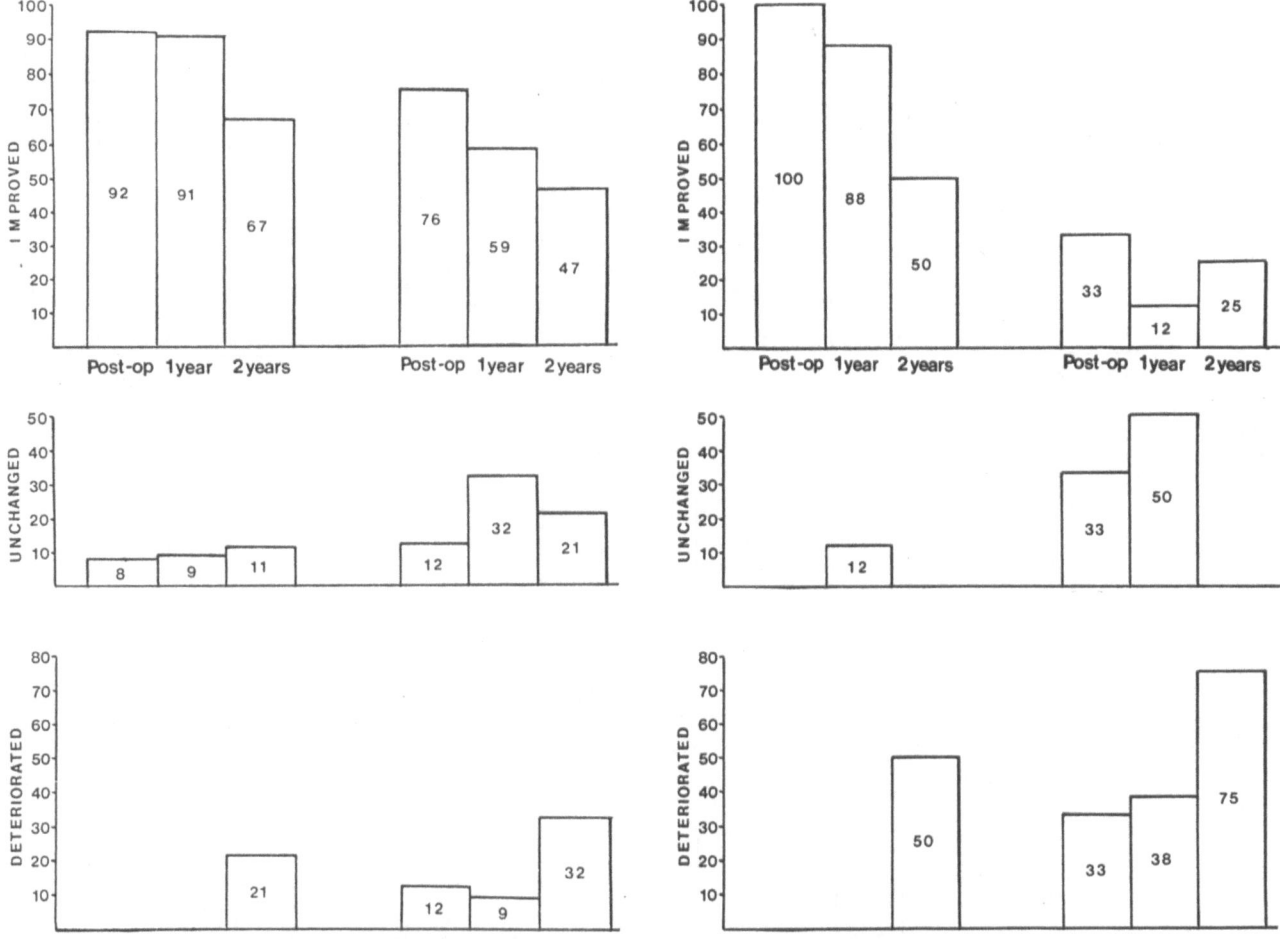

Fig. 3. Parkinsonian and essential tremor   Fig. 4. Demyelination

ently positive association with increasing age and a likelihood of functional improvement.

Discussion

Van Manen (1967)[5] selected 35 symptoms in patients with parkinsonism and evaluated each on a four-point scale and using factor analysis divided the clinical picture into four groups. Tremor and rigidity improved in 90% of patients after thalamotomy with less improvement in hypokinesia, gait and posture and little or no improvement in speech. His results were based on a three-month assessment in the belief that most recurrences occurred before that time and that thereafter the chance of recurrence was small. Riechert and Mundinger also devised on assessment scheme for patients with parkinsonism but they noted early recurrence of symptoms in the immediate postoperative period and further

recurrence months or years after the first operation. Using this scheme Mundinger (1969)[2] reported 500 subthalamotomies for parkinsonism. Improvements in posture were noted in 68% and in gait in 86%. It was also noted that the more severe the tremor the more effective the operation.

Our results suggest that for movement disorders as a whole the functional outcome is proportional to the degree of impairment existing preoperatively.

Shima, Nakagaki and Kitamura[4] examined the influence of age on outcome following stereotactic surgery for tremor in 22 patients followed-up for an average of 2.9 years. 45.8% of the 11 thalamotomies in older patients produced complete abolition of tremor, 45.5% resulted in a marked decrease and only minor improvement resulted in 9%. In 11 patients in the age group below 65 complete tremor abolition was secured

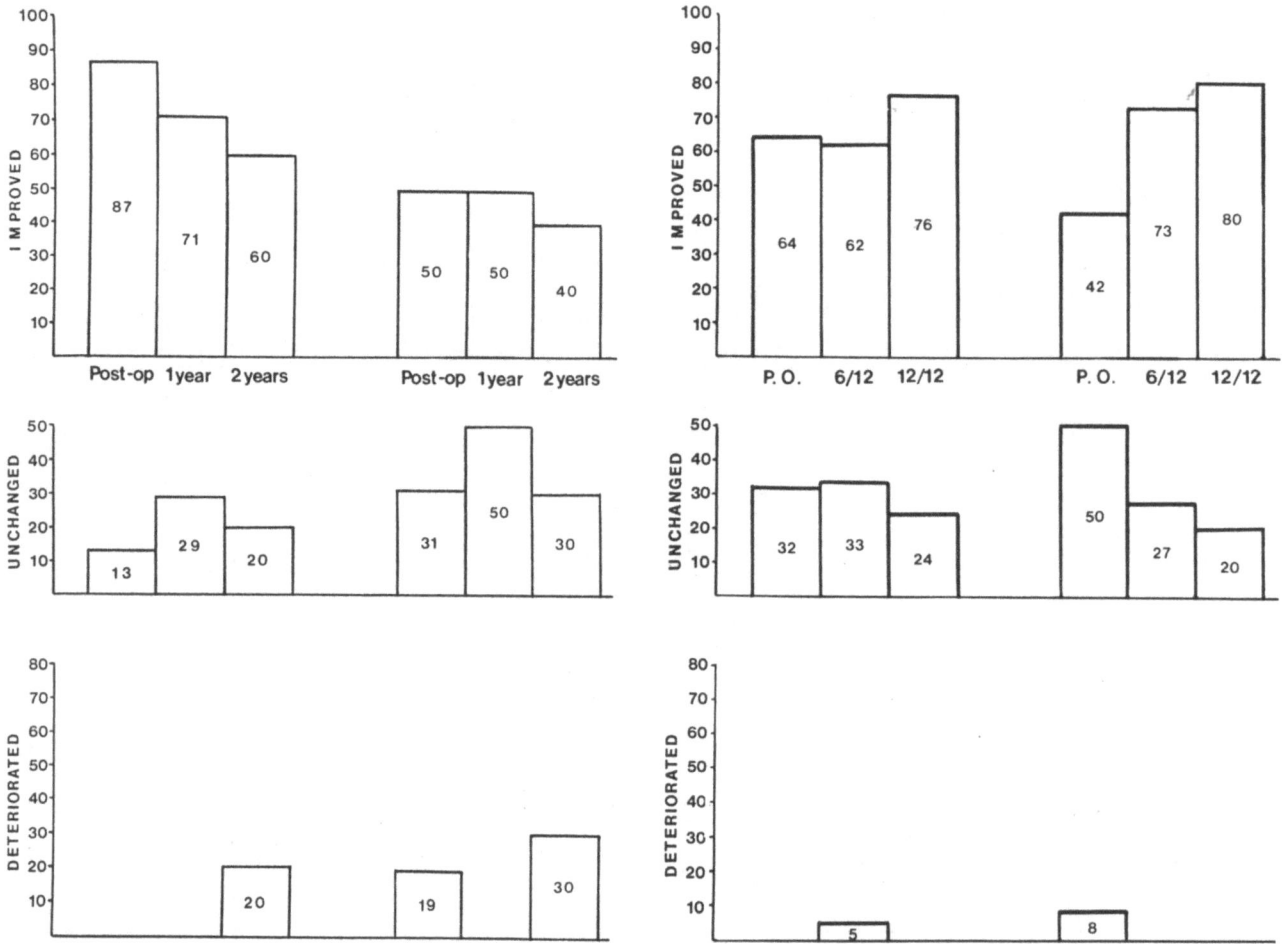

Fig. 5. Dystonia, athetosis, chorea

Fig. 6. Parkinsonian and essential tremors. Dystonia, athetosis, chorea

in 64.3%, a marked decrease in 21.4% and a minor improvement in 14.3%. The authors concluded that the long-term results showed that the procedure was almost equally effective in both aged and younger patients although slightly better for the younger. Our results for the total group of movement disorders suggest that functional improvement is somewhat better in the older age group although this may be because more refractory conditions, such as dystonia and cerebellar tremors, tend to occur in younger patients.

Hoehn and Yahr (1969)[1] examined the relationship of improvement of extremity tremor to general functional ability and reviewed 150 patients undergoing 215 operations with a mean follow-up of 4.4 years. After unilateral thalamotomy 48% had good to excellent relief of tremor, 35% mild to moderate relief and in 16% the tremor was the same or worse. Assessments of functional ability showed 31.5% had improvement,

44% were the same and 24.5% were worse. They also noted that functional ability continued to deteriorate despite maintenance of the improvement in tremor.

Ore (1969)[3] classified patients into four functional groups based on degree of dependance and ability to work. Examining the results of 609 chemothalamectomies he concluded that in the majority of cases the procedure only produced a one-step improvement in the functional capacity. In patients preoperatively in groups three and four improvement was brief and after six months to two years there was a slow regression which he attributed to progressive akinesia.

A variety of sometimes complex and time-consuming assessments have shown that symptomatic improvement following thalamotomy for Parkinson's disease is not accompanied by functional improvement. The simpler assessment scheme described here has confirmed these earlier findings not only for Parkinson's disease but also for other movement disorders. Up to 12 months postoperatively functional improvement is maintained or in the case of dystonia, increased. Thereafter there is a gradual loss of both symptomatic and functional improvement at a rate dependent upon the underlying pathology.

The World Health Organization classification[6] of disability in terms of impairment and its effect on performance (disability) and the results of such disability in fulfilling a role (handicap) attempts a universal classification but like many "complete" schemes the coding is very time consuming. We present our scheme as the minimum that should be followed in presenting results for stereotactic surgery for movement disorder. It is rapid and easy to use yet still able to give a very general evaluation. For an exploration of the effectiveness of particular lesions a more detailed survey is necessary but for comparison of the results of different treatments a generally accepted simple assessment scheme is needed. Perhaps the society should try to agree such a scheme which would aid us in assessing the effectiveness of our surgical procedures.

References

1. Hoehn MM, Yahr MD (1969) Evaluation of the long term results of surgical therapy. In: Gillingham FJ, Donaldson IML (eds) Third Symposium on Parkinson's Disease. Livingstone, Edinburgh, pp 274–280
2. Mundinger F (1969) Results of 500 subthalamotomies in the region of the zona incerta. In: Gillingham FJ, Donaldson IML (eds) Third Symposium on Parkinson's Disease. Livingstone, Edinburgh, pp 261–265
3. Ore GD (1969) Long term results of surgical treatment. In: Gillingham FJ, Donaldson IML (eds) Third Symposium on Parkinson's Disease. Livingstone, Edinburgh, pp 283–284
4. Shima F, Nakagaki H, Kitamura K (1983) Stereotactic treatment of tremor in aged patients. Appl Neurophysiol 316
5. Van Manen J (1967) Stereotactic methods and their applications in disorders of the motor system. Van Gorcum, Assen
6. World Health Organization (1980) International Classification of Impairments, Disabilities and Handicaps. WHO, Geneva

Acta Neurochirurgica, Suppl. 39, 66–69 (1987)

Stereotaxis and Abnormal Movements

F. Frank*, A. P. Fabrizi, R. Frank-Ricci, and G. Gaist

Division of Neurosurgery, Bellaria Hospital, Bologna, Italy

Summary

A series is presented of 106 patients with extrapyramidal syndromes treated by stereotaxy. The importance of preoperative screening to assure favorable results is stressed. Surgical contraindications are elderly patients with dementia and akinesia. Axial dystonia and spasmodic torticollis respond poorly to stereotaxy; and bilateral interventions should be avoided.

Keywords: Abnormal movements; stereotaxis; hyperkinetic syndromes.

Introduction

After a boom in the 60's, stereotactic surgery, performed at times without proper criteria, was almost abandoned; while there was a rebirth in the early 70's of medical therapy due to the discovery of a new drug, levodopa. When temporary efficacy and numerous side-effects arose due to L-dopa, in the late 70's there was a slow revival of surgery in abnormal movements, which presented with more refined techniques and more precise criteria for the patient selection. The authors present their experience in the surgery of abnormal movements, not because of large numbers; but, because the series stresses the precise criteria in patient selection and surgical technique, which influence the quality of the results.

Materials and Method

From 1978 to 1985 106 patients with abnormal movements underwent surgery. 8 patients were treated bilaterally. 75 cases presented with parkinsonism, and 31 cases had other forms of movement disorders (Table 1).

---

* F. Frank, M.D., Division of Neurosurgery, Bellaria Hospital, Bologna, Italy.

Table 1. *Abnormal Movements-Personal Surgical Experience: 106 patients (Bologna 1978–1985)*

75 patients	parkinsonism		
31 patients	other abnormal movements	choreoathetosis	12
		dystonia m. d.	6
		action myoclonus	4
		cerebellar tremor	6
		torticollis	3

Table 2. *Immediate Postoperative Results. Parkinsonism*

Unilateral interventions (67 patients)			Bilateral interventions (8 patients)		
Good	Fair	Poor	Good	Fair	Poor
63 (94%)	4* (7%)	—	2 (25%)	—	6** (75%)

* Relapse after 1 month.
** 1 patient deceased after surgery.

Results

The immediate postoperative results are presented in Tables 2 and 3. The surgical outcome was assessed: 1 good, a complete disappearance of the symptoms, without side-effects, and complete social reintegration; 2 fair, a partial or complete disappearance of the symptoms, with few side-effects, and partial social reintegration; 3 poor, when symptoms persisted, with grave side-effects, and no social reintegration. The best results were in hyperkinetic syndromes after unilateral intervention (82.7%). Good results were also obtained in rigid parkinsonism; but in dystonic syndromes, including spasmodic torticollis the results were poor. Bilateral interventions, even if performed after long intervals from the first and second operation (minimum time elapsed: 6 months), gave a high percentage (75%) of poor results. The symptoms disappeared, but the patients were burdened by grave side-effects (dementia, incontinence and aphasia) (Tables 2 and 3).

Table 4 presents the long-term follow-up of the patients with good initial results. The percentage of recurrence is low (4.9%), and the good results remained stable.

The targets selected, according to the literature[3, 5, 8, 9], are presented in Table 5.

The Mundinger-Riechert stereotactic frame* was used, and the target coordinates were calculated after radioopaque ventriculography (using 3 cm^3 of iopamiro**).

Coagulations were performed by a monopolar electrode with an extractable lateral probe*. To avoid recurrence of the symptoms, 4–5 coagulations at 70 °C for 30 seconds each were made, giving at 0.33 cm^3 lesion (seen on CT image controls). The coagulations were made only after precise neurophysiological responses were obtained with electrical stimulation.

* Fischer Inc., Freiburg i. Br., Federal Republic of Germany.
** Bracco Pharmaceutical Co., Milan, Italy.

Table 3. *Immediate Postoperative Results. Other Abnormal Movements (31 Patients)*

	Number of patients	Results		
		Good	Fair	Poor
Choreoathetosis	12	7	4	1
Dystonia m. d.	6	1	2	3
Action myoclonus	4	4	—	—
Cerebellar tremor	6	6	—	—
Torticollis	3	—	—	3
Total		18 (58.1%)	6 (19.4%)	7 (22.5%)

Table 4. *Long-Term Follow-Up (81 Patients). Only Patients with Good Initial Results (6 Months–7 Years)*

	Recurrence	Contralateral symptoms
63 patients parkinsonism	3 (4.8%)	21 (33.3%)
18 patients other abnormal movements	1 (5.5%)	1 (5.5%)

Table 5. *Targets Selected in Relationship to Symptoms*

Hyperkinetic	resting tremor intentional (or cerebellar) tremor action myoclonus	VIM
Rigidity Choreoathetosis Dystonia musculorum deformans	zona incerta	
Torticollis		VOI

Discussion

Parkinsonism is a disease with an incidence of 20–30 cases per 1,000 individuals of age 50–70 years[1], presenting a grave social burden, and where a correct therapy is imperative. Less frequent are the dystonic syndromes, both major and minor (torticollis) (1 patient per 100,000)[1], and the same is true for choreoathetosis. The cerebellar hyperkinetic syndromes (posttraumatic or multiple sclerotic) can not be statistically evaluated for occurrence. It is certain however that very few of these disorders can be treated by neurosurgical means (due to spontaneous resolution, grave progression of the disease, etc).

To obtain good results, an accurate screening of the patients is mandatory. The disease must be present for at least 2 years, and show signs of deterioration, even when under adequate medical therapy. The patient must also be less than 70 years old. Surgery is contraindicated in all cases presenting dementia or rapid psychological decay (severe cortical atrophy on CT). Also contraindicated are patients with marked akinesia.

Other authors[3, 5, 8, 10] have pointed out that surgery worsens preexisting psychic disorders, precipitating situations in precarious psychic equilibrium, and noticeably worsening the kinesia.

Others[3, 7] have stated that bilateral interventions in extrapyramidal disorders do not have great risk; whereas our limited experience with bilateral procedures, shows that even if the interventions are performed 6 months apart, numerous and grave side-effects appear. Although the initial symptoms are overcome, for example tremor, 75% of the cases are burdened with grave psychic disorders, aphasia and incontinence. These disturbances have caused us to abandon bilateral procedures.

Even our experiences with major and minor dystonia has shown little success. The dystonic manifestations are frequently bilateral, and involve the axial musculature. Therefore, a unilateral intervention is only palliative, and unsatisfactory. Contrary to what was pointed out by Bertrand[2], spasmotic torticollis is a minor dystonia that does not respond to surgery.

Choreoathetosis[4, 5], if lateralized and in young patients that do not have either Wilson's or Huntington's disease, responds satisfactorily to stereotactic surgery. These patients must be carefully screened with a good clinical history, specific laboratory examinations, and psychological evaluation for a suitable outcome.

As confirmed by many others[3, 5, 8, 10], stereotactic lesions should never be performed without obtaining suitable response during electrical stimulation of elective targets. For the VIM thalamic nucleus, the elective target for hyperkinetic Parkinson and cerebellar syndromes, it is necessary, according to Tasker[8], to obtain, during stimulation, a slight vibratory sensation of the lips and tongue, and accompanied by arrest of tremor at frequencies of 30–40 Hz. At times, an acceptable stimulatory response may alter the rhythm and amplitude of the tremor. If the target is the zona incerta, elective for choreoathetosis, along with a slowing down of the choreic movements and a reduction of the amplitude at low frequencies (20–30 Hz), one can observe a rhythmic reduction of pupillary diameter, due to stimulation of sympathetic fibers, that cross the zona incerta and H and H_1 Forel fields[7]. When the target is the VOI thalamic nucleus, the stimulation may produce either no response or ocular nystagmus. This target, according to Bertrand[2, 6] is elective for axial dystonia and spasmodic torticollis.

Conclusions

On the whole, our experience in the treatment of abnormal movements is good. Along with an accurate screening of the surgical candidates, precise neurophysiological information prior to the stereotaxic lesion is mandatory. Although treatment can produce spectacular results we have had an appreciable morbidity (10.4%) and mortality (0.9%).

References

1. Barbeau A (1976) The nonsurgical treatment of Parkinson's disease: a personal view. From: Current controversies in neurosurgery. Saunders, Philadelphia, Pa., pp 419–427
2. Bertrand C (1976) The treatment of spasmodic torticollis with particular reference to thalamotomy. From: Current controversies in neurosurgery. Saunders, Philadelphia, Pa., pp 445–460
3. Bertrand C, Martinez SN, Hardy J, Molina-Negro P, Velasco F (1973) Stereotactic surgery for parkinsonism. In: Progress in neurological surgery. S Karger, Basel, pp 79–112
4. Cooper IS (1982) Dystonia. From: Stereotaxy of the human brain. G Thieme, Stuttgart New York, pp 545–561
5. Guiot G, Derome P (1982) The principles of stereotaxic thalamotomy. In: Correlative neurosurgery. Ch C Thomas, Springfield, Ill, pp 481–505
6. Hassler R, Dieckman G (1982) Stereotaxic treatment for spasmodic torticollis. From: Stereotaxy of the human brain. G Thieme, Stuttgart New York, pp 522–531

7. Hassler R, Riechert T, Mundinger F, Umbach W, Gangelberger JA (1960) Physiological observations in stereotaxic operations in extrapyramidal motor disturbances. Brain 83: 337–350

8. Tasker RR (1976) Surgery for Parkinson's disease. From: Current controversies in neurosurgery. Saunders, Philadelphia, Pa. pp 411–418

9. Tasker RR, Organ LW, Hawrylyshyn PA (1982) Extrapyramidal Thalamus. From: The thalamus and midbrain of man. ChC Thomas, Springfield, Ill, pp 275–323

10. Walker AE (1982) Stereotaxic surgery for tremor. From: Stereotaxy of the human brain. G Thieme, Stuttgart New York, pp 515–521

Acta Neurochirurgica, Suppl. 39, 70–72 (1987)

Thalamotomy for Tremor After a Vascular Brain Stem Lesion

J. van Manen* and **J. D. Speelman**

Academisch Medisch Centrum, Amsterdam, The Netherlands

Summary

Three patients with a postural tremor caused by a thrombosis in the area of the thalamo-geniculate arteries are presented. The tremor disappeared completely after ventrolateral thalamotomy. The result persisted also after a follow up of more than 5 years. The location of the lesion causing the tremor is discussed.

Keywords: Cerebrovascular brain stem; tremor; stereotactic thalamotomy.

Introduction

Ischemic vascular lesions in the mesodiencephalic area of the brain may lead to a variety of unvoluntary movement disorders as myoclonias, choreas, tremor and athetosis. These symptoms arise in various combinations with hemisensory disturbances, pain, hemiparesis, spasticity, hemianopia and psychiatric disturbances as delusions and hallucinations, confusion or loss of consciousness. Migraine and the use of oral contraceptives is frequently found as a risk factor in the history of young adults suffering from cerebral thrombosis[1]. The infarction is quite frequently localized in the area supplied by the posterior cerebral artery in these patients. We saw three patients in whom a postural and intention tremor developed some weeks to months after the stroke. The other symptoms had subsided for the greater part. Poirier[2] localized the lesions causing a postural tremor in the ventromedial tegmentum of the mesencephalon. This localization did not seem likely in at least two of our patients.

One can distinguish three areas of thalamic vascularization with a different symptomatology in case of thrombosis. The anterior part of the thalamus is supplied from the posterior communicating artery by the polar thalamic artery penetrating the thalamus along the bundle of Vicq d'Azyr. Bilateral thrombosis of this vessel leads to a softening of the anterior thalamus up to the VOA nucleus and causes a Korsakow symptomatology. This was recently described by von Cramon *et al.*[3]. The top of the basilar artery and the basilar communicating artery or mesencephalic artery supplies the medial thalamus and mesencephalon by a number of paramedian thalamic and mesencephalic perforating arteries. Thrombosis in this region leads to oculomotor paresis, peduncular hallucinosis, disturbances of consciousness, myoclonias and myorhythmias[4]. The softening may involve the grey substance around the aqueduct, the oculomotor nuclei, the brachium conjuctivum, the red nucleus and the medial thalamus up to the centromedian nucleus.

The posterolateral thalamus is supplied by branches of the posterior cerebral artery originating behind the junction with the posterior communicating artery. These are the thalamo-geniculate and posterior choroidal medial and lateral arteries. They supply the ventrolateral thalamus, the ventroposterior thalamus and pulvinar, and medially reach the lateral part of the centromedian nucleus. Also a part of the internal capsule and the field of Wernicke are supplied[5], the region where the thalamic syndrome originates. Motor symptoms, choreoathetosis, as well as tremors are described. Sigwald[6] described a patient with tremor and ataxia caused by a lesion in this vascular supply area. He assumed the tremor to be caused by a lesion in the lateral part of the centromedian nucleus and the

---

* J. van Manen, M.D., Academisch Medisch Centrum, Department of Neurology, Meibergdreef 9, NL-1105 AZ Amsterdam, The Netherlands.

dentato-thalamic fibers medially and behind the VOP and VIM nuclei. Lapresle[7] summarized the literature about this subject in 1973.

Case Histories

Patient A: a 35-year-old woman with a history of migraine attacks, heavy smoking and the use of oral contraceptives. In December 1978 a right-sided transient paresis and hemianopia developed. Followed four months later by a left-sided transient paresis, hemianopia, ataxia and disturbances of proprioception with transient clouding of consciousness. During recovery she had a psychoorganic syndrome for a time. One month after the second stroke a left-sided postural and intention tremor developed. The CT scan showed an area of softening in the territory of the left and right posterior cerebral artery and the basilar artery (Fig. 1). One year after the first stroke a right-sided ventrolateral thalamotomy was performed which abolished the tremor completely. The postoperative course was characterized by speech and swallowing difficulties, apathy, and a worsening of the left-sided paresis. These symptoms recovered well apart from a transient psychotic period. After a long period of rehabilitation the patient again became independent for the activities of daily life. She returned to her home where she is now living alone, independent for the greater part, though with markedly limited visual fields. There is no tremor and the left hand is quite usable.

hemiparesis; some ataxia persisted. After one year there was no tremor, but the hand was slightly clumsy and slow. The tendon reflexes were somewhat enhanced on the right side. After four years there was still no tremor but some dystonic posturing of the hand, proprioception was slightly diminished. 11 years postoperatively: no tremor, dystonia agravated, functionally deteriorated.

Patient C: a man of 21 years old suffered from a right-sided hemiparesis, hemihypesthesia, hemianopia and a short period of clouding of consciousness in 1972. Hemiparesis recovered, some ataxia persisted as did the hemianopia. After some weeks a right-sided tremor developed, which persisted despite medical treatment. Six years later we found a postural and intention tremor of the right hand, slight hypesthesia and minimal paresis. Hemianopia on the right side. The CT scan showed a softening of the caudal part of the left thalamus (Fig. 2). A left thalamotomy was performed and the tremor disappeared completely. Ten months later there were slight oscillations of the right arm during intentional movements, writing was clumsy as were fine movements of the hand. Gross movements were performed well with a normal force. Hemianopia proved unchanged. This situation persisted for 7 years postoperatively.

The stereotactic target in the three patients was chosen 15 mm behind the foramen of Monro at the level of the commissural line and 14–15 mm lateral from the median plane of the brain.

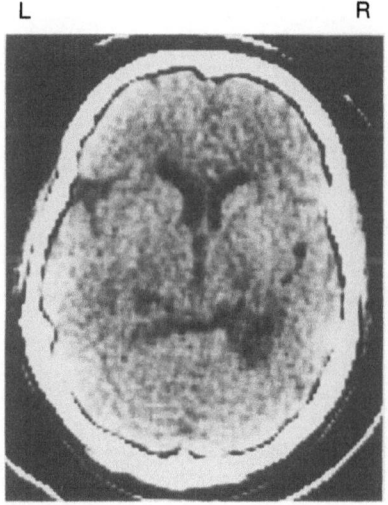

L   R

Fig. 1

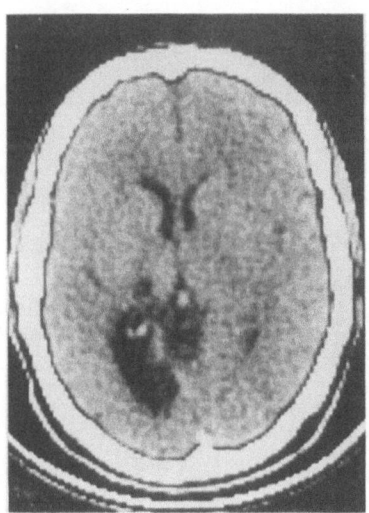

Fig. 2

Patient B: a woman of 24 years old, with a history of migraine and the use of oral contraceptives. In December 1972 after a short period of right-sided visual disturbances and unsteady gait, a transient right-sided paresis developed with tingling of the right arm. Right-sided hemianopia, hemiparesis, hemihypesthesia and a Babinski sign were found. After recovery 6 weeks later a postural and intention tremor of the right arm developed. This tremor persisted after medical treatment and slight hypesthesia persisted. A diagnosis of thrombosis in the proximal branches of the posterior cerebral artery was made. Thirteen months after the stroke a left ventrolateral thalamotomy was performed. Tremor was abolished at the cost of a slight

Discussion

In all three patients a softening in the posterior and posterolateral thalamus caused by a thrombosis of the thalamo-geniculate arteries was the most likely diagnosis, because of the hemianopia, hemisensory disturbances and transient hemiparesis.

Postural and intention tremor and ataxia may occur but infrequently in these kind of cases. The vascular lesion must reach quite medially and destroy the lateral part of the centromedian nucleus and adjacent dentato-thalamic radiation sparing of the VOP and VIM nuclei. This lesion, described by Sigwald and Lapresle, seems more probable in our patients than a lesion in the ventromedial tegmentum of the mesencephalon. (In patient A however it is possible that the mesencephalon was also involved.) It is astonishing that this region is not frequently damaged during stereotactic thalamotomy for tremor when the lesion is placed far medially, or during thalamotomy for intractable pain when lesions in the VPM and centro median nuclei are performed. Castaigne[8] described such a case.

These three patients confirm that VOP/VIM thalamotomy can abolish or greatly improve postural and intention tremor, when the lesion causing the tremor is not in the mesencephalon or dentate nucleus but at thalamic level.

References

1. Spaccavento LJ, Solomon GD (1984) Migraine as an etiology of stroke in young adults. Headache 24: 19–22
2. Poirier LJ, Bouvier G, Bedard P, Boucher R, Larochelle L, Olivier A, Singh P (1969) Essai sur les circuits neuronaux impliqués dans le tremblement postural et l'hypokinesie. Rev Neurol 120: 15–40
3. Cramon DY von, Hebel N, Schuri U (1985) A contribution to the anatomical basis of thalamic amnesia. Brain 108: 993–1008
4. Graff-Radford NR, Damasio H, Yamada T, Eslinger PJ, Damasio AR (1985) Nonhaemorrhagic thalamic infarction: clinical, neuropsychological and electrophysiological findings in four anatomical groups defined by computerized tomography. Brain 108: 485–516
5. Percheron G (1973) The anatomy of the arterial supply of the human thalamus and its use for the interpretation of the thalamic vascular pathology. Z Neurol 205: 1–13
6. Sigwald MM, Monnier M (1936) Syndrome thalamo-hypothalamique avec hemitremblement (Ramollissement du territoire arterial thalamo-perforé). Rev Neurol 66: 616–631
7. Lapresle J, Haguenau M (1973) Anatomico-clinical correlation in focal thalamic lesions. Z Neurol 205: 29–46
8. Castaigne P, Cambier J, Escourolle R (1968) Les myoclonies d'intention et d'action. Rev Neurol 119: 107–120

Acta Neurochirurgica, Suppl. 39, 73–76 (1987)

Stereotactic Neurosurgery in the Treatment of Tremor

G. Broggi*, C. Giorgi, and **D. Servello**

Department of Neurosurgery, Istituto Neurologico "C. Besta", Milano, Italy

Summary

The results of stereotactical thalamotomy in 40 adult patients suffering from tremor of different etiology are presented. A combination of lesions of VOA-VOP-ZI seems to be optimal. Early results have been excellent in 63%, good in 23%, fair in 6% and poor in 8%. Long-term follow-up showed a negative shift with 40% excellent, 6% good and 54% poor results.

In Parkinson disease with predominant tremor relief of this invalidating symptom can be achieved. But L-dopa therapy must be continued and surgical treatment does not stop the general disease progression.

Keywords: Tremor; stereotactic thalamotomy; computer graphics; neurofunctional data base.

Introduction

For years surgery was the only treatment possible for Parkinson disease[4]. The discovery that dopamine deficit was the cause of the disease and the introduction of L-dopa, meant that the indication for surgery all but disappeared[1]. Nevertheless, the problem of tremor has not been solved, so that, in some neurosurgical centers stereotactic thalamotomy is still proposed as definitive therapy for tremor[7, 13, 19]. We present our experience in the last ten years.

Material and Method

From 1975 to 1985 at Istituto Neurologico C. Besta, 40 adult patients suffering from tremor of different etiology, underwent stereotactic thalamotomy. Of these 28 suffered from Parkinson disease a predominant tremor. In four cases surgery was also performed contralaterally after more than a year. In five patients the operation has been repeated again on the same side: on the whole 36 thalamotomies have been performed.

---

* G. Broggi, M.D., Department of Neurosurgery, Istituto Neurologico "C. Besta", Milano, Italy.

There were 15 males and 13 females, aged from 45 to 72 years, the average age for males being 58.5 and for females 58 (Table 1).

Table 1

Male	24 (15 PK)
Female	16 (13 PK)
Total	40 patients
Bilateral procedure	4 cases
Repeated procedure	5 cases
Age ranging from	24 to 73 years
Mean	56.2 (male 55.1, female 57.8)
Follow-up ranging from	2 to 9 years

Table 2

Results	Immediate	Short-term (12 months)	Long-term
Excellent	31 (63.2%)	23 (57.5%)	14 (40.0%)
Good	11 (22.5%)	6 (15.0%)	2 (5.7%)
Fair	3 (6.1%)	0	0
Poor	4 (8.2%)	11 (27.5%)	19 (54.3%)
Total	49	40	35

The symptomatology appeared from 1 to 15 years prior to the surgery, with an average of 7.5 years. From 1977 all the patients have had CT with or without contrast enhancement.

The surgical technique adopted was the stereotactic one using the Riechert-Mundiger apparatus[10]. In the patients operated upon in 1975–76 air encephalography by suboccipital puncture was performed immediately before surgery. Subsequently, using the data supplied by CT scan, volumetric reconstruction of the ventricular system, and the help of bone landmarks[20], stereotactic ventriculography with positive watersoluble contrast medium (Amipaque®, Jopamiro®) has been used. After 1985 also somatosensory-evoked potentials have been used to locate thalamic structures.

The following targets have been chosen: a) the nuclei ventralis oralis anterior (VOA) and ventralis oralis posterior (VOP) of the thalamus; b) the zona incerta; c) the nucleus ventralis intermedius; d) the fasciculus dentato thalamicus. The exact dimension of the thalamus, its relations with the internal capsule, and the exact positioning of the electrode in the nuclei were checked by radiological and neurophysiological assessment. The lesion was produced by radiofrequency current heating (Radionics Inc., Burlington, Massachusetts) to 70–75 °C with a straight thermocouple electrode or a side-protruding electrode (Fischer Co., Freiburg, Germany).

The theoretical volume of the lesion calculated on the atlas[18] was always corrected following the intraoperative clinical and neurophysiological controls. The aim was to improve the results with small (1 to 2 mm in diameter) adjacent lesions instead of concentrically enlarging the radiofrequency lesions since the thalamic nuclei do not have a spherical shape, but have a more complex spatial configuration[18]. Computed tomographic scanning was also done in 15 cases pre- and postoperatively to check the dimensions of the lesion[3, 15].

The anatomo-physiological basis of these targets is as follows. Lateral thalamotomy was chosen when the predominant symptom was hyperkinesia (large or small excursion) or tremor. The nuclei of the lateral thalamus act as an interface between the pallidal and cerebellar inputs[17] and so constitute the principal centre of sensorimotor integration. The VOA nucleus receives fibers from the medial and lateral globus pallidus, outputs of the striatal system, through the ansa lenticularis. These fibers[13] having crossed the internal capsule and running medially along the superior contour of the subthalamic nucleus, enter Forel's fields, demarcating the zona incerta. The cerebellar afferents to the thalamus originate from the contralateral dentate nucleus: the fibers, having passed into the brachium conjunctivum, cross the red nucleus to end in the VOP. The neurones of these nuclei, distinguishable only in humans, receive a monosynaptic input converging from the globus pallidus and from the dentate nucleus[5, 6]. These nuclei were chosen as surgical targets because their destruction induces an alteration of the input to the motor cortex areas, 6α and 4γ, which ends by reducing the small excursion hyperkinesias[12]. The lesion made in the subthalamus further diminishes the pallidal input to the thalamus and thus inhibits the large excursion hyperkinesias. Lesions of the VOP was chosen also for the correction of intention and postural tremor because this nucleus, via the 4γ cortex, influences the spinal motoneurons on which the striatoreticulospinal, nigrospinal, and rubrospinal inhibitory influences have been diminished by the disease[11]. The nucleus ventralis intermedius was chosen as alternative target in case

Table 3

Targets	Excellent	Good	Fair	Poor	Total
VOA, VOP, ZI, VIM	10	2	0	2	14
VOA, VOP, ZI, FDTH	1	1	0	0	2
VOA, VOP, VIM, FDTH	0	1	0	0	1
VOA, VOP, ZI	6	2	1	0	9
VOA, VOP	5	1	0	0	6
VOP, ZI, VIM	1	0	0	0	1
VOP, ZI, FDTH	2	0	0	0	2
VOP, ZI	5	0	1	2	8
ZI	1	3	0	0	4
VOP	0	1	1	0	2
Total	31 (63.3%)	11 (22.5%)	3 (6.1%)	4 (8.1%)	49

Table 4

Targets	Excellent	Good	Fair	Poor	Total
VOA, VOP, ZI, VIM	8	2	0	0	10
VOA, VOP, ZI, FDTH	1	1	0	0	2
VOA, VOP, VIM, FDTH	0	1	0	0	1
VOA, VOP, ZI	5	2	1	0	8
VOA, VOP	3	0	0	0	3
VOP, ZI, VIM	1	0	0	0	1
VOP, ZI, FDTH	2	0	0	0	2
VOP, ZI	3	1	0	2	6
ZI	1	0	0	0	1
VOP	0	1	1	0	2
Total	24 (66.6%)	8 (22.2%)	2 (5.6%)	2 (5.6%)	36

of distal tremor of upper arm[14] and fasciculus dentato-thalamicus was chosen when a "cerebellar" component of the tremor was also present.

Results

The immediate results have been very satisfying (Table 2). Results have been judged optimal with complete disappearance of tremor, good when there has been a remission up to 70%, fair for remission of 70 to 50%, and poor with lower than 50%. The mortality has been nil. Morbidity consisted in one case of thalamic hematoma which caused a mild but handicapping hemiparesis and a case of neglect[16].

In Table 3 the relevance of different nuclei lesions is matched with the cases result, whilst in Table 4 the analysis concerns the parkinsonism patients group. The

optimal combination resulted from VOA-VOP-ZI lesions.

The position of the probe with respect to different thalamic nuclei has been plotted using a program developed at our department[8, 9], which stores neurophysiological observations collected during successive procedures, within an anatomical frame of reference.

Although the process of "normalization" of individual cerebral anatomies (allowed by the program on the basic of CT and MRI data) could not be satisfactorily carried out because all patients were operated on the basis of ventriculography only, the images seem to confirm that best results were achieved in cases in which a combination of targets was used (Fig. 1).

Fig. 1. This sketch outlines the output of a color monitor, not suitable for black and white reproduction. The picture refers to left side results only. X, Y and Z axes represent the atlas frame of reference, X lying in the sagittal plane, pointing anteriorly, Z and Y being vertical and lateral, respectively. The continuous outlines represent the borders of ZI and VOA-VOP nuclei of the thalamus, derived from the horizontal sections of the Schaltenbrand-Wahren atlas, digitized and processed through a 3-d computer program (see reference in the text). Graphic symbols show the position of probes at different operations, plotted after "normalization". Squares and circles represent good results at short- and long-term follow-up; solid circles symbolize poor long-term results, and seem to be located outside the borders of the nuclear complex. They frequently coincide with sites where prolonged coagulations were performed to stop tremor (stars). Solid triangles represent sites where side effects were observed

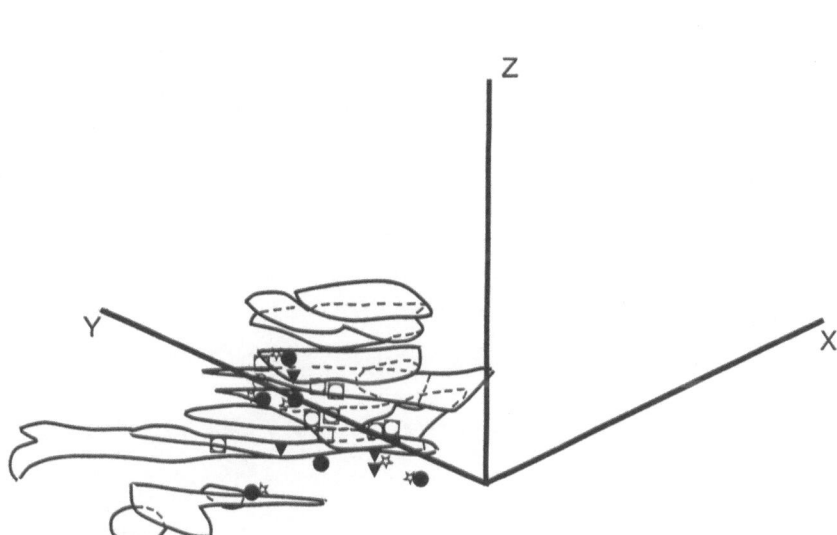

Conclusions

From these results it appears clear that since a drug with a specific action against tremor has not yet been found, the surgical therapy of stereotactic ventrolateral thalamotomy and subthalamotomy, gives reliable results small risk surgery is fundamental in the treatment of one of the major invalidating symptoms of the Parkinson disease.

The side-effects that do appear are generally temporary[2], while permanent relief of tremor can be achieved. Alternative medical therapy based on L-dopa must be continued and the timing of surgery is indi-

vidual for each patients. Surgical treatment does not stop the disease progression.

References

1. Birkmayer W, Hornykiewicz O (1961) Der 3,4-Dioxyphenylalanin (= L-Dopa-)Effekt bei der Parkinson-Akinese. Wien Klin Wschr 73: 787–788
2. Broggi G, Angelini L, Giorgi C (1980) Neurological and psychological side effects after stereotactic thalamotomy in patients with cerebral palsy. Neurosurg 7: 127–134
3. Cala LA, Mastaglia FL, Vaugham RJ (1976) Localization of stereotactic radiofrequency thalamic lesions by computerized axial tomography. Lancet 2: 1133–1134

4. Cooper IS (1961) Parkinsonism. Its medical and surgical therapy. Ch C Thomas, Springfield, Ill

5. Desiraju T, Broggi G, Prelevic S, Santini M, Purpura DP (1969) Inhibitory synaptic pathways linking specific and nonspecific thalamic nuclei. Brain Res 15: 542–543

6. Desiraju T, Purpura DP (1969) Synaptic convergence of cerebellar and lenticular projections to the thalamus. Brain Res 15: 544–547

7. Gillingham FJ, Donaldson IMC (eds) (1969) Third symposium on Parkinson's disease. Livingstone Publ, Edinburgh, London

8. Giorgi C, Cerchiari U, Broggi G, Birk P, Struppler A (1985) Digital image processing to handle neuroanatomical and neurophysiological data. Appl Neurophysiol 48: 30–33

9. Giorgi C, Cerchiari U, Broggi G, Contardi N, Birk P, Struppler A, An intraoperative interactive method to monitor stereotactic functional procedures (in this volume)

10. Hassler R (1959) Stereotactic brain surgery for extrapyramidal motor disturbances. In: Schaltenbrandt G, Bailey P (eds) Einführung in die stereotaktischen Operationen. G Thieme, Stuttgart, pp 472–488

11. Hassler R (1967) Funktionelle Neuroanatomie und Psychiatrie. In: Gruhle HW *et al* (eds) Psychiatrie der Gegenwart. Springer, Berlin Heidelberg New York, pp 152–285

12. Hassler R (1972) Physiopathology of rigidity. In: Siegfried J (ed) Parkinson's disease. H Huber, Bern, pp 20–45

13. Hassler R, Mundinger F, Riechert T (1979) Stereotaxis in Parkinson syndrome. Springer, Berlin Heidelberg New York, pp 1–315

14. Narabayashi H, Ohye C (1978) Parkinsonian tremor and nucleus ventralis intermedius (V.im) of the human thalamus. In: Desmedt JE (ed) Progress in clinical neurophysiology. 5: Physiological tremor, pathological tremors and clonus. Karger, Basel, pp 165–172

15. Passerini A, Broggi G, Giorgi C, Savoiardo M (1962) CT studies in patients operated with stereotaxic thalamotomies. Neuroradiology 16: 364–367

16. Perani D, Nardocci N, Broggi G (1982) Neglect after unilateral thalamotomy. Ital J Neurol Sci 1: 61–64

17. Purpura DP (1970) Operations and process in thalamic and synaptically related subsystems. In: Schmitt FO (ed) The Neurosciences: The second study program. Rockefeller University Press, New York, 1970

18. Schaltenbrand G, Wahren W (1977) Atlas for stereotaxy of the human brain. 2nd ed. G Thieme, Stuttgart

19. Siegfried J (ed) (1972) Parkinson's disease. H Huber, Bern Stuttgart Wien, pp 1–219

20. Siegfried J, Brandli-Graber S (1981) Repéragr radiologique simple du trou de Monro sur les radiographies craniennes à vide. Neurochirurgie 27: 146–148

Acta Neurochirurgica, Suppl. 39, 77–79 (1987)

Severity in Movement Disorders: a Quantitative Approach

R. T. Shann[*, 1], **R. H. Lye**[2], and **G. W. Rogers**[1]

[1] Department of Medical Illustration and [2] Department of Neurosurgery, University of Manchester, U.K.

Summary

A method for quantitative assessment of severity of movement disorders using video recordings and computer analysis is presented. It allows quantified comparison of pre- and post-treatment conditions.

Keywords: Torticollis; intention tremor; TV

Introduction

The assessment of severity of movement disorders is notoriously subjective. In most instances a trained observer is needed in order to minimize the inconsistencies of the clinical examination[1]. There is therefore a need for quantification of the severity of a disorder so that a more objective evaluation of the response to different therapeutic regimes may be obtained. We describe a method whereby pre- and post-treatment video recordings of patients suffering from movement disorders may be quantified for the purpose of direct comparison. We present our preliminary results following analyses of simulation of the movement disorder of spasmodic torticollis and of the intention tremor of patients suffering from multiple sclerosis.

Method

A light emitting diode (LED) attached to the subject was used as a reference marker. When assessing patients suffering from spasmodic torticollis, the LED was rigidly attached to a hat. Mechanical amplification was obtained by fixing the LED to the hat by a 200 mm length of wire. An overhead TV camera tracked the LED, the flashing of the latter being synchronized at the start of each scan of the TV field (*i.e.*, every 0.02 seconds) in order to avoid blurring of the image.

In patients with intention tremor the LED was fixed to a stick held in the hand. The patients were asked to touch a target at arm's length with the LED. The video camera signal was converted into the coordinates of the LED "spot" within the field of view of the camera using a small circuit designed and built in our hospital workshop[**] interposed between the camera and computer[*]. Two minutes of continuous video recording was obtainable at any one time. Programs were written to enable calibration, analysis of data and also to provide an "preview" facility so that the computer could replay in real time the movement observed. This enabled adjustments to the camera aperture and triggering level of the detection circuit as required. The accuracy of the system was verified using a model of damped simple harmonic motion comprising an oscillating flexible pointer fixed at one end and with the LED at the free tip. Least squares curve fitting techniques demonstrated a high degree of accuracy (up to derivatives of the fourth order) when measuring transverse motion, although the accuracy was marginally reduced when measuring the small range of longitudinal motion.

Results

Many parameters may be measured for a given movement. Besides the various components of velocity and acceleration, there are frequency spectra, momenta, angular speeds and impulses. The "sharpness" and frequency distribution of the peaks of any of these parameters can be calculated. In order to establish which parameters were relevant we wrote a general suite of programs. These enabled samples of data to be treated to a wide range of statistical analyses. Graphical displays assisted in identifying an appropriate characterization.

* R. T. Shann, D. Phil., Department of Medical Illustration, University of Manchester, U.K.

* BBC Model B Microcomputer manufactured by Acorn Ltd., Cambridge CB1 4JN.
** Details available on request.

To test our system, several neurosurgeons familiar with the movement disorder of torticollis mimicked the condition with varying degrees of severity. Simultaneous recordings on both the computer and videotape were then available for comparison between the clinical judgement of severity and the computer's analysis of the same motion. One of us (R. H. Lye) rated the video recordings on a 0–4 scale as described by Burke *et al.*[1] according to the severity of the movement disorder. Such assessment would correspond to that of the "untrained examiner" in that study. Typical results are shown in Figs. 1 and 2. On our small samples there was apparently better correlation between the mean angulation of the head and the Marsden-Fahn severity factor than when one considered mean angular speed of motion.

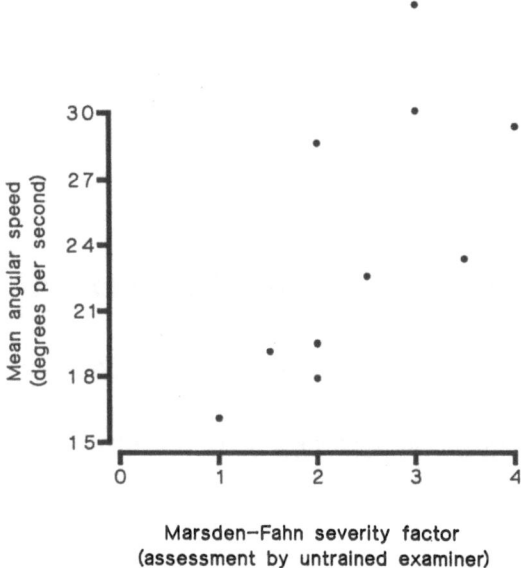

Fig. 2. Comparison between mean angular speed of the head and the Marsden-Fahn severity score. Untrained examiner. Simulation of torticollis

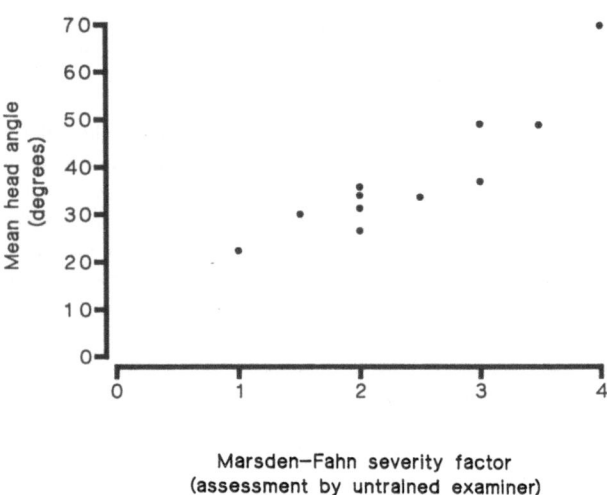

Fig. 1. Comparison between mean head angle and the Marsden-Fahn[1] severity factor as assessed by an untrained examiner. Simulation of torticollis

In patients demonstrating intention tremor secondary to multiple sclerosis, one parameter of interest is the average speed in the direction perpendicular to the intended motion on "finger-nose" testing. We found this speed was only a few millimeters per second in a normal control but in a severe case of intention tremor (Fig. 3 a) this corrective speed could average 250 mm/second. Successful surgery reduced this by a factor of 10 (Fig. 3 b).

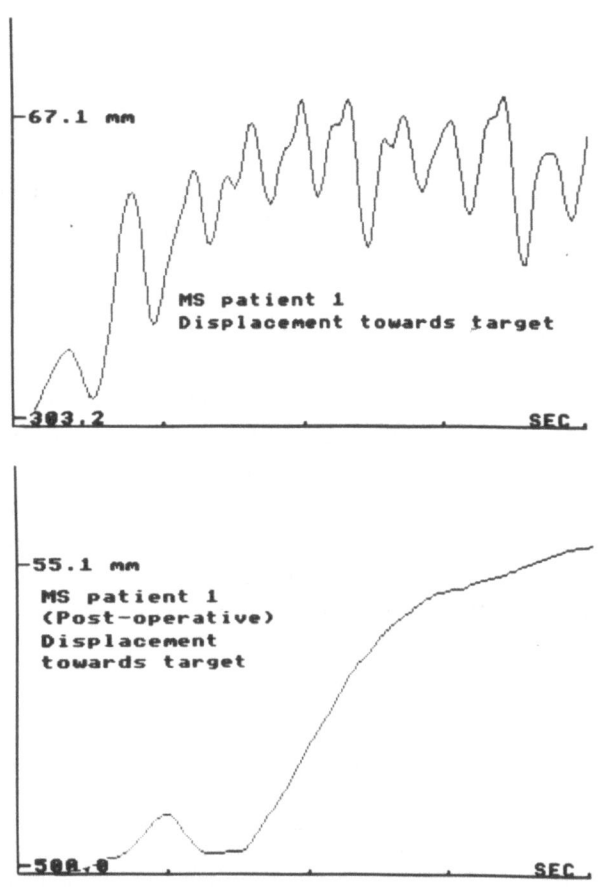

Fig. 3. Intention tremor in a patient with multiple sclerosis. Displacement towards target against time. a) Before, b) after stereotactic thalamotomy

Conclusions

The method described is simple to use. Apart from an electronic circuit which is inexpensive and easily constructed, the equipment required is readily available to most audiovisual services. We believe it provides an objective assessment of movement disorders to supplement existing scoring systems based on clinical judgement. It should prove of value in the management of patients suffering from movement disorders by enabling an objective evaluation of the response to treatment and in comparing the effectiveness of different therapies.

Acknowledgements

This research was supported by Manchester Central District Research Grants Committee, Research grant no. 096. The authors wish to express their appreciation to the staff of the University Department of Medical Illustration and the Department of Medical Physics, Manchester Royal Infirmary for technical assistance. The authors would also like to thank Miss Pamela Brown of the University Department of Neurosurgery for secretarial assistance.

Reference

1. Burke RE, Fahn S, Marsden CD, Bressman SB, Moskowitz C, Friedman J (1985) Validity and reliability of a rating scale for the primary torsion dystonias. Neurol 35: 73–77

Acta Neurochirurgica, Suppl. 39, 80–84 (1987)

EMG Investigations in Patients with Torticollis

S. Price*, J. E. Fox, and **E. R. Hitchcock**

Department of Neurosurgery, University of Birmingham, U.K.

Summary

EMGs have been performed on patients suffering from organic torticollis, hysterical torticollis and on normal control subjects. The EMG activity of the sternomastoid muscles during head rotation in control subjects and those with hysterical torticollis showed similar characteristics and neither group showed a response to body tilt. Subjects suffering from organic torticollis, however, did show a response to tilt.

The results suggest that the response to backward tilt might aid in distinguishing the organic and hysterical forms of torticollis.

Keywords: Torticollis; EMG; organic torticollis; hysterical torticollis.

Introduction

Torticollis, an involuntary movement disorder characterized by spasmodic or sustained deviation of the head, was first described by Wepfer in 1727[8]. It has been well documented since[2, 4, 5, 7] and yet the pathophysiology of the disorder remains obscure[5, 6]. It is believed that the condition may be organic[1, 3] or psychogenic[9, 10] in origin. The distinction, on clinical grounds, is very difficult as both forms exhibit similar symptoms and it is made more so by the fact that long-standing cases of the organic form may develop secondary psychiatric features[1, 3].

In the present study, EMG and certain reflex responses have been investigated in patients presenting with this disorder in an effort to develop *tests* which might help distinguish the organic from the hysterical form.

Method

Results are reported from investigations on twenty-one subjects. Nine were suffering from organic torticollis alone, three had torticollis as a concomitant to dystonia and three were thought to be suffering from the hysterical form of the disorder on the basis of the effectiveness of placebo nerve block and the response to psychotropic drug therapy. Six normal control subjects were also investigated.

All subjects were tested while reclining on a couch. The EMG was recorded from the sternomastoid muscles, bilaterally, using surface electrodes: activity was amplified, displayed and photographed in the conventional way (Medelec MS 6). Recordings were taken at rest, during rotation of the head both to the right and left (unopposed and against resistance) and finally during a tilting manoeuvre. The latter procedure consisted of tilting the couch backwards from the resting (neutral) position, through an angle of 32°, at a rate of approximately $20° \, sec^{-1}$; the couch was held in this (*down*) position until EMG activity had again stabilized and was then returned to *neutral*.

Results

a) Controls: Results showed normal reciprocal activation of the sternomastoid muscles during head rotation, and no change in the EMG activity during tilting (see Figs. 1 and 2).

b) Torticollis: Approximately 50% of the subjects with torticollis had head deviation to the right and 50% to the left; the data has been pooled (the *ipsilateral* sternomastoid was considered the muscle on the side to which the head deviated pathologically).

Rotation of the head to the side of the deviation resulted in activity in the contralateral muscle. Rotation to the opposite side was generally performed with great difficulty and often resulted in coactivation of ipsilateral and contralateral sternomastoids.

All subjects with organic totricollis showed a response to tilting, although the response was not the same in all patients. In some the activity of the

---

* S. Price, Department of Neurosurgery, University of Birmingham, BI5 2TJ U.K.

contralateral muscle showed a transient decrease when the couch was tilted backwards and a transient increase on return to the neutral position, while in others the contralateral muscle showed a transient increase in activity both on downward tilt and on return to neutral. Activity in the ipsilateral muscle was unchanged throughout the manoeuvre in all patients (examples are shown in Figs. 3 and 4).

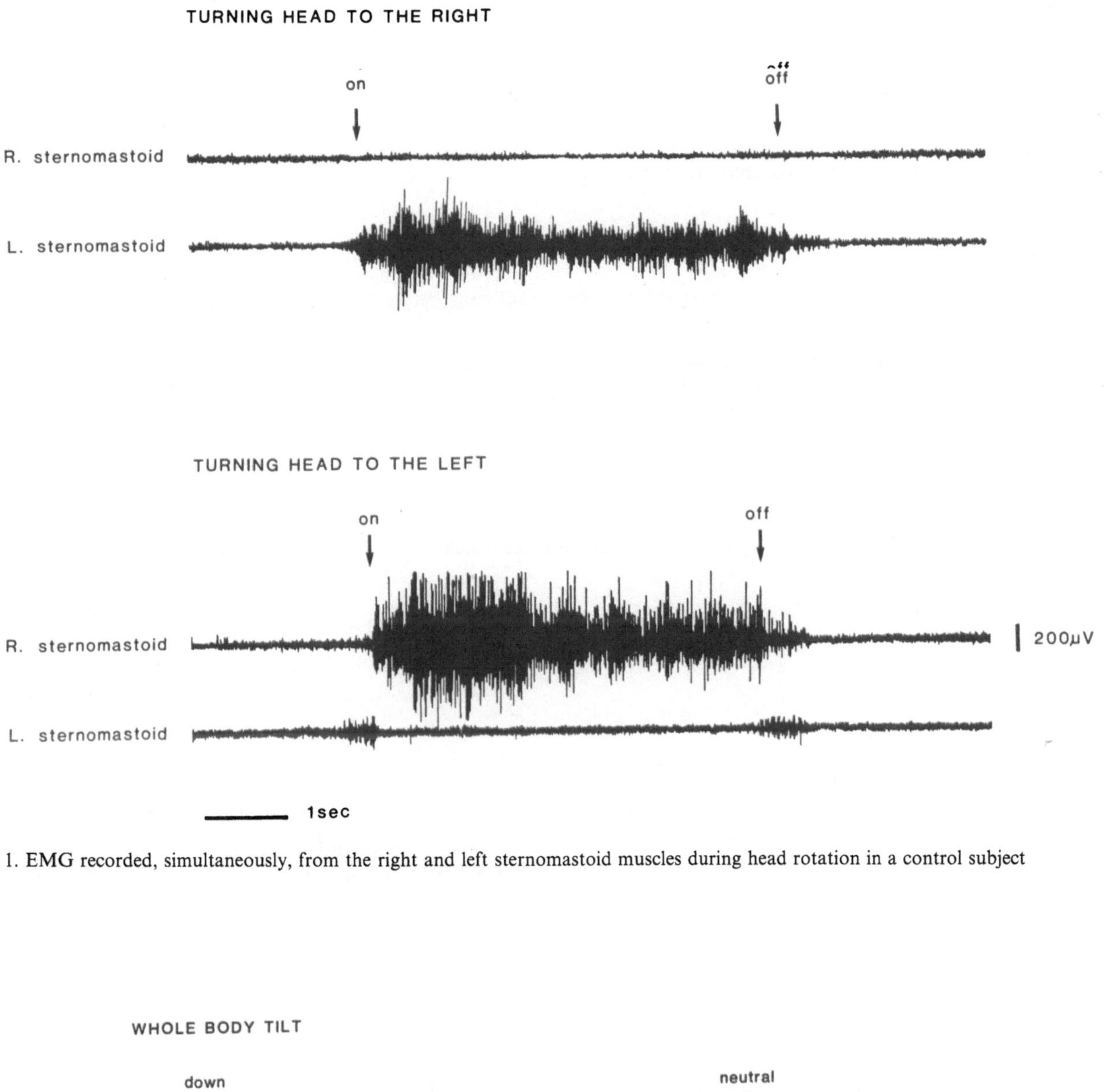

TURNING HEAD TO THE RIGHT

on
off

R. sternomastoid

L. sternomastoid

TURNING HEAD TO THE LEFT

on
off

R. sternomastoid

L. sternomastoid

| 200μV

1sec

Fig. 1. EMG recorded, simultaneously, from the right and left sternomastoid muscles during head rotation in a control subject

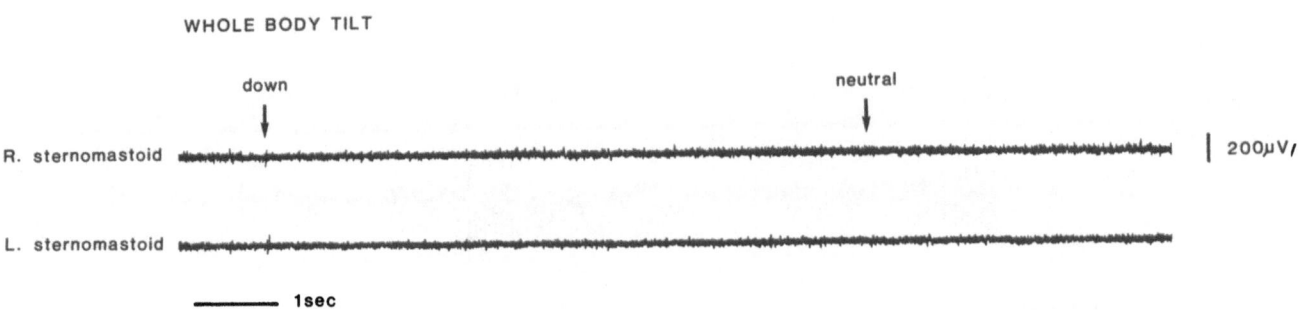

WHOLE BODY TILT

down
neutral

R. sternomastoid

| 200μV/

L. sternomastoid

1sec

Fig. 2. EMG recorded from the sternomastoid muscles during the body tilt maneuver in a control subject

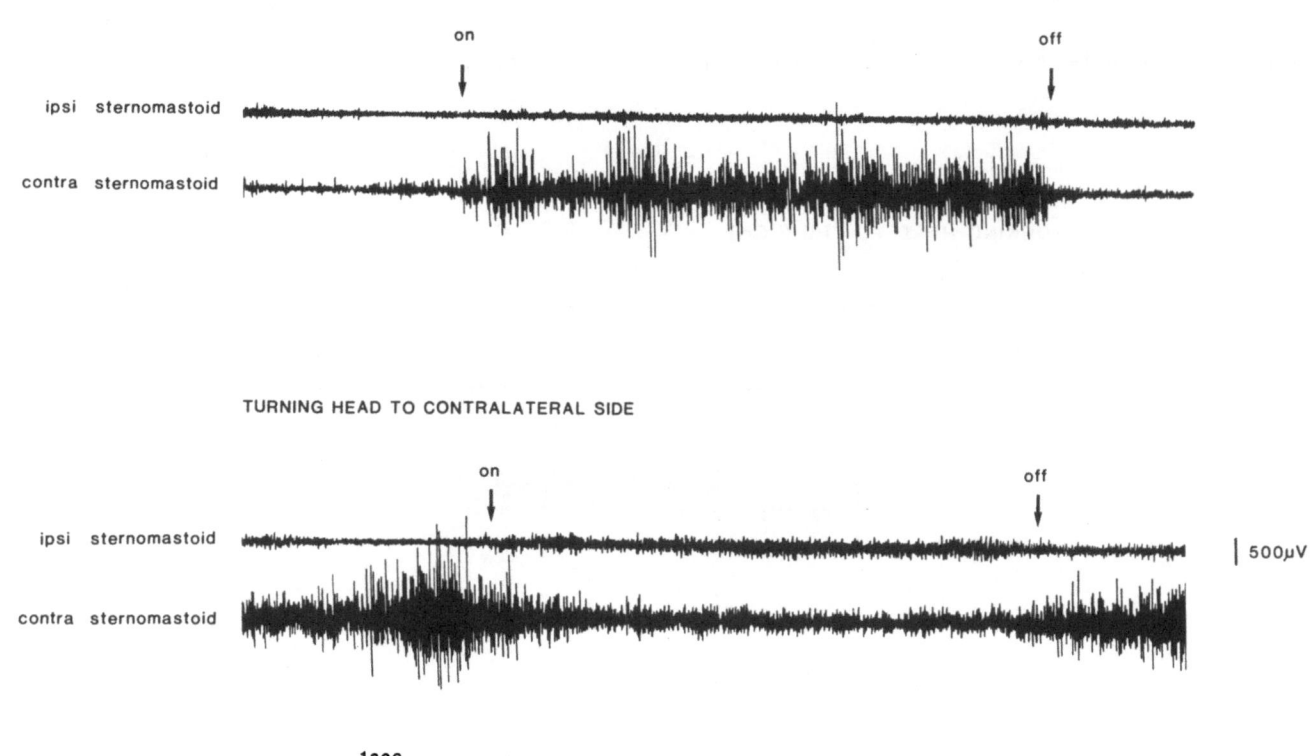

Fig. 3. EMG recorded, simultaneously, from the right and left sternomastoid muscles during head rotation in a patient with organic torticollis

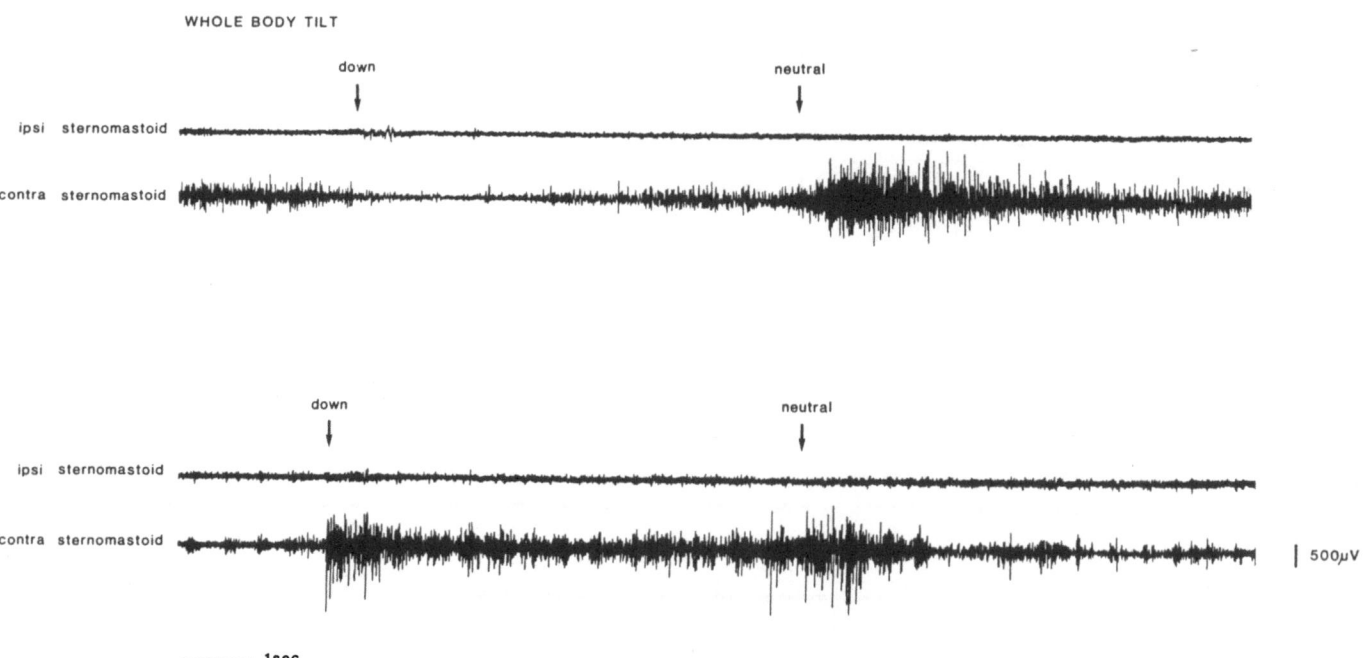

Fig. 4. EMG recorded during the body tilt maneuver in two patients with organic torticollis

c) Hysterical: Rotation of the head to the side of the pathological deviation was associated with activity in the contralateral muscle. Rotation towards the opposite side resulted in activation of the ipsilateral muscle. No change in EMG activity was observed during tilting (see Figs. 5 and 6).

Conclusion

The EMG activity of the sternomastoid muscles during head rotation in control subjects and those with hysterical torticollis showed similar characteristics and neither group showed a response to body tilting. Subjects with organic torticollis were found to have greater variability in EMG activity during head rotation and coactivation was often observed. In addition, the EMG from these patients showed a transient response to tilting backwards and a further response on return to the neutral position.

The results suggest that the response to backward tilt might aid in distinguishing the organic and hysterical forms of torticollis.

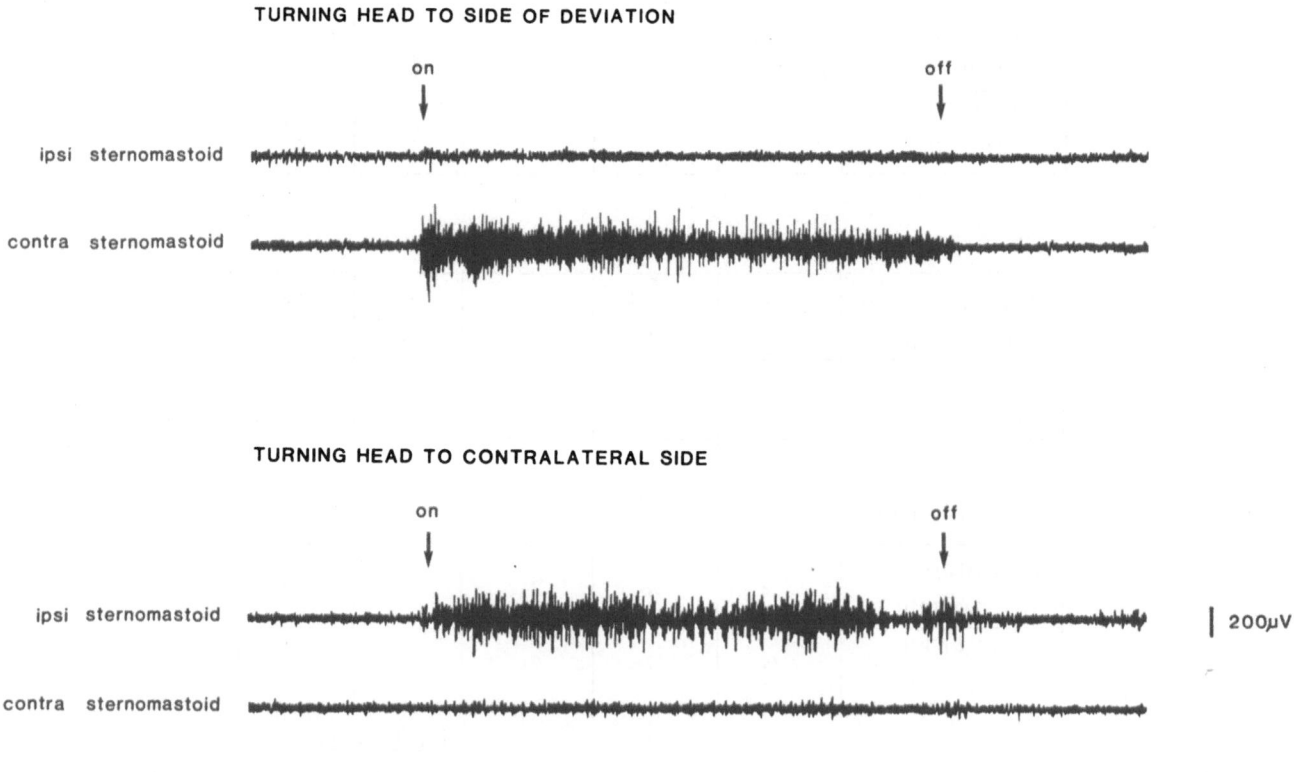

Fig. 5. EMG recorded from the right and left sternomastoid muscles during head rotation ina patient with hysterical torticollis

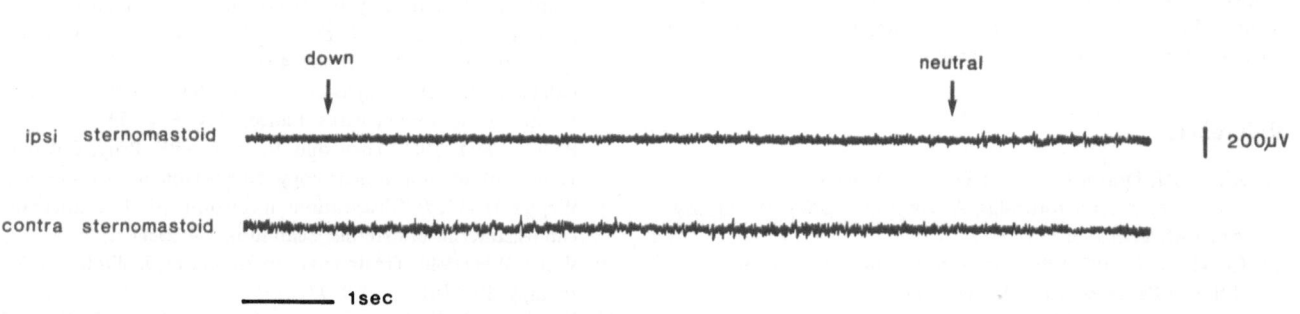

Fig. 6. EMG recorded during the body tilt maneuver in a patient with hysterical torticollis

CONTROLS (n = 6)

| | HEAD TURNING | | | WHOLE BODY TILT | |
	TO RIGHT	TO LEFT	FORWARD	BACK	RETURN
RIGHT STERNOMASTOID	6φ	6+	6+	6φ	6φ
LEFT STERNOMASTOID	6+	6φ	6+	6φ	6φ

ORGANIC TORTICOLLIS (n = 9)

| | HEAD TURNING | | | WHOLE BODY TILT | |
	IPSI	CONTRA	FORWARD	BACK	RETURN
IPSILATERAL STERNOMASTOID	7φ 1+	6+ 1-	3+ 1φ	7φ 1+ 1-	5φ 2+
CONTRALATERAL STERNOMASTOID	7+ 1-	3+ 3- 1φ	3++ 1+	4+ 4- 1φ	5+ 1φ 1-

TORTICOLLIS WITH DYSTONIA (n = 3)

IPSILATERAL STERNOMASTOID	2+ 1φ	2++ 1+	1++	1φ 1+	2φ
CONTRALATERAL STERNOMASTOID	2++ 1+	2+ 1--	1++	1++ 1--	1φ 1--

HYSTERICAL TORTICOLLIS (n = 3)

IPSILATERAL STERNOMASTOID	3φ	2+	2+	3φ	3φ
CONTRALATERAL STERNOMASTOID	3+	2φ	2+	3φ	3φ

ACTIVITY SCALE

φ = No Change
+ = Increase
++ = Very Marked Increase
- = Depression
-- = Very Marked Depression

Fig. 7. Table of results

Acknowledgements

We would like to thank Barbara Carter and Karen Hébert, Department of Physiology, and Alistair Rose, Sandwell District General Hospital, for help with the photography.

References

1. Alpers BJ, Drayer CS (1937) The organic background of some cases of spasmodic torticollis: A report of a case with autopsy. Am J Med Sci 193: 378–384
2. Cockburn JJ (1971) Spasmodic torticollis: a psychogenic condition. J Psychosom Res 15: 471–477
3. Foester O (1933) Mobile spasm of the neck muscles and its pathological basis. J Comp Neurol 58: 725–735
4. Horton PC, Miller I (1972) The etiology of spasmodic torticollis. Dis Nerv System 33: 273–275
5. Matthews WB, Beasley P, Parry-Jones E, Garland G (1978) Spasmodic torticollis: a combined clinical study. J Neurol Neurosurg Psychiatry 41: 485–492
6. Paterson MT (1945) Spasmodic torticollis: results of psychotherapy in twenty-one cases. Lancet Nov: 556–559
7. Podovinsky F (1968) Torticollis. In: Vinken PK, Bruyn GW (eds) Handbook of clinical neurology. North-Holland, Amsterdam
8. Wepfer JJ (1727) Observations medico-practical de affections capitis, Internis et externis, Scaphusii. J A Ziegleri
9. Whiles WH (1940) Treatment of spasmodic torticollis by psychotherapy. Br Med J 1: 969–971
10. Yashkin JC (1935) Treatment of spasmodic torticollis with special reference to psychotherapy. J Nerv Ment Dis 81: 299–309

Acta Neurochirurgica, Suppl. 39, 85–87 (1987)
© by Springer-Verlag 1987

The Foerster-Dandy Operation for the Treatment of Spasmodic Torticollis

J. D. Speelman*, [1], J. van Manen[1], K. Jacz[2], and G. T. van Beusekom[2]

[1] Neurological Department, Academic Medical Centre, Amsterdam, The Netherlands, [2] Neurosurgical Department, Sophia Ziekenhuis, Zwolle, The Netherlands

Summary

The short-term results of the Foerster-Dandy operation in 9 patients with spasmodic torticollis had been studied retrospectively. A good result was obtained for dyskinetic movement, but less for head tilt. Neckpains improved in 4 patients, but developed in 2 other patients postoperatively. Limitation of neck motility and abduction of the arms was seen in all patients. One patient died following a cerebral sinus thrombosis.

Other surgical procedures are discussed and the results of studies of anterior rhizotomy are compared.

The Foerster-Dandy procedure is indicated for severe disability or pain of more than 2 years duration not responding to other treatments.

Keywords: Torticollis; rhizotomy.

Introduction

Spasmodic torticollis can be a disabling and painful disorder. Although spontaneous remissions have been described in 10–35%[10, 12, 13], some permanently, for the remaining patients nonoperative treatment is mostly unsuccessfull[12].

Few neurologists are familiar with the operations for spasmodic torticollis. We present a short review of operative procedures and discuss the results of a retrospective study of our patients with spasmodic torticollis referred for a Foerster-Dandy operation.

---

* J. D. Speelman, Neurological Department, Academic Medical Centre, Meibergdreef 9, NL-1105 AZ, Amsterdam, The Netherlands.

Methods and Patients

The Foerster-Dandy operation is a bilateral ventral rhizotomy of C_1–C_3, and intradural bilateral section of the spinal nervus accessorius followed by the sectioning of the nervus accessorius at the entry of the m. sterno-cleido-mastoideus, bilaterally.

The scoring of the seriousness of spasmodic torticollis uses part of the Fahn and Marsden scoring scale for primary torsion dystonia[1]: grade 0—no dystonia; grade 1—slight torticollis; occasional pulling; grade 2—obvious torticollis, but mild; grade 3—moderate pulling; grade 4—extreme pulling.

This study concerns 9 patients out of a group of 22 patients with spasmodic torticollis, seen in our clinic in the period 1976–1983. It was performed in the period November 1984–February 1985. The mean postoperative follow up was 3 years (1–5 years). This group of 9 patients consists of 4 females and 5 males; mean age at operation, 55 years (36–79 years); mean duration of illness before the operation, 4.4 years (2,5–9 years). For a grading of the seriousness of the spasmodic torticollis: see Table 1. Four patients complained of pain due to cervical spondylosis.

Treatments before the operation: see Table 2. Every patient had tried at least 2 types of drugs.

Results

The results are based on the data of 8 patients, because 1 patient expired the ninth postoperative day (see below):

Improvement of spasmodic torticollis: *The dyskinetic component*, see Fig. 1 a. Seven patients are without dyskinetic movements of the neck and 1 patient improved considerably.

Abnormal posture of the neck, see Fig. 1 b. In 3 patients no improvement of the head tilt was been obtained. Three patients have normal position of the head and 2 patients improved.

Table 1. *The Seriousness of Spasmodic Torticollis*

Grade 4	3 patients
Grade 3	5 patients
Grade 2	1 patient

Table 2. *Treatments Before Operation*

1. Pharmacotherapy:	
anticholinergics	9 patients
neuroleptics	6 patients
benzodiazepines	5 patients
L-dopa/dopamine agonists	4 patients
propranolol	4 patients
antiepileptics	2 patients
amantadine	2 patients
2. A form of behavior therapy	5 patients
3. Myo-feedback	2 patients

Table 3. *Effects on Social Functioning and/or Profession*

No improvement	3 patients
Good improvement	4 patients
Excellent	1 patient

Table 4. *Opinion Concerning the Result of the Operation*

	Patients	Neurologist
Worsening	1	—
No change	—	—
Slight improvement	2	2
Good improvement	3	5
Very good improvement	2	1

The *pain* due to the cervical spondylosis disappeared in the 4 patients, but 2 others complain of neckpains since the operation and are still using a soft collar most of the day due to weakness of the neck muscles.

For *effects on social functioning and/or profession*, see Table 3.

For the *subjective opinion of the patients* concerning the result of the operation and that of the neurologist, see Table 4.

Complications

1. One patient died on the ninth postoperative from brain oedema and pulmonary emboli, due to cerebral sinus thrombosis, despite prophylaxis.

2. One patient developed a slight Brown-Séquard syndrome.

3. All patients developed limited neckmotility and arm abduction, due to the denervation of neck muscles, m. levator scapulae and cervical part of the m. trapezius.

4. Two patients still have an "unstable" neck and pain, despite vigorous physiotherapy and have to wear a soft collar.

Discussion

Bilateral stereotactic thalamotomy has no role in the treatment of spasmodic torticollis, because of the lack of lasting benefit and side-effects[3, 9]. In our opinion unilateral stereotactic thalamotomy combined with cervical rhizotomy[1] may only be considered if the torticollis is associated with a disabling tremor or hemidystonia. Myotomy of selected muscles[2] and denervation of a limited group of muscles have both a high risk of failure or a later recurrence of the symptoms[11].

Bilateral rhizotomy, as a Foerster-Dandy procedure, is a disputed treatment. Favorable reports mention beneficial results in 70–80% of the cases, although side-effects are described as transient and include instability of the head, difficulties in swallowing and, rarely, ischemic complications affecting the spinal cord or medulla[8, 9, 14–16]. However Meares[13] concluded that the unoperated cases in his series fared better than those receiving bilateral cervical rhizotomy. In our study we see a favorable effect on dystonic movements of the neck, but less on the posturing of the head. In all patients there is a loss of normal functioning of the neck and of abduction of the arms, a situation preferred by 7 of the 8 patients.

Torticollis can be the cause of a painful cervical spondylosis as in 4 of our patients. The operation had a favorable effect on this complaint for these patients, but was the cause of neckpains in 2 other patients. One patient died in the immediate postoperative period and another had a slight Brown-Séquard syndrome, probably due to an ischemic lesion of the spinal cord. One patient had an anomaly of the PICA which looped around the right accessory nerve associated with most spasms of the left sterno-cleido-mastoid muscle. This

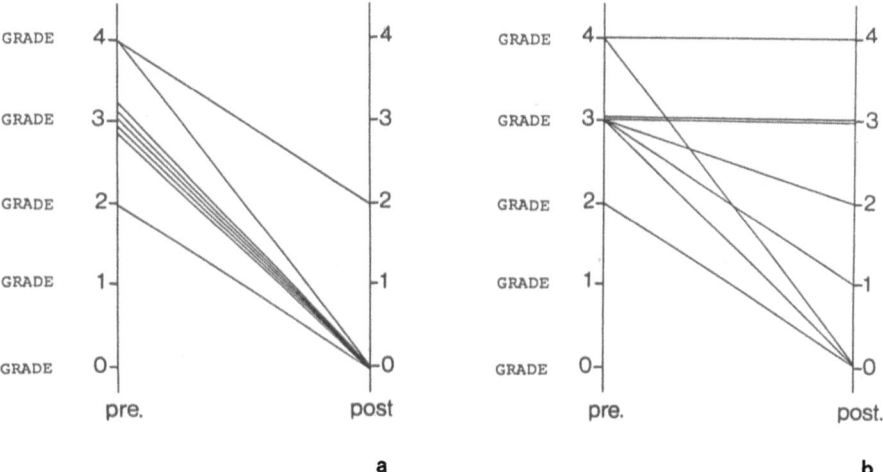

Fig. 1. Results of the operation on spasmodic torticollis. a) Effects on the dyskinetic movement, b) effects on the posturing of the head

case indicates that a proven vascular anomaly around an accessory nerve or cervical root does not automatically mean a positive relation with the torticollis. In our opinion the extension of the Foerster-Dandy operation with section of the dorsal roots C_1, C_2 and C_3 at the same time, for the treatment of cervical pain as described by Siegfried[15], is not justified. Four patients with pain due to cervical spondylosis were without pain after the operation.

The position of cervical spinal cord stimulation is still disputed[5, 7]. A problem is that a case of severe spasmodic torticollis requires a partial laminectomy to fix the electrode to the dura.

In our opinion, the application of botulism toxin is not indicated because of the harzards due to the amount of toxin that has to applied repeatedly. The appearance of antibodies in 3 patients has been reported[6].

In conclusion, despite this short terms follow up of mean 3 years, the Foerster-Dandy operation should still be considered for spasmodic torticollis if the following criteria are satisfied: a severe disabling state, grades 3 and 4; existence of symptoms for more than 2 years; no improvement with other therapies, pharmacotherapy, behavior therapy and spinal cord stimulation.

References

1. Bertrand CM (1982) Peripheral versus central surgical approach for the treatment of spasmodic torticollis. In: Marsden CD, Fahn S (eds) Movement disorders. Butterworths Scientific, London, pp 315–321

2. Chen Xinkang (1981) Selective resection and denervation of cervical muscles in the treatment of spasmodic torticollis; results in 60 cases. Neurosurg 8: 680–688

3. Cooper IS (1977) Neurosurgical treatment of dyskinesias. Clin Neurosurg 24: 367–390

4. Fahn S, Jankovic J (1984) Practical managements of dystonia. In: Jankovic J (ed) Neurologic clinics, vol 2. W.S. Saunders Co n Phil, Tokyo, pp 555–569

5. Fahn S (1985) Lack of benefit from cervical cord stimulation for dystonia. N Engl J Med 313: 1229

6. Fahn S (1981) Rigidity/Dystonia: Clinical aspects and treatments. 1986 Congress of International Medical Society of Motor α disturbances. Abstract and pers. communication

7. Gildenberg PL (1981) Comprehensive management of spasmodic torticollis. Appl Neurophysiol 44: 233–243

8. Hamby WB, Schiffer S (1970) Spasmodic torticollis; results after cervical rhizotomy in 80 cases. Clin Neurosurg 17: 28–37

9. Hernesniemi J, Laitinen L (1977) Résultats tardifs de la chirurgie dans le torticollis spasmodique. Neuro-chirurgie 23: 123–131

10. Jayne D, Lees AJ, Stern GM (1984) Remission in spasmodic torticollis. J Neurol Neurosurg Psychiatry 47: 1236–1237

11. Maccabe JJ (1982) Surgical treatment of spasmodic torticollis, In: Marsden CD, Fahn S (eds) Movement disorders. Butterworths Scientific, London, pp 308–314

12. Matthews WB, Beasley P, Parry-Jones W, Garland G (1978) Spasmodic torticollis: a combined clinical study. J Neurol Neurosurg Psychiatry 41: 485–492

13. Meares M (1971) Natural history of spasmodic torticollis; and effect of surgery. Lancet 2: 149–150

14. Schürman K (1957) Die Chirurgie der extrapyramidalen Hyperkinesen. In: Olivecrona H, Tönnis W (eds) Handbuch der Neurologie, Bd 6. Springer, Berlin Göttingen Heidelberg, pp 116–121

15. Siegfried J, Hood T (1983) Current status of functional neurosurgery. In: Krayenbühl H et al (eds) Advances and technical standards in neurosurgery, vol 10. Springer, Wien New York

16. Sorensen BF, Hamby WB (1966) Spasmodic torticollis. Results in 71 surgically treated patients. Neurol 16: 867–878

Section II

Movement Disorder and Spasticity

B. Spasticity

Acta Neurochirurgica, Suppl. 39, 91–95 (1987)
© by Springer-Verlag 1987

Spasticity: from Pathophysiology to Therapy

P. J. Delwaide*

Section of Neurology and Clinical Neurophysiology, University of Liège, Liège, Belgium

Summary

To be a fully effective, a treatment against spasticity has to match as closely as possible the major pathophysiological disorders responsible for increased tone and hyperreflexia. It is possible to analyze those existing at the spinal cord level by techniques of clinical neurophysiology. Among the functional modifications, increased excitability of alpha motoneurones, reduction in presynaptic inhibition and changes in interneuronal excitability are the most clearly documented. In fact, spasticity results from a combination in various proportions of these different pathophysiological troubles. As the myorelaxants have a specific mode of action, drug selected should be me which is the more able to correct the predominant trouble; this latter can easily be known by clinical neurophysiologic analysis.

Introduction

Spasticity—which is one of the most frequently encountered neurological conditions—is not synonymous with the upper motor neurone (UMN) syndrome but is just one of its components. It has been defined as follows: Spasticity is a motor disorder characterized by a velocity dependent increase in tonic stretch reflex ("muscle tone") with exaggerated tendon jerks, resulting from hyperexcitability of the stretch reflex as one component of the upper motor neuron syndrome[18].

Once developed, spasticity frequently persists with eventual changes and, becoming chronic, leaves the patient with varying degrees of handicap. It is not yet well established to what extent spasticity functionally disables a patient. A certain degree of spasticity in a lower limb is sometimes useful making it possible for the patient to use his leg as a firm stick. If spasticity is too intense, it impedes residual voluntary movements and can be an important problem if complicated by clonus. So, a careful clinical study is always mandatory before attempts are made to modify spasticity intensity.

Spasticity is of a great interest for at least 2 reasons: first, it can be interpreted in physiological terms from data derived from animal studies: it reflects hyperexcitability of the myotatic reflex arc whose control mechanisms are reasonably well-known. Second, it is at present the only manifestation of the UMN syndrome which can be modified by therapeutic measures, of either physical, surgical or drug type. It is clear that therapeutic effectiveness will be the greater, the more the treatment used matchs the pathophysiological disorders. For these reasons, the state of the art in interpretation of spasticity deserves consideration.

Possible Mechanisms for Hyperexcitability of the Myotatic Reflex and Experimental Studies in Man

The stretch reflex has been extensively been studied in animals and various control mechanisms of its excitability have been demonstrated[2, 22]. Table 1 lists all the *possible* troubles which could account for increased tone and stretch reflexes in man.

Due to progresses in clinical neurophysiology, many of the functions which could theoretically explain spasticity are now open to experimental testing in man. Table 2 indicates some of the available techniques and the mechanism they explore. The various techniques applied in man lend to quantitative results so that it is possible to define normal values. In a group of spastic patients if the results of a given test differ from normal values, it can be inferred that the neurophysiological mechanism under study is abnormal.

* P. J. Delwaide, M.D., Section of Neurology and Clinical Neurophysiology, University of Liège, Liège, Belgium.

Table 1. *Possible Mechanisms of Hyperreflexia and Spasticity*

Alpha motoneuron hyperactivity

Gamma motoneuron hyperactivity

Excitatory IA interneuron hyperexcitability

Reduction of presynaptic inhibition

Reduction of recurrent inhibition (Renshaw)

Reduction of autogenic inhibition (IB and IA)

Reduction of reciprocal inhibition

---

a) Alpha Motoneurone (MN) Hyperexcitability: The excitability of alpha MN can be measured by the ratio H max/M max where H max is the maximum amplitude of the electrically induced monosynaptic reflex and M max, the maximum amplitude of the direct motor response. This ratio reflects the proportion of the MN which can be activated reflexly: the higher the excitability of the motor nucleus, the greater the number of activated motoneurons. In a group of spastic subjects, the H max/M max ratio is statistically higher than in normal controls although the values may be dispersed in patients. This result indicates an increased excitability of the motor nucleus[1, 6]. A similar conclusion is reached using another test: the F wave amplitude. The F wave is the late response which appears in a muscle after supramaximal stimulation of its motor nerve. It is due to the backfiring of the motor nucleus after antidromic stimulation. Here also, the response amplitude will reflect the degree of excitability of the motor nucleus. When performed bilaterally in hemiplegic patients, the F wave amplitude is regularly higher on the diseased side provided spasticity is present[19]. In the early stages after a stroke, there may be no difference or the response can be unevokable in a hypotonic limb[12].

It can thus be concluded that spasticity is characterized by an increased excitability of alpha motoneurons.

b) Gamma System Hyperactivity: Once the indirect route of alpha MN facilitation was demonstrated, speculations about the pathophysiological meaning of the gamma system have been florishing. A specific increase of gamma motoneuron has been advocated as the only cause of hyperreflexia[11, 24]. However, the experimental data put forward to support this hypothesis have been scare and disputable. On the one hand, a major argument was based on the comparison between mechanically and electrically induced reflexes. In spastic patients, the tendon jerk is much more enhanced than the electrically induced reflex (H reflex)[3]. This comparison has been proposed to assess gamma system activity on the assumption that the only difference was in the fact that the electrically induced reflex (H reflex) short-circuits the neuromuscular spindles and is thus independent of its tone[21]. However, it has been shown that these two reflexes differ in many more respects than involvement of the spindles and Burke (1985)[4] has listed at least 6 differences. So, a higher tendon jerk does not necessarily mean a gamma system hyperexcitability. On the other hand, procainization of the motor nerve may induce a selective palsy of the smaller gamma fibers[20]. Procainization also abolishes spasticity. Here also, criticisms have been presented and it has been shown that selectivity of anesthesia is far from being absolute. It could be that anesthesia also affects other fibers, namely the IA fibers. Despite these restrictions, the hypothesis of gamma system hyperactivity has enjoyed a wide popularity and is still sometimes presented as being the mechanism responsible for spasticity.

A more direct approach of the gamma system was developed in the seventies[27]. With a tungsten microelectrode, it is possible to record single IA fiber discharges whose frequency reflects the neuromuscular spindle tone. Applied to spastic patients, this technique failed to disclose more discharges in IA fibers as anticipated from an increase in gamma system hyperactivity. Although the results are difficult to quantify and the experience is rather limited, it appears that the hypothesis of gamma system hyperexcitability is no more supported by reliable experimental data. It can be concluded that there is no argument at present favoring that mechanism.

c) Presynaptic Inhibition: This mechanism, well known in spinal cord animal physiology has not been taken into account in human studies before the introduction of the vibratory stimulation[14, 17]. Vibration of a muscle sets off discharges in IA afferents and thus submits the spinal cord to a barrage of excitatory influences. It thus appeared as a paradox that, when elicited during vibratory stimulation, the monosynaptic reflex was not enhanced as expected but on the contrary, markedly inhibited. Animal as well human experiments have shown that the inhibition is due to a mechanism of presynaptic inhibition acting on the IA terminals and triggered by homonymous IA afferents[6, ±, 13]. The measure of vibratory inhibition—if achieved in standardized conditions—reflects intensity of presynaptic inhibition. In spastics, the vibratory inhibition of monosynaptic reflexes is statistically reduced and a reduction in presynaptic inhibition can thus be considered constitutive of the pathophysiological troubles of spasticity.

This conclusion is supported by results obtained using another technique: the spinal cord evoked potentials. When responses are averaged over the Th 12 spinal process after stimulation of the tibial nerve at the popliteal fossa, the evoked potential is complex. It is made of a large negative wave (S wave) followed by a longer lasting positive wave (P 2). This latter reflects spinal cord presynaptic inhibition and is stastically reduced in spasticity[10]

At present, reduction in presynaptic inhibition as the cause correlates best with spasticity as shown by factorial analysis[15].

d) Reciprocal Inhibition: In addition to activation of the MN of the muscle they originate from, IA afferents also inhibit the antagonist motoneuron pool via an interposed interneuron, namend IA interneuron. On this interneurone converge many afferents both segmentary and suprasegmentary. The IA interneurone can be tested in man by conditioning a monosynaptic reflex with stimulation of a nerve coming from an antagonist[26]. This technique is applicable both to lower and upper limb[5] where the inhibitory effects are more marked. In spastic patients, the inhibition observed at rest in reduced and even less when the subject is asked to voluntarily contract his tibialis anterior[8, 9]. It can thus be taken for granted that reciprocal inhibition is less active in spasticity. The functional consequences of this observation are clear. When a voluntary isotonic movement is performed, the passively stretched antagonist muscle may be the site of a myotatic reflex which hinders the active contraction. The role of reduced reciprocal inhibition is probably important to account for the handicap brought about by spasticity.

e) Recurrent Inhibition: Motor axon collaterals activate inhibitory cells (Renshaw's cells) projecting to the motoneurone pool. This recurrent inhibition is influenced by descending pathways and, if reduced, might explain alpha motoneuron hyperexcitability. The functioning of this particular interneurone can be studied in man through a sophisticated technique proposed by Pierrot-Deseilligny *et al.* (1976). The data show no difference at rest between normal and hemiplegic patients[16]. This means that hyperexcitability of the myotatic arc cannot be explained by a reduced activity of the Renshaw's cells. However, in some paraparetic or paraplegic patients, clear differences

with control subjects may appear[9]. Moreover, during contraction, the efficacy of Renshaw's cell is reduced. It can however be concluded that a reduction in recurrent inhibition is not required to explain myotatic arc hyperexcitability in spastic patients at rest even if it plays a role in certain circumstances.

f) Interneuronal Pool: Many other interneurons both facilitatory and inhibitory intervene to adjust the degree of excitability of the motoneurons. They cannot yet be studied specifically in man but an indirect reflection of their functioning is obtained by the study of polysynaptic reflexes. For example, the interneurones involved in the flexor reflex can be easily studied. It appears that the stimulation threshold for flexor reflexes is lowered in spasticity, fact which indicates that transmission within polysynaptic pathways is not reduced[25]. Moreover, an exteroceptive stimulation may evoke responses in muscles which normally do not respond. So, the channeling of information is altered, probably because inhibitory interneurons are less active.

From what has just been presented, it is clear that spasticity does not result from a single abnormality as was believed some 15 years ago when a controversy was going on to decide whether spasticity was due or not to gamma system hyperactivity. On the contrary, it is clear that many pathophysiological troubles do coexist in spasticity. Table 2 lists them but it is possible that new dysfunctions will still be discovered once new techniques will be made available to test additional neurophysiologic mechanisms in man.

Is Spasticity Correlated with a Specific Pathophysiological Trouble?

It is not surprising that many pathophysiological changes can be demonstrated in spasticity as several descending pathways which project onto various spinal cell layers with distinct functions are generally lesioned. However, it could be that only one mechanism accounts for hyperexcitability of the myotatic arc. To test this possibility, we looked for a correlation between spasticity intensity and abnormal results found in the various tests.

Spasticity intensity was assessed by the Ashworth's scale. In a group of 88 spastic patients, there was no good correlation between the scores in this scale and changes in H/M ratio (measure of alpha motoneuron excitability), reduction in vibratory inhibition (presynaptic inhibition) and reciprocal inhibition. Nor was there any relationship with changes in flexor reflexes. The same degree of spasticity can be observed with high or low H/M ratio, with very marked or slight reduction in vibratory inhibition or reciprocal inhibition. The clinical picture does not permit to predict which type of modifications will appear from the electrophysiological testing. From these results, it can be concluded that spasticity intensity is not a function of one specific trouble. It is more likely to consider that it reflects the combined effect of the various troubles.

Table 2. *Segmentary Spinal Functions Studied in Man*

1. Excitability of the MN pool	H/M ratio
2. Gamma system activity	microneurography
3. Interneuronal activity	
global	threshold of flexor reflex
selective	
presynaptic inhibition	vibratory inhibition of the MS reflex (Delwaide 1971)
excitatory IA interneurone	tonic vibration reflex
inhibitory IA interneurone (reciprocal inhibition)	Tanaka's technique (1974)
Renshaw's inhibition (recurrent inhibition)	Pierrot-Deseilligny's technique (1976)
IB autogenic inhibition	Pierrot-Deseilligny's technique (1979)

Table 3. *Pathophysiological Analysis of Spasticity. Abnormalities at Spinal Segmentary Level*

Increased montoneuronal excitability

Decreased excitability of interneurones responsible for:
 presynaptic inhibition
 reciprocal inhibition
 IA facilitation

Modified excitability of interneurones responsible for polysynaptic reflexes (flexor reflexes)

Renshaw cells function normally at rest

No evidence for fusimotor (gamma) hyperactivity

Moreover, the abnormalities in the different tests are not linked and any combination of changes in electrophysiological analysis can be found. So, it can be added that the various troubles contribute in variable proportion to produce clinical spasticity.

Toward a Therapy Based on Pathophysiology

What are the practical implications of these conclusions? It seems unlikely that spasticity could be treated by correcting only one abnormality. To be fully effective, a logical treatment should be multifactorial. In other words, a good myorelaxant drug or a good surgical procedure have not to be extremely specific. Moreover, as the contribution of the various troubles is variable, the same treatment may lead to unpredictable results, sometimes excellent and sometimes disappointing. To reduce uncertainty, there are however prospects. A pathophysiological analysis can now been easily performed by Clinical Neurophysiology. The techniques used to achieve an analysis of the various pathological troubles and to assess their relative contribution are non invasive, painless but time consuming. It is more than likely that better therapeutic results could be expected from a therapeutic approach based on pathophysiology than from only clinical data. Pathophysiology does not appear to be linked with etiology or intensity but varies largely from patient to patient. Time has come when clinicians may obtain pertinent data concerning the individual pathophysiological profile of their patients, enabling the development of therapeutic strategies on logical background.

References

1. Angel RW, Hoffmann WW (1963) The H reflex in normal, spastic and rigid subjects. Arch Neurol 8: 591–596
2. Baldissera F, Hultborn H, Illert M (1981) Integration in spinal neuronal system. In: Brooks VB (ed) Handbook of physiology, Sect I, The nervous system, vol II, Motor control. Am Phys, Soc, Bethesda, pp 509–595
3. Buller A, Dornhorst AC (1957) The reinforcement of tendon reflexes. Lancet ii: 1260–1262
4. Burke D (1985) Mechanisms underlying the tendon jerk and H-reflex. In: Delwaide PJ, Young RR (eds) Clinical neurophysiology in spasticity. Restorative neurology, vol 1. Elsevier, Amsterdam, pp 55–61
5. Day BL, Rothwell JC, Marsden CD (1983) Transmission in the spinal reciprocal IA inhibitory pathway preceding willed movements of the human wrist. Neurosci Lett 37: 245–250
6. Delwaide PJ (1971) Étude expérimentale de l'hyperréflexie tendineuse en clinique neurologique. Éditions Arscia, Bruxelles, 324 p
7. Delwaide PJ (1973) Human monosynaptic reflexes and presynaptic inhibition. In: Desmedt JE (ed) New developments in EMG and clinical neurophysiology. Karger, Basel, pp 508–522
8. Delwaide PJ (1984) Contribution of human reflex studies to the

understanding and management of the pyramidal syndrome. In: Shahani B (ed) Electromyography in CNS disorders: Central EMG. Butterworths, Boston, pp 77–109

9. Delwaide PJ (1985) Electrophysiological testing of spastic patients: its potential usefulness and limitations. In: Delwaide PJ, Young RR (eds) Clinical neurophysiology in spasticity. Restorative neurology, vol 1. Elsevier, Amsterdam, pp 185–203

10. Delwaide PJ, Schoenen J, Pasqua V de (1985) Lumbosacral spinal evoked potentials in patients with multiple sclerosis. Neurol 35: 174–179

11. Dietrichson P (1971) The role of the fusimotor system in spasticity and parkinsonian rigidity, Oslo, pp 101

12. Fisher MA, Shahani BT, Young RR (1978) Assessing segmental excitability after acute rostral lesions. I. The F response. Neurology 28: 1265–1271

13. Gillies D, Lance JW, Neilson PD, Tassinari CA (1969) Presynaptic inhibition of the monosynaptic reflex by vibration. J Physiol (Lond) 205: 329–339

14. Hagbarth KE, Eklund G (1966) Motor effects of vibratory stimuli in man. In: Granit R (ed) Muscular afferents and motor control. Nobel Symposium I. Almqvist & Wiksell, 466 p

15. Hayat A (1979) Factorial analysis of specific parameters in spasticity. Electromyogr Clin Neurophysiol 19: 541–553

16. Katz R, Pierrot-Deseilligny E (1982) Recurrent inhibition of alpha-motoneurons in patients with upper motor neuron lesions. Brain 105: 103–124

17. Lance JW (1966) The reflex effects of muscle vibration. Proc Aust Assoc Neurol 4: 49–56

18. Lance JW (1980) Symposium synopsis. In: Feldman RG, Young RR, Koella WP (eds) Spasticity: Disordered motor control. Year Book, Chicago, pp 485–494

19. Liberson WT (1976) Averaged late reflex responses in hand muscles of spastic patients. In: International symposium on reflexes and motor disorders, Brussels, pp 118

20. Matthews PB, Rushworth G (1957) The selective effect of procaine on the stretch reflex and tendon jerk of soleus muscle when applied to its nerve. J Physiol (Lond) 135: 245–262

21. Paillard J (1955) Réflexes et régulations d'origine proprioceptive chez l'Homme. Thèse Fac. des Sciences de Paris (Série A. n 2858). Arnette, Paris, 293 p

22. Pierrot-Deseilligny E, Mazieres L (1985) Spinal mechanisms underlying spasticity. In: Delwaide PJ, Young RR (eds) Clinical neurophysiology in spasticity. Restorative neurology, vol 1. Elsevier, Amsterdam, pp 63–76

23. Pierrot-Deseilligny E, Bussel B, Held JP, Katz R (1976) Excitability of human motoneurones after discharges in a conditioning reflex. Electroenceph Clin Neurophysiol 40: 279–287

24. Rushworth G (1960) Spasticity and rigidity: an experimental study and review. J Neurol Neurosurg Psychiatry 23: 99–117

25. Shahani BT, Young RR (1971) Human flexor reflexes. J Neurol Neurosurg Psychiatry 34: 616–627

26. Tanaka R (1974) Reciprocal Ia inhibition during voluntary movements in man. Exp Brain Res 21: 529–540

27. Vallbo AB, Hagbarth KE, Torebjörk HE, Wallin BG (1979) Somatosensory, proprioceptive, and sympathetic activity in human peripheral nerves. Physiol Rev 59: 919–957

Acta Neurochirurgica, Suppl. 39, 96–98 (1987)

Percutaneous Selective RF Neurotomy in Spasticity

Y. Kanpolat*, C. Cağlar, E. Akiş, A. Ertürk, and **H. Uluğ**

University of Ankara, İbni Sina Medical Center, Department of Neurosurgery, Ankara, Turkey

Summary

Percutaneous selective radiofrequency (RF) neurotomy is described in the treatment of spasticity dominant in femoral adductor and gastrocnemius muscles. The effectiveness of neurotomy has been tested by means of infiltration of nerves with local anesthetic agents before neurotomy in one case. A temperature monitoring needle electrode system with 5 mm noninsulated tip has been used for localization of the nerves by stimulation. Percutaneous obturator neurotomy was performed in the most proximal portion accesible. Posterior tibial neurotomy was performed in two branches of the nerve which innervates the medial and lateral heads of gastrocnemius. Four lesions lasting 60 seconds were made in four different directions between 65° and 70°.

Keywords: Spasticity; neurotomy.

Introduction

One of the most important problems in spasticity related to cerebral palsy is spasticity of the femoral adductor and gastrocnemius muscles. Spasticity of the femoral adductor muscles leads to the adduction, inner rotation and flexion of the thigh. Spasticity of the gastrocnemius muscle causes knee flexion and equinus deformity of the foot.

Since the beginning of 20th century, peripherial applications have been attempted to lessen the spasticity of femoral adductor and gastrocnemius muscles. The adductor muscles spasticity was reduced by lesions of the Obturator nerve. Infiltration of the nerve with local anesthetics, application of neurolytic agents to the nerve (phenol, alcohol) and section of the nerve can be cited[3, 5]. Spasticity of the gastrocnemius muscle has been reduced by resection of Achilleus tendon or section of medial and lateral head of the muscle[4]. Partial denervation of the gastrocnemius muscle is performed by means of partial popliteal neurotomy[2].

Method

Obturator neurotomy has been used in cases where spasticity was dominant in femoral adductor and gastrocnemius muscles at lower extremities. The effectiveness of this procedure has been tested by the infiltration of local anesthetics into the obturator nerve, and two divisons of the posterior tibial nerve for each of the defined points. Here, 2 ml of local anesthetics was injected. Following injection, a decrease of spasticity indicated that a definitive procedure might be successful.

Obturator nerve: The procedure is done under neurolept-anesthesia. The patient is put in a supine position. A 5 mm uninsulated tip, temperature monitoring needle electrode system is used. With the needle electrode a perpendicular puncture is made to a depth of 2.5–3.5 cm to 1.5 cm lateral to the pubic tubercle. At this point the inferior ramus of os pubis can be felt. The needle electrode is withdrawn gently and pushed several mm laterally and caudally. This region is approximately the entrance of obturator foramen. The electrode system is connected to the stimulator, with an average of 0.5–1 volt of stimulation. During the procedure the indifferent electrode is placed into the contralateral thigh (Fig. 1). When contact is made on the obturator nerve adduction and the inner rotation of the thigh is observed synchronically at each stimulation. The procedure is terminated having made four lesions of 65–70 °C with a duration 60 seconds, in four different positions (cephal, caudal, medial and lateral).

Posterior tibial nerve: The patient is put in to a prone position. The puncture is made with a 30° angle towards the medial head of the gastrocnemius muscle through the middle of popliteal fossa and one cm below the popliteal line. The needle electrode system is introduced and having been pushed to 1.5–2 cm depth, it is withdrawn during of

---

* Dr. Y. Kanpolat, University of Ankara, İbni Sina Medical Center, Department of Neurosurgery, Ankara, Turkey.

0.5–1 volt of stimulation. The region where motor synchronous response is produced is the ideal point. Here four lesions are made, at 65–70 °C for 60-second duration at four different sites (medial, lateral, ventral and dorsal). The needle electrode is withdrawn this time from the same point of puncture and introduction is made toward the lateral head of the gastrocnemius muscle at a 30° inclination up to a depth of 1.5–2 cm. At the region where motor response is received synchronous with 0.5–1 volt stimulation, the

same type of four lesions are made. The indifferent electrode is placed on the contralateral side (Fig. 2).

The method has been used on an 11-year-old boy who had been rehabilitated continuously for ten years. He was able to walk after this procedure at the end of a 6 weeks rehabilitation period. Subsequent to one year follow up period, it was observed that the patient did not need any other further applications (pre- and postoperative conditions of the patient are presented in Table 1).

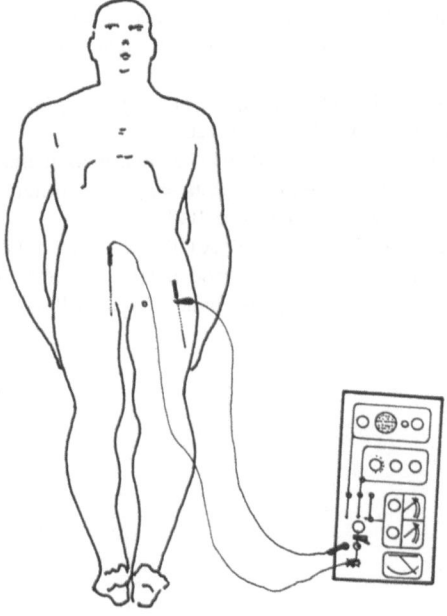

Fig. 1. Percutaneous selective RF obturator neurotomy. 1.5 cm lateral to the pubic tubercle, 2.5–3.5 cm depth, 0.5–1 volt stimulation, 65–70 °C, 4 lesions

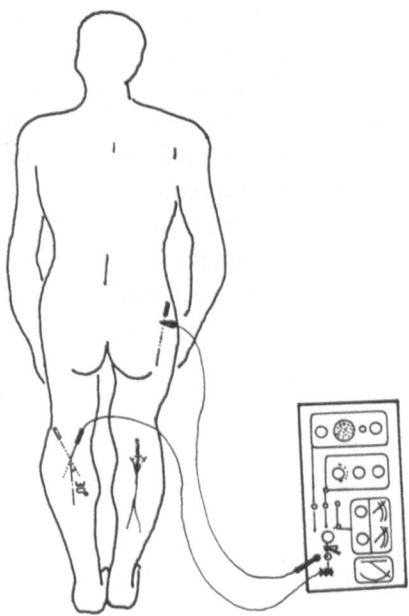

Fig. 2. Percutaneous selective RF posterior tibial neurotomy. Middle of poplitea, 1 cm below the poplitea line, 30° angle, medial or lateral head of gastrocnemius muscle, 0.5–1 volt stimulation, 65–70 °C, 4 lesions

Table 1. *Preoperative and Postoperative Conditions of the Patient*

The position of the patient	Preoperative	Postoperative
Coming to standing position	plenty of assistance is needed	little assistance is needed
Standing	plenty of assistance is needed hips in flexion and adduction knees in flexion heels are raised up from the floor (spontaneous plantar flexion)	little assistance is needed hips in extension, slight adduction knees in extension heels touch the floor
Kneeling and half kneeling	plenty of assistance is needed (hips tend to flex)	the patient can take the position with little assistance, keeps the position without support (hip flexion do not pull the trunk forward)
Hip abduction range in supine position	right: 15° left: 25°	right: 45° left: 45°
Walking	not possible	1st week: between parallel bars, the patient can walk without assistance using hip flexors, without the interference of adductors and without the heels rising up from the floor 6th week: can walk without assistance

Discussion

In the treatment of spasticity, peripherial denervation is generally provided by neurolytic agents or direct neurectomy.

With percutaneous RF neurotomy, it is possible to reach the peripherial nerves to be denervated without any surgical exploration. The effectivenss of the procedure can be tested by preliminary infiltration of local anesthetic. Some advantages of this procedure are; simplicity of the method, percutaneous application, possibility of localizing the nerve with the help of stimulation and its reproducibility when needed. Another advantage is immediate rehabilitation of the patient just after the procedure.

In conclusion, percutaneous selective RF neurotomy is a noninvasive simple and effective procedure in order to denervate the spastic muscle or muscle groups where contracture has not developed. The merit of this procedure will prove itself when applied on wide series of patient groups whose long-term follow up will be possible in the future.

References

1. Brunelli G, Brunelli F (1983) Partial selective denervation in spastic palsies (hyponeurotization). Microsurg 4: 221–224
2. Gross C (1979) Spasticity-Clinical classification and surgical treatment. In: Krayenbühl H et al (eds) Advances and technical standards in neurosurgery, vol 6. Springer, Wien New York, pp 55–97
3. Felsentahl G (1979) Spasticity modified by obturatory nerve block. MD State Med J: 123–126
4. Silver CM, Simon SD (1959) Gastrocnemius muscle resection (Silverskiold operation) for spastic-equinus deformity in cerebral palsy. J Bone Joint Surg 41 A: 1021–1028
5. Tardieu G, Got C, Lespargot A (1975) Indication d'un nouveau type d'infiltration au point moteur (alcool a 96°). Application clinique d'une étude expérimentale. Ann Méd Physique 18: 539–557

Acta Neurochirurgica, Suppl. 39, 99–102 (1987)

Microsurgical Selective Posterior Rhizotomy in the Dorsal Root Entry Zone for Treatment of Limb Spasticity

M. Sindou*, **J. J. Mifsud**, **Ch. Rosati**, and **D. Boisson**

Department of Neurosurgery, Hôpital Neurologique, Lyon, France

Summary

SPR in the DREZ, which was introduced in 1972 by the senior author on the bases of anatomical studies in humans, selectively interrupts the (lateral) nociceptive and the (central) myotatic fibers, while sparing the (medial) lemniscal fibers. In addition it enhances the inhibitory mechanisms of the Lissauer's tract and dorsal horn. The procedure was effective—with a follow-up ranging from 1 to 14 years—in 93% of the paraplegic patients with flexion-adduction postures (50 cases) or severe hypertextension (3 cases) and in 89% of the hemiplegic patients with irreducible flexion of the upper extremity (23 cases) or lower limb (5 cases).

Keywords: Dorsal root entry zone lesions; hemiplegia; pain; paraplegia; selective posterior rhizotomy; spasticity.

Introduction

When spasticity cannot be successfully modified by physical and pharmacological therapy, functional neurosurgery can provide some alternative solutions. Electrical neurostimulation can only alleviate moderate spasticity. For more severe hypertonic disorders with irreducible abnormal postures in the limbs, one must use destructive procedures at the level of peripheral nerves, spinal roots and spinal cord.

Our interest in the surgical treatment of spasticity was elicited by an observation made after performing selective microsurgical lesions in the posterior root-spinal cord junction—the so-called dorsal root entry zone (DREZ)—for pain treatment and reported in 1972 in the medical thesis of the senior author[1]. On this occasion, unwanted hypotonia was produced and the myotatic reflexes were abolished in all the territories corresponding to the operated painful segments. Therefore the same year we started to apply the same method to improve spasticity in both paraplegic and hemiplegic patients[2, 5–7].

Anatomical Bases

This technique—selective posterior rhizotomy (SPR)—was developed after a careful anatomical study of the DREZ was performed in man[1, 3, 4, 8] and a topographical segregation of the afferent fibers according to their size, destination and function was revealed (Fig. 1).

It consists of microsurgical coagulations and incisions performed ventrolaterally at the entrance of the rootlets in the posterolateral sulcus of the spinal cord. The incision is 2 mm deep into the internal part of Lissauer's tract and is internally orientated at 45° angle.

Such lesions aims at interrupting the small (nociceptive) and the large (myotatic) fibers situated respectively in a lateral and a central position in the DREZ, whilst sparing most of the large (lemniscal) fibers regrouped medially to reach the dorsal column.

On these anatomical backgrounds, the procedure was presumed to selectively suppress the afferent pathways of the polysynaptic nociceptive as well as the monosynaptic stretch reflexes, which in spastic syndromes are separated from their suprasegmental inhibitory control whilst preserving the inhibitory anatomical structures of the DREZ. The surgical lesions try to preserve: 1. the lemniscal fibers and their recurrent collaterals to the dorsal horn shown as inhibitory for the segmental electrophysiological mechanisms, 2. the substantia gelatinosa interconnecting fibers running through the external part of the Lissauer's tract, demonstrated as inhibitory for the intersegmental modulating system which regulates the metameric excitability from the adjacent levels.

* M. Sindou, M.D., Department of Neurosurgery, Hôpital Neurologique, 59 bd Pinel, F-69003 Lyon, France.

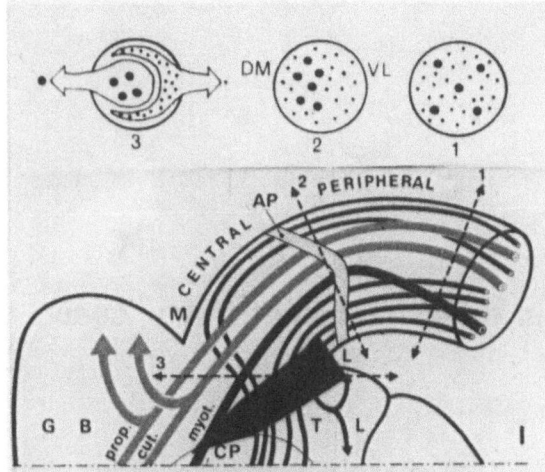

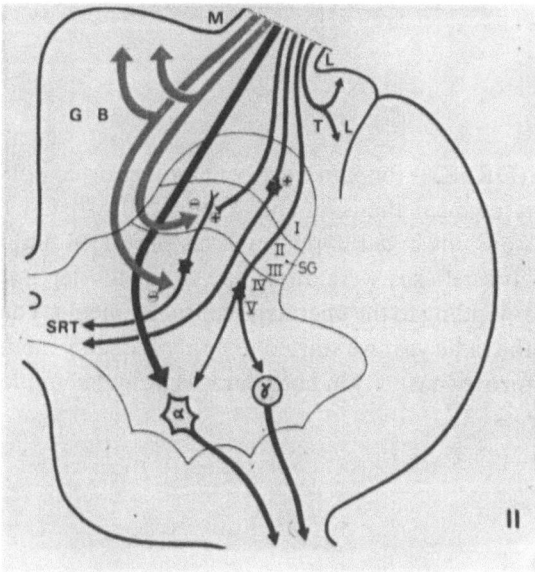

Fig. 1. Organization of fibers at the posterior rootlet-spinal cord junction in humans. I) Each rootlet has a peripheral and a central segment. The junction between the two constitutes the pial ring (*AP*). Peripherally the fibers have no organization. In the neighborhood of the pial ring, the small fibers are situated on the rootlet surface, predominantly on its lateral side. In the central segment, they regroup laterally to enter the tract of Lissauer (*TL*). The large fibers are situated centrally for the myotatic fibers and medially for the lemniscal ones. Selective posterior rhizotomy affects the black triangle. II) The small fibers terminate on the spino-reticulothalamic cells (SRT) which they activate, and on polysynaptic arcs to the motoneurons (in particular those to the flexor muscles). The large myotatic fibers project onto the anterior horn cells and form the myotatic arcs. The short collaterals of the large lemniscal fibers (of cutaneous or proprioceptive origin) terminate at the SRT cells which they inhibit

Operative Technique

The procedure, which has been detailed elsewhere[1, 3, 4] is illustrated in Fig. 2.

For paraplegia, the lumbosacral segments are approached through a T 10–L 2 laminectomy, while for the upper limb in hemiplegic patients, a C 4–C 7 hemilaminectomy with conservation of the spinous processes is sufficient to reach the C 5–T 1 segments.

Identification of the accurate metameric levels—*i.e.*, those selected as supporting the harmful tonic mechanisms—is achieved by studying the muscle responses to bipolar electrical stimulation of the accessible anterior and/or dorsal roots. Then under the microscope, the ventrolateral aspect of the posterior root-spinal cord junction is exposed by retracting backwards all the rootlets of each selected root. The lesions consist of a continous coagulation made at the lateral edge of the posterolateral sulcus using a sharp bipolar microforceps and afterwards a 2 mm deep incision with a microbistoury orientated at 45° angle, performed into the lateral part of the DREZ and the internal part of the Lissauer's tract.

Results

The results for harmful spasticity and abnormal postures are summarized in Tables 1 and 2. Pain was relieved in 90% of the patients affected with severe painful manifestations, which were present 25 times in the paraplegic and 17 times in the hemiplegic group. In 39 patients, there was reappearance of or improvement in some useful motility, after surgery.

Discussion

As demonstrated in the series, SPR in the DREZ—provided that patient selection is correct—was most often useful in 1. suppressing the excess of spasticity, 2. correcting the abnormal postures and flexion spasms when present (with complementary lengthening of tendons and capsulolysis of joints, if they were irreducibly retracted), 3. relieving the frequently associated pain.

In the most severe situations (so-called "comfort" indications), these effects allow nursing and physiotherapy to be resumed and sometimes reappearance of some useful voluntary movements. In the less affected patients (so-called "functional" indications), correction of the localized harmful spasticity facilitates the remaining functional capacities.

Table 1. *Results in Spastic Paraplegia (59 Cases)*

Etiology	*Results (1–14-year follow-up)*	
M.S. (31), trauma (15), misc. (13)	1. Hyperflexion group: success in 52 of 56 cases	
Clinical status	2. Hyperextension group: success in 3 of 3 cases	93% effective
hyperspasticity		
abnormal flexion (or hyperextension) postures	*Complications*	
pain	death due to pulmonary insufficiency (2)	
	wound infection (2)	
	motor aggravation (2)	

Table 2. *Results in Spastic Hemiplegia (28 Cases)*

Etiology	*Results (1–13-year follow-up)*	
Trauma (16), vasc. (10), misc. (2)	1. Uper limb group: success in 21 of 23	
Clinical status	2. Lower limb group: success in 4 of 5 cases	89% effective
hyperspasticity		
abnormal flexion postures	*Complications*	
pain	motor aggravation (1)	
	paresthesias (1)	

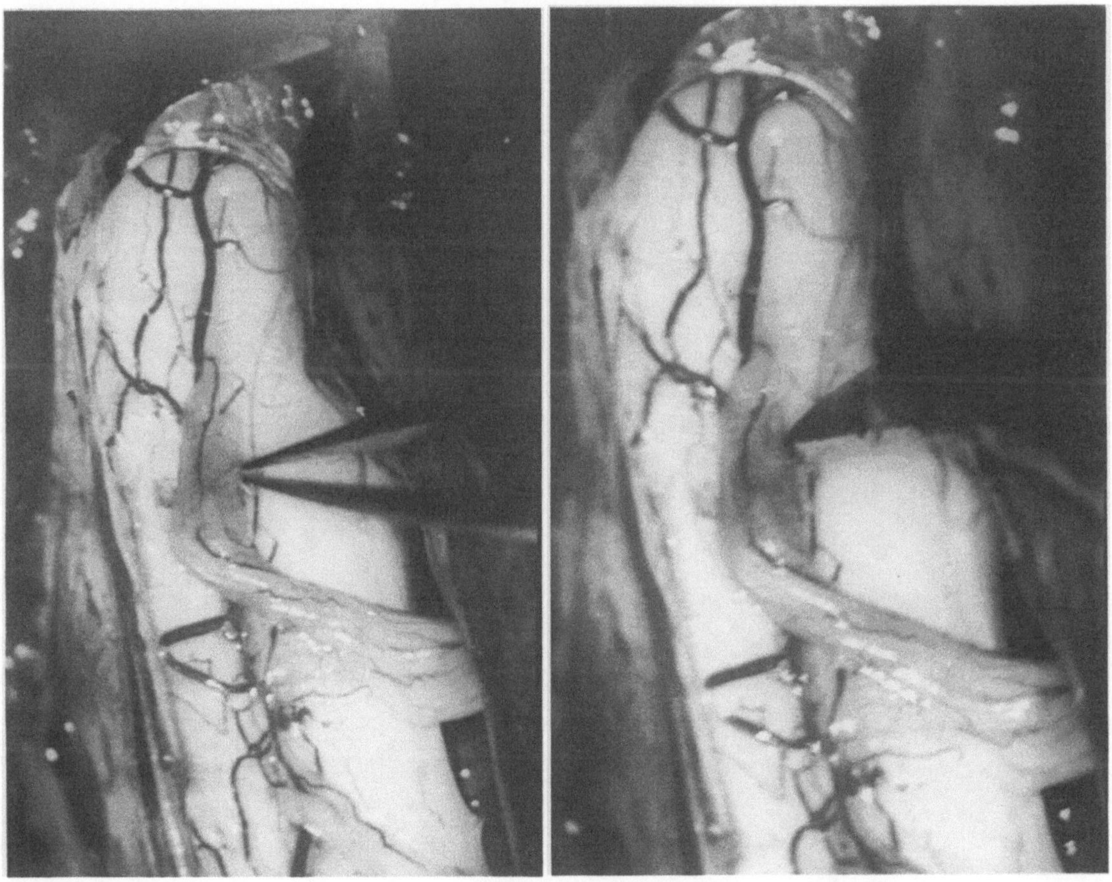

Fig. 2. Selective posterior rhizotomy in the DREZ, at the cervical level. Posterolateral aspect of the spinal cord under the microscope during selective posterior rhizotomy at the cervical level. Left: the right C 6-root (which has 6 rootlets) has been retracted toward the midline to make the lateral region of the spinal cord-rootlet junction accessible. This lateral region is the site of small pial vessels which are coagulated by means of a bipolar forceps. Right: the lateral selective incision is made along the 6 rootlets in the lateral part of the posterolateral sulcus using a small piece of a razor blade. The incision is 2 mm deep in the posterior rootlet-spinal cord junction at an angle of 45° with the posterior plane of the dorsal column

References

1. Sindou M (1972) Study of the dorsal root entry zone. The selective posterior rhizotomy for pain surgery. M.D. Thesis N° 173, Lyon
2. Sindou M, Fischer G, Goutelle A, Schott B, Mansuy L (1974) La radicellotomie postérieure sélective dans le traitement de la spasticité. Rev Neurol 130: 201–215
3. Sindou M, Fischer G, Mansuy L (1976) Posterior spinal rhizotomy and selective posterior rhizidiotomy. Prog. Neurol Surg 7: 201–250
4. Sindou M, Goutelle A (1983) Surgical posterior rhizotomies for the treatment of pain. In: Krayenbühl H *et al* (eds) Advances and technical standards in neurosurgery, vol 10. Springer, Wien New York, pp 147–185
5. Sindou M, Millet MF, Mortamais J, Eyssette M (1982) Results of selective posterior rhizotomy in the treatment of painful and spastic paraplegia secondary to multiple sclerosis. Appl Neurophysiol 45: 335–340
6. Sindou M, Mifsud JJ, Boisson D, Goutelle A (1986) Selective posterior rhizotomy in the dorsal root entry zone for treatment of spasticity in the hemiplegic upper limb. Neurosurg (in press)
7. Sindou M, Pregelj R, Boisson D, Eyssette M, Goutelle A (1985) Surgical selective lesions of nerve fibers and myelotomies for modifying muscle hypertonia. In: Eccles, Dimitrijevic (eds) Recent achievements in restorative neurology 1: Upper motor neuron functions and dysfunctions. Karger, Basel, pp 10–26
8. Sindou M, Quoex C, Baleydier C (1974) Fiber organization at the posterior spinal cord-rootlet junction in man. J Comp Neurol 153: 15–26

Acta Neurochirurgica, Suppl. 39, 103–105 (1987)

Spinal Cord Stimulation Improves Motor Performances in Hemiplegics: Clinical and Neurophysiological Study*

B. Cioni** and **M. Meglio**

Istituto di Neurochirurgia, Università Cattolica, Roma, Italy

Summary

We studied the effects of spinal cord stimulation (SCS) on motor performances in patients with spastic hemiparesis due to cerebrovascular ischemic accident.

11 patients were evaluated before and after 7 days of SCS by means of the Albert's motor scale and a surface polyelectromyography. SCS significantly improved motor performances in 63% of the patients. It reduced agonist-antagonist coactivation and clonus. Such an effect was particularly evident during voluntary movements and gait.

Keywords: Spinal cord stimulation; hemiplegia.

Introduction

Spinal cord stimulation (SCS) has been used to manage spastic motor disorders since 1973[5]. A large literature reports on its effectiveness in ameliorating abnormal motor control of spinal or brain-stem origin[2–4, 6–8, 10]. Few reports deal with the modifications of motor disturbances from cerebrovascular accidents[9, 11].

The purpose of this study was to check the effectiveness of SCS in patients with spastic hemiplegia from ischemic stroke and to quantitative the modifications of motor performances by means of clinical and neurophysiological tests.

Material and Method

11 patients were selected for the study. They met the following requirements: 1. spastic hemiparesis from ischemic accident in the territory of the middle cerebral artery, documented by CT scan; 2. stabilized motor dysfunction (time from the ictus to SCS: 3 months—4 years); 3. no antispastic medication for a period of two weeks.

SCS was performed via a percutaneously implanted epidural electrode facing the dorsal aspect of the cervical cord. Rectangular pulses of 0.2 msec duration, 80 c/sec and intensity threshold for paresthesia were delivered continously. All patients were studied before and after 5 days of SCS according to the following protocol: 1. clinical evaluation utilizing the Albert's motor scale[1]; 2. neurophysiological evaluation: surface polyelectromyography of the upper and lower extremities paretic muscles recorded (time constant: 0.03 msec, paper speed: 15 mm/sec, amplification: 100 µv/cm) during spontaneous relaxation, voluntary and passive specific movements, provoked clonus, phasic stretch reflexes and Babinski's response and free gait.

Results

The mean total score of the Albert's motor scale changed from 81.9 to 91.7, following SCS. It was significantly increased in 7 patients, in 1 was unchanged, but in no case was decreased. This result was mainly due to the improvement in the quality and endurance of upper and lower extremity voluntary movements. The analysis of the surface EMG showed that SCS broke down pathological patterns of muscle coactivation during voluntary movements in $^8/_{11}$ patients (Fig. 1). Clonus was greatly reduced in $^8/_{10}$ patients. The phasic stretch reflexes were unchanged as far as amplitude was concerned, while the after discharge was reduced in $^7/_9$ patients (Fig. 2). The EMG activity provoked by passive movements was decreased

---

* Supported by MPI 71365.

** B. Cioni M.D., Istituto di Neurochirurgia, Università Cattolica, Largo A. Gemelli 8, I-00168 Roma, Italy.

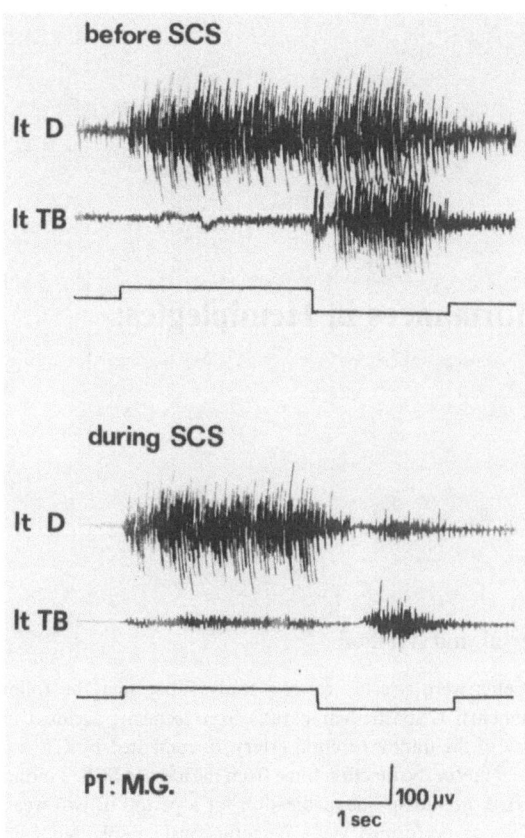

Fig. 1. EMG recording from deltoid (*D*) and triceps brachii (*TB*) of a left (*lt*) hemiparetic patient during voluntary shoulder lifting (black marker up) and lowering (marker down). A better agonist-antagonist coordination was evident during SCS

in 3 cases, unchanged in the remaining 8. No significant modification of the Babinski's response was noticed. Gait was improved in $^8/_{11}$ patients. The EMG recordings showed a more phasic activity with the appearance of the silent period and a better agonist-antagonist coordination. Clonus in triceps surae during the stance phase was markedly reduced, when present (4 cases).

Conclusions

Our study demonstrates that SCS can improve motor performance in patients with hemiparesis due to cerebrovascular ischemic accident. The clinical effectiveness of SCS was confirmed by EMG recordings. In $^7/_{11}$ patients the considerable amelioration led us to implant a permanent system for SCS.

Our data, though very preliminary, seem to suggest a possible role of SCS in the rehabilitation of poststroke hemiparetics.

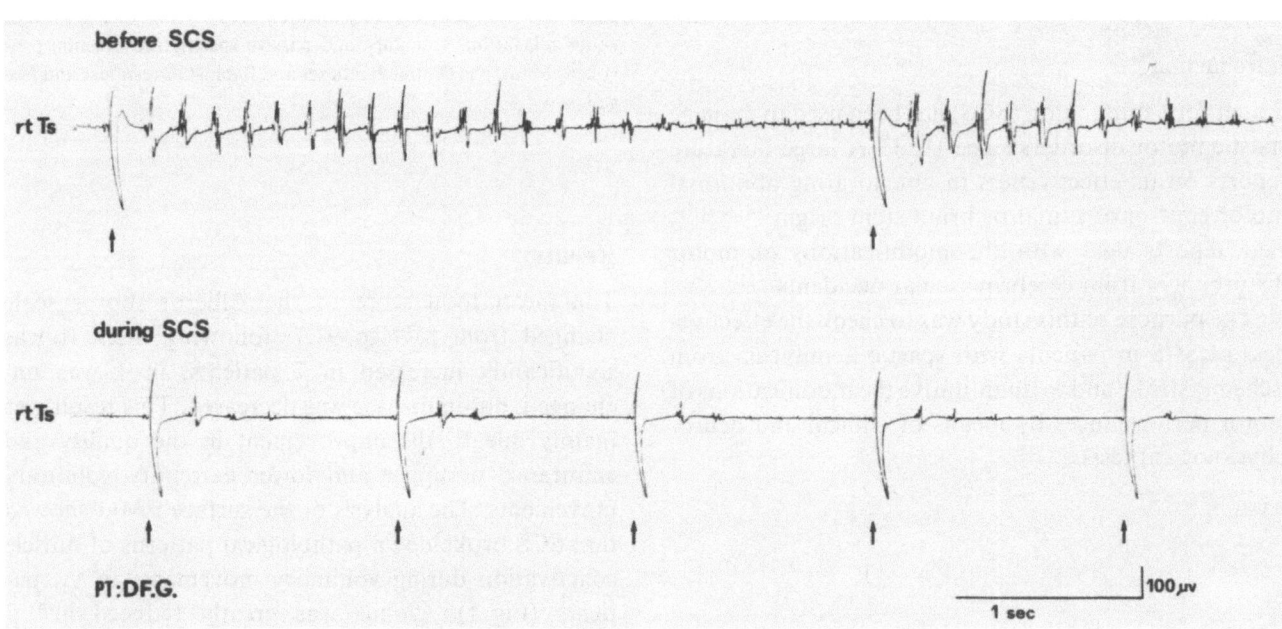

Fig. 2. Phasic stretch reflexes mechanically evoked by tapping the Achilles tendon with a hammer (arrow). EMG recording from right triceps surae (*rt TS*). The after discharge was markedly reduced during SCS

References

1. Albert A (1965) Bilan fonctionnel de l'hémiplégique adulte. Kinésithérapie 94: 13–33
2. Barolat-Romana R, Myklebust JB, Hemmy DC, Myklebust B, Wenninger W (1985) Immediate effects of spinal cord stimulation in spinal spasticity. J Neurosurg 62: 558–562
3. Campos RJ, Dimitrijevic MM, Dimitrijevic MR, Sharkey PL (1979) Suppression of decerebrate rigidity in patients with brainstem lesions by continuous spinal cord stimulation. Acta Neurol Scand 60 [Suppl] 73: 224
4. Campos RJ, Dimitrijevic MM, Faganel J, Sharkey PC (1981) Clinical evaluation of the effects of spinal cord stimulation on motor performances in patients with upper motor neuron lesions. Appl Neurophysiol 44: 141–151
5. Cook AW, Weinstein SP (1973) Chronic dorsal column stimulation in multiple sclerosis. NY State J Med 73: 2868–2872
6. Davis R, Gray E, Kudza J (1981) Beneficial augmentation following dorsal column stimulation in some neurological diseases. Appl Neurophysiol 44: 37–49
7. Dooley DM (1977) Demyelinating, degenerative and vascular diseases. Neurosurg 1: 220–224
8. Illis LS, Sedgwick EM, Tallis R (1980) Spinal cord stimulation in multiple sclerosis: clinical results. J Neurol Neurosurg Psychiatry 43: 1–14
9. Nakamura S, Tsubokawa T (1985) Evaluation of spinal cord stimulation for postapoplectic spastic hemiplegia. Neurosurg 17: 253–259
10. Siegfried J, Lazorthes Y, Broggi G (1981) Electrical spinal cord stimulation for spastic movement disorders. Appl Neurophysiol 44: 77–92
11. Waltz JM, Pani KC (1978) Spinal cord stimulation in disorders of the motor system. In: Advances in external control of human extremities. Yugoslav Comitee for Electronics and Automation, Belgrade, pp 545–556

Acta Neurochirurgica, Suppl. 39, 106–111 (1987)
© by Springer-Verlag 1987

High-Frequency Cervical Spinal Cord Stimulation in Spasticity and Motor Disorders*

J. Broseta**,[1], G. Garcia-March[1], M. J. Sánchez-Ledesma[1], J. Barberá[2], and J. González-Darder[2]

[1] Departamento de Neurocirugía, Hospital Clínico Universitario, Salamanca, Spain, [2] Departamento de Neurocirugía, Hospital Mora, Cádiz, Spain

Summary

High frequency stimulation of the cervical spinal cord was used in an attempt to moderate motor disorders in 10 patients, 5 spastics, 3 dystonics and 2 with spasmodic torticollis. Through a C_4 flavectomy a quadripolar flat electrode was introduced into the epidural space, placing the first terminal at C_2. Electrical parameters for stimulation were established at 0.05–0.10 msec pulse width, rate of 200–1,400 Hz and amplitude to paresthesia threshold. Stimulation was delivered 1 hour on and 1 hour off during daytime. Subjective, clinical and/or neurophysiological initial improvement were initially observed in 9 cases, in which the stimulation system was permanently implanted. After a mean follow-up of 41.4 months all patients but one had a clinical and neurophysiological condition almost similar to the prestimulation one.

To investigate the effect of the so-called high-frequency stimulation on the nervous system, 9 dogs were implanted following the same surgical technique and stimulated with analogous electrical parameters than in human practice. After follow-up of 8 months the pathological studies of the animals did not show any local or suprasegmental alterations on nervous structures except a local huge dural scarring reaction.

Keywords: Cerebral palsy; dystonia; spasmodic torticollis; spinal cord stimulation.

Introduction

Based on the preliminary results reported in early studies using spinal cord stimulation in the treatment of spasticity[1, 2] and other motor disorders[3, 4], Waltz[9] designed an investigational protocol on this problem, obtaining in the first large series very satisfactory results and concluding that the higher clinical benefits occurred in cases with the electrode placed in the cervical area between C_2 and C_4 using in stimulation rates of 1,200 to 1,400 Hz.

Encouraged by this positive experience, five years ago our team began to use high-frequency cervical spinal cord stimulation attempting to control motor disturbances in a pilot group of 10 patients, whose long-term results are here reported. To investigate the possible effect of this type of stimulation on nervous tissue an experimental study in dogs using similar surgical technique an electrical parameters than in human practice was done.

Materials and Methods

Patient Population: Table 1 summarizes the general data, diagnosis, neurological condition and evaluation of the group. Spasticity was attributed to multiple sclerosis in 3 cases, stroke in 1 and cerebral palsy in 1; axial dystonia was present in 3 cases, in 2 associated with abnormal movements and in 1 with retrocollis, and spasmodic torticollis was evident in 2 cases. All patients were chronically resistent to every pharmacological approach and case 4 prior to stimulation underwent a VL thalamotomy and stereotactic bilateral dentatectomy without any effect. The neurological condition was evaluated scoring the functional group disturbances (pyramidal, cerebellar, brain stem, sensory, visceral, visual and mental functions), neurological examination, associated involuntary movements, some skilled acts (dressing, eating, bathing, cutting meat, buttoning, brush teeth, comb hair, tie shoe, turning electrical appliances, swallowing, speech quality, handwritting, etc) and in spastics the Disability Status Scale[5]. During the neurophysiological study conventional electromyographic recording, H reflex, H max/M max ratio, T wave amplitude, H max/T max ratio, clonus recording and silent period were measured in every patient in the different controls.

---

  * This study was supported by a grant (DGPC-270882) from the Ministry of Education and Science of Spain.

  ** J. Broseta, M.D., Departamento de Neurocirugía, Hospital Clínico Universitario, E-37007 Salamanca, Spain.

Table 1. *Global Clinical Data*

Case	Age, sex	Diagnosis	Neurological condition	Neurological evaluation		
				Functional groups	Kurtzke's scale	Skilled acts
1	43, male	multiple sclerosis	spastic paraparesia, spasms, neurogenic bladder, walk with canes	6	6	19
2	45, female	multiple sclerosis	spastic quadriparesia, cerebellar and brain stem symptoms, bed restricted	10	7	23
3	26, male	poststroke spasticity	moderate spastic hemiplegia	3	3	21
4	51, female	dystonia	axial dystonia, abnormal movements and tremor in upper limbs	0		30
5	55, female	spasmodic torticollis	spasmodic torticollis	0		19
6	42, female	spasmodic torticollis	spasmodic torticollis	0		17
7	39, male	multiple sclerosis	spastic paraplegia, spasms, wheel chair	9	7	21
8	32, male	dystonia	axial dystonia	0		24
9	24, male	cerebral palsy	spastic quadriparesia, dystonia, abnormal movements, cerebellar and brain stem symptoms	9	7	36
10	32, female	dystonia	axial dystonia, retrocollis	0		22

Table 2. *Results*

Case	Global early results	Follow up	Neurological late results			Neurophysiological late results	Global late results
			Functional groups	Kurtzke's scale	Skilled acts		
1	subjective improvement	57 months	6	6	19	no change	no change
2	no change, system was not implanted						
3	neurophysiological improvement	55 months	3	3	20	no change	slightly improved
4	subjective improvement	54 months	0		30	no change	no change
5	clinical improvement	54 months	0		11	improved	improved
6	subjective improvement	22 months*	0		17	no change	no change
7	neurophysiological improvement	50 months	9	7	23	no change	no change
8	subjective improvement	49 months	0		24	no change	no change
9	neurophysiological and clinical improvement	26 months**	9	7	35	no change	no change
10	subjective improvement	37 months	0		17	no change	retrocollis slightly improved

* In this case a VOI thalamotomy was performed after 22 months of follow-up.

** The stimulation device was removed due to late local infection on the connector.

Surgical Technique and Stimulation Procedure: For spinal cord stimulation a flat electrode with four contacts was introduced in the epidural space through a C_4 flavectomy (Fig. 1 A), performed under general anesthesia and the patient in sitting position. Then the electrode was manipulated under fluoroscopy until the first terminal was located at C_2 and midline (Fig. 1 B), bringing out through the skin the percutaneous extensions for temporal stimulation. During trial period the most benefitial combination between contacts and effective electric parameters would be found. The former generally coincided with those combination reproducing electric paresthesias in the affected motor area; the latter used to be constant on pulse duration (0.05 to 0.10 msec) and amplitude (paresthesia threshold) were not uniform respecting rates. Patient's response to frequency was variable, without confirming a better effect using 1,200 to 1,400 Hz, but in 3 cases. In this period stimulation was used 1 hour on and 1 hour off during the daytime. When patients showed some clinical, neurophysiological and/or subjective improvement the system was totally implanted, adapting the ideal combination of contacts to a radiofrequency receiver, placed subcutaneously, and delivering the stimulation by means an external transmitter*.

Experimental Material and Methods: Using the same surgical technique this quadripolar electrode was implanted in 9 dogs. Similar electrical parameters were programed, delivered by an external transmitter. In the animals high-frequency cervical spinal cord stimulation was continuous. After 8 months of stimulation they were sacrified and the nervous specimens referred for pathological study.

Results

Clinical Results: Table 2 summarizes the early and late results of the series after a mean follow-up of 41.4 months. In 1 out of the 10 cases the system was not definitively implanted. The others initially showed some clinical amelioration, consisting in an improvement of spasmodic torticollis in 1 case and spasticity with a better axial stability in an other case; or neurophysiological changes in 3 cases. Otherwise the majority of patients presented a subjective improvement, muscular relaxation and decrease of spasms, occasionally associated some increase in the movement ranks, but without any clinical correlation. In these early controls involuntary movements were unchanged. Later 6 out of the 9 cases remain unchanged, with no variation in the final scoring of functional groups, skilled acts or Kurtzke's scale. Only 1 case of spasmodic torticollis is almost symptom-free following 54 months of using stimulation (Fig. 2); an other case with dystonia is partially alleviated from the retrocollis component; and finally an other case with poststroke spasticity showed a slight improvement in walk ability.

---

* Stimulation equipment was provided by Medtronic Inc., Minneapolis, Minn.

During the first year follow-up 4 out of the 5 spastics showed some changes in repetitive neurophysiological studies, observing a temporal decrease of H max/ M max ratio and a 4 mV mean reduction in T wave amplitude. As illustrated in Fig. 3 these changes were transitory returning to the previous values. Moreover these neurophysiological fluctuations were not related with an objective clinical amelioration.

There were two adverse effects due to the technique, a moderate and transitory spinal cord compression after electrode implantation with a further favorable course and a late local infection on the connector with extrusion of the system and its removal, that was not replaced considering the lack of effect.

Experimental Results: Gross examination of the cervical area revealed a constant finding of a large fibrous reaction around the electrode which involved the local dura and caused a substantial cord compression (Fig. 4 A). Histologic study of this scarring reaction showed proliferation of collagenic and reticulin fibers with abundant fibroblasts and local lymphocyte infiltration (Fig. 4 B). Spinal cord, brain stem, cerebellum and thalamic structures showed no other changes.

Discussion

Waltz[10] in a large series using this technique for motor disorders reported an impressive percentage of excellent long-term results, observing major amelioration in 85% of cases with spasticity resulting from cerebral palsy, in 65% from multiple sclerosis, in 74% from trauma, in 31% from stroke, in 66% of cases with dystonia and in 60% of cases with spasmodic torticollis. Oakley and Reynolds[7] also presented satisfactory results in 6 cases with diverse motor disturbances treated with this type of stimulation. Unfortunately our results do not confirm these improvement, when only 1 out of 10 patients has had some relief in her spasmodic torticollis. Scerrati *et al.*[8] found a 20% reduction of H/M ratio in 3 cases with spasticity using spinal cord stimulation with rates between 50 and 120 Hz. In our spastic cases this neurophysiological improvement was not constantly and not permanently observed.

In our experience the results obtained with high-frequency cervical spinal cord stimulation to moderate spasticity and other motor disorders has been poor and there are few indications for its use. Moreover, the use of stimulation with high rates of 1,000 to 1,400 Hz in our cases did not produce clinical benefit, and did not

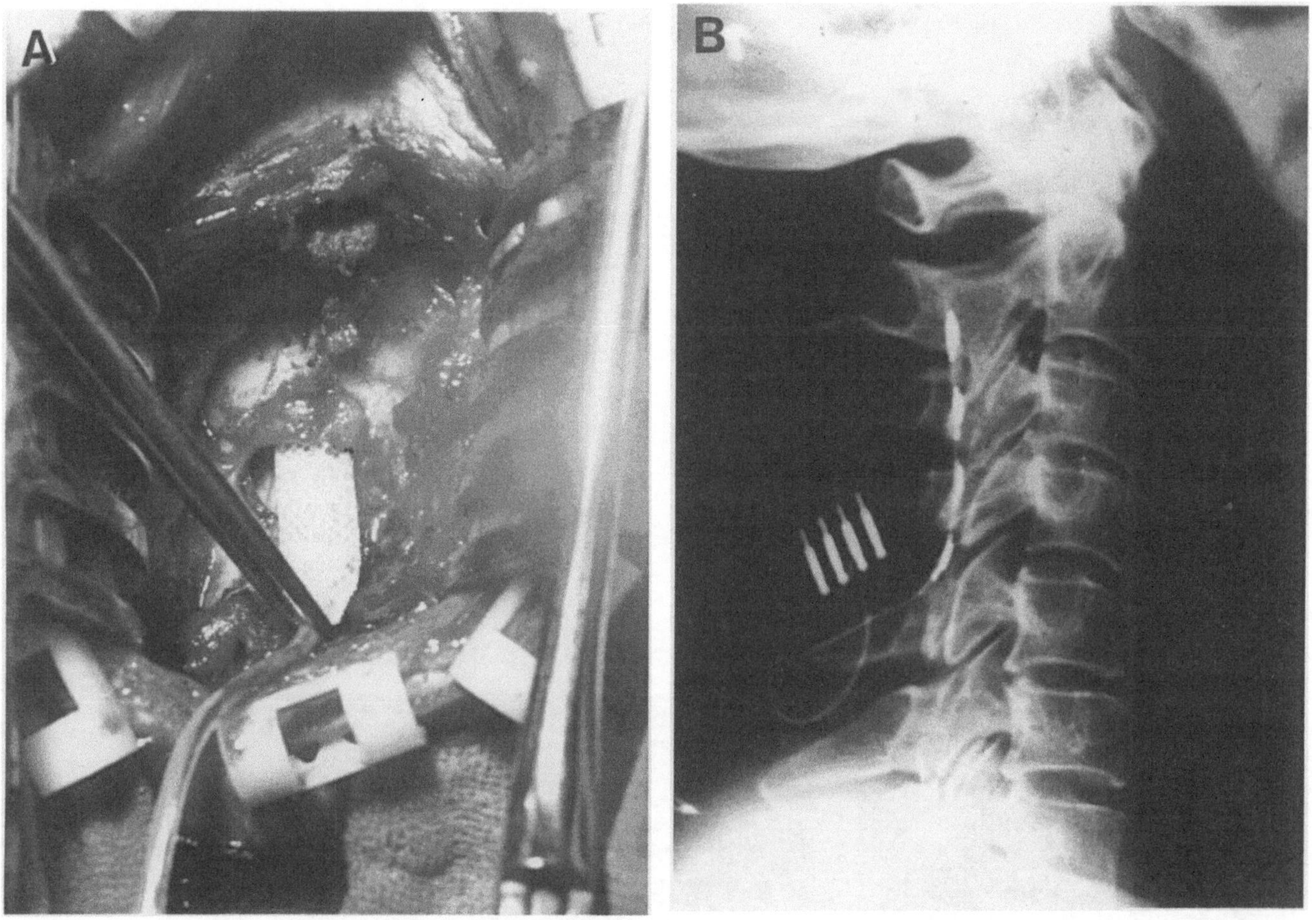

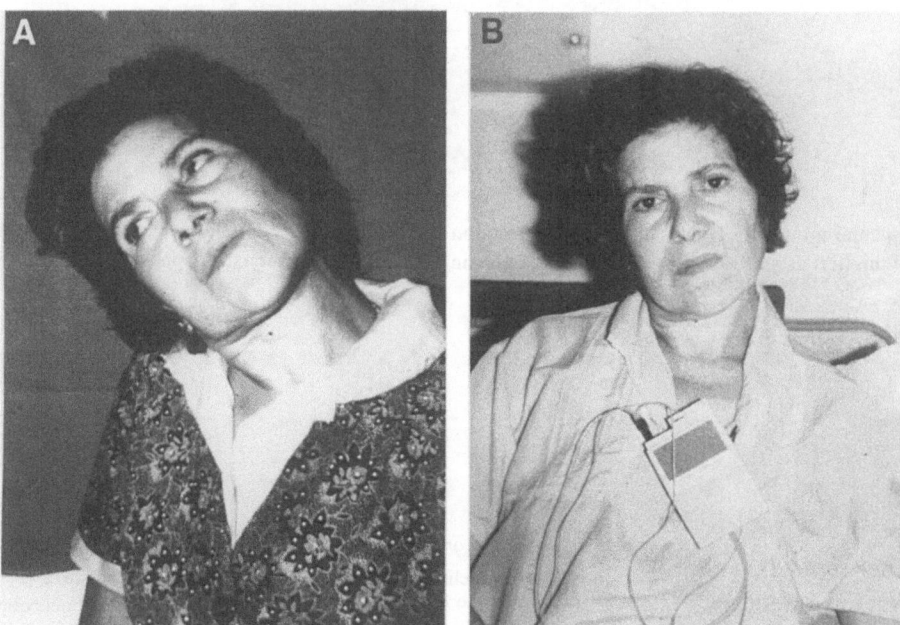

Fig. 1. A) Introduction of the quadripolar flat electrode in the epidural space through a C$_4$ flavectomy. B) Radiological control of the electrode position at the cervical area with the first terminal placed at C$_2$

Fig. 2. Case 5 presenting spasmodic torticollis. A) Prestimulation clinical condition. B) Poststimulation control

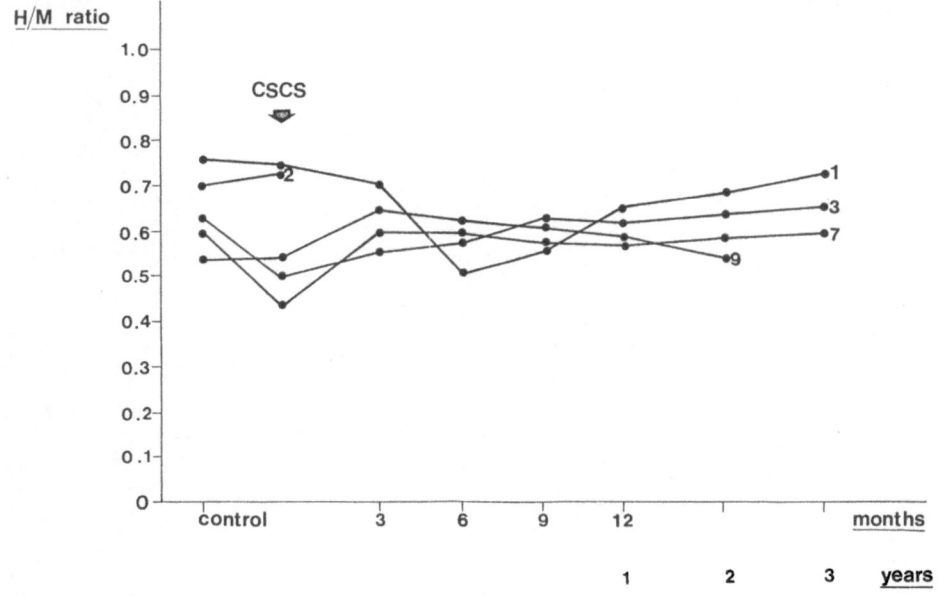

Fig. 3. Evolutive changes of H max/M max ratio of patients with spasticity during the follow-up with stimulation in use (*CSCS:* cervical spinal cord stimulation)

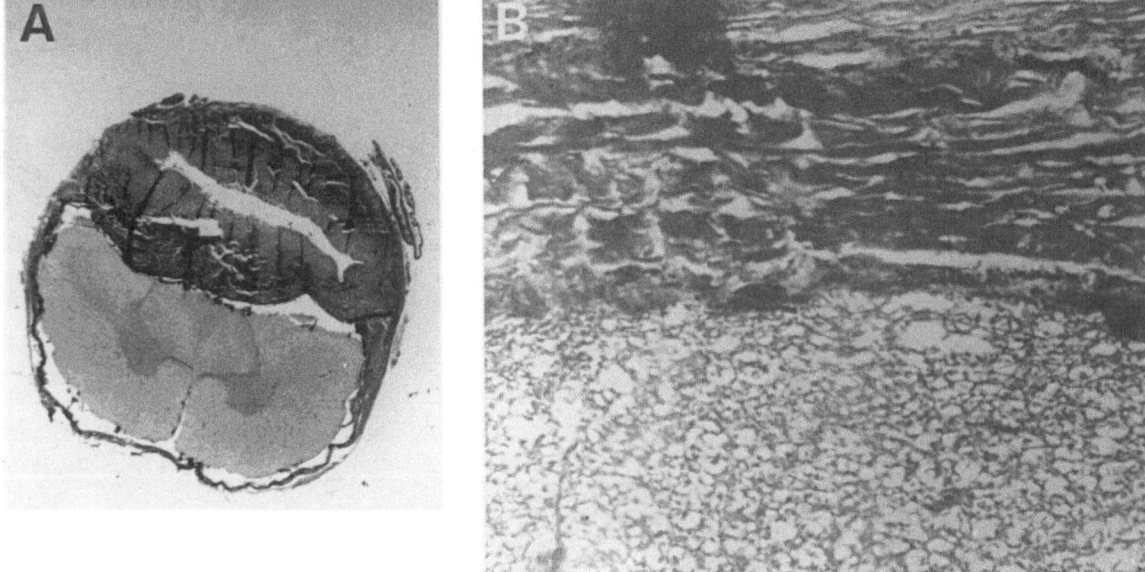

Fig. 4. Pathological findings in the experimental group: A) Huge dural scarring reaction around the electrode that caused a spinal cord compression. B) Hystological appearance of this fibrous reaction with collagenic and reticulin proliferation, in contact with posterior spinal cord pathways

improve the results reported in other series that proposed for these motor disturbances a low frequency of 100 to 120 Hz[6].

References

1. Cook AW (1973) Chronic dorsal column stimulation in multiple sclerosis: Preliminary report. NY St J Med 73: 2868–2872
2. Cook AW (1974) Stimulation of the spinal cord in motor neurone disease. Lancet 2: 230–231

3. Dooley DM (1976) Electrical stimulation of the spinal cord in patients with demyelinating and degenerative diseases of the central nervous system. J Flo Med Ass 63: 905–909

4. Gildenberg PL (1976) Treatment of spasmodic torticollis by dorsal column stimulation. Rev Inst Nal Neurologia 10: 11–15

5. Kurtzke JF (1955) New scale for evaluating disability in multiple sclerosis. Neurology 5: 580

6. Lazorthes Y, Siegfried J, Broggi G (1981) Electrical spinal cord stimulation for spastic motor disorders in demyelinating disease. A cooperative study. In: Hosobuchi Y, Corbin T (eds) Indications for spinal cord stimulation. Excerpta Medica, Amsterdam, pp 48–57

7. Oakley J, Reynolds A (1982) High frequency cervical epidural stimulation for spasticity. Appl Neurophysiol 45: 93–97

8. Scerrati M, Onofri M, Pola P (1982) Effect of spinal cord stimulation on spasticity: H-reflex study. Appl Neurophysiol 45: 62–67

9. Waltz JM (1978) Spinal cord stimulation in disorders of the motor system. In: Proceedings of the Sixth International Symposium on External Control of Human Extremities, Belgrado, pp 545–555

10. Waltz JM (1982) Computerized percutaneous multilevel spinal cord stimulation in motor disorders. Appl Neurophysiol 45: 73–92

Acta Neurochirurgica, Suppl. 39, 112–116 (1987)

Application of SCS for Movement Disorders and Spasticity

A. Koulousakis*, U. Buchhaas, and **K. Nittner**

Department of Stereotaxis, University of Cologne, Cologne, Federal Republic of Germany

Summary

20 patients with movement disorders and spasticity were treated with SCS to decrease tonus and improve impaired motor function. 12 patients with multiple sclerosis had gait disorders and slight up to considerable increase in tonus. We observed quantitative changes in diminution of spasticity, voluntary motor function, bladder function, lessening of ataxia and pain relief. During a follow-up of up to 4 years no deterioration in the patients' condition could be observed. 11 patients reported a more fluent gait in 6 cases accompanied by a decrease in tonus. Increase in tonus and deterioration of gait pattern were observed after break down of the stimulation. Although the indications for electrostimulation still remains uncertain in some cases with definite diagnosis without any progressive character of the disease and without severe neurological disturbances improvement can be obtained.

Keywords: SCS; movement disorders; spasticity.

Introduction

First reports[1–3, 5] about the application of SCS in patients suffering from multiple sclerosis or other types of degenerative or demyelinic diseases or movement disorders were received with scepticism because one could not explain the effect on movements by stimulating the dorsal column[6, 9]. In 1981 Siegfried *et al.*[7] authoritatively reported about 164 of his own and 800 patients from the literature between 1973 and 1980. Some authors reported very bad results, others quite satisfying ones[2, 6]. Klinger *et al.*[4] reported an improvement of movement disorders of 20–30%, however they pointed out that it was difficult to impartially analyse and quantify the success of SCS.

Since 1983 multiprogrammable pulse generators (Itrel) are used—particularly in handicapped patients—with more combinations of stimulation parameters, and better control of the type and duration of the stimulation; this has made us critically reconsider the results of SCS in patients with spinal spasticity.

Patients and Methods

In the last 4 years 20 patients with spinal spasticity received treatment by means of electrodes implanted into the spinal canal (Fig. 1). 12 patients suffered from multiple sclerosis, 4 of them with a quadriparesis and 8 with paraplegia. 3 patients suffered from myelopathy, 3 patients had degenerative or vascular diseases of the spinal cord and 2 patients had cerebral palsy.

According to international reports[5] we implanted multipolar resume electrodes by means of laminectomy. During the last 3 years we preferred Sigma electrodes which are percutaneously inserted, in patients with parapareses in the area of the lumbar enlargement and in the case of quadriparesis at the cervical region. The electrode was attached to the pulse-generator when a more fluent and steadier gait or a decrease in tonus were observed. Previously the electrode was mostly attached to a subcutaneously implanted RF receiver, but since 1983 we have mostly used ITREL generators. Out of 20 implantations 14 electrodes were attached to an ITREL generator.

Patients with multiple sclerosis were only treated by means of implanted electrodes to improve their residual motor function if the disease had not shown any progression in the course of the last 2 years. In all patients diagnosis was confirmed by immunological or radiological test.

The patients were regularly examined after 3, 6, and 12 months and thereafter annually. We then particularly noted tonus and gait pattern. The patients and their relatives as well as the nursing staff, were asked about special abilities concerning writing, sitting, dressing and undressing, standing, and walking unaided or with canes. The tonus was regularly noted although we know that it may change spontaneously during the day. We agree with Siegfried[7] that subjective and clinically objective improvement of movement disorders may be a better functional evaluation of the method than for example an electromyography.

---

* A. Koulousakis, Department of Stereotaxis, University of Cologne, D-5000 Köln 41, Federal Republic of Germany.

Total number (n=20) of cases with S.C.S.

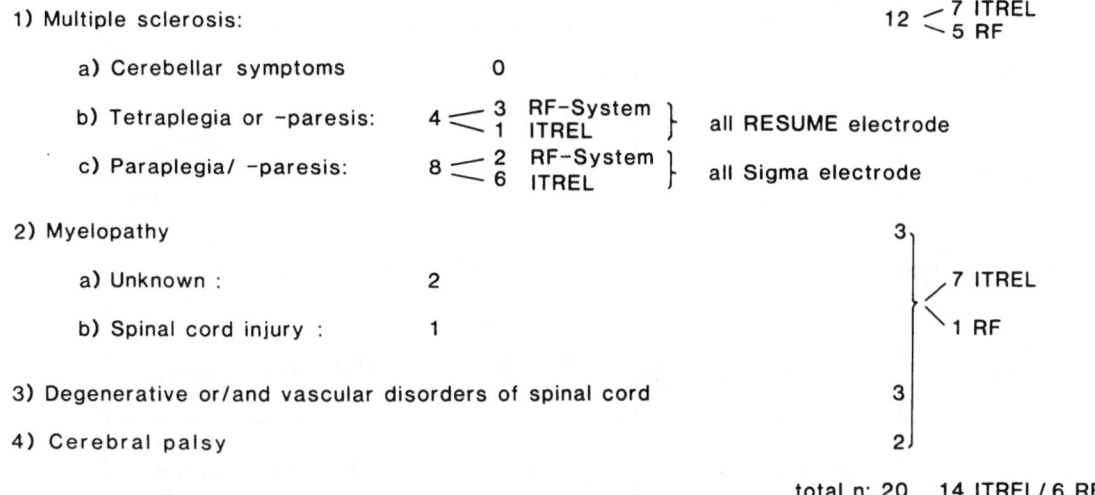

1) Multiple sclerosis: 12 ⟨ 7 ITREL / 5 RF

    a) Cerebellar symptoms   0

    b) Tetraplegia or –paresis: 4 ⟨ 3 RF-System / 1 ITREL   } all RESUME electrode

    c) Paraplegia/ –paresis: 8 ⟨ 2 RF-System / 6 ITREL   } all Sigma electrode

2) Myelopathy   3

    a) Unknown :   2   7 ITREL

    b) Spinal cord injury :   1   1 RF

3) Degenerative or/and vascular disorders of spinal cord   3

4) Cerebral palsy   2

                                 total n: 20   14 ITREL/ 6 RF

Fig. 1. Patients with SCS and movement disorders (n = 20)

Pat.	Elec.	pain		spasticity		motor function		coordination		standing		walking ability		improved life style
O., G. ♀ 52 y.	C2·C5 R RF	∅	∅	↑↑		↓(↓)		↓		∅		wheel- chair		
P., H. ♂ 36 y.	C2·C5 R ITREL	∅	∅	↑↑(↑)	↑↑(↑)	↓↓↓	↓↓↓	∅	∅	∅	∅	bed – ridden	bed – ridden	NO
E., M. ♀ 61 y.	C2·C5 R RF	∅	∅	(↑)	(↓)	↓↓	↓↓	↓(↓)	↓(↓)	(↑)	(↑)	walker	one cane auto- nomous	YES	walking standing bladder nursing
P., A. ♂ 39 y.	C2·C4 R RF	∅	∅	↑↑(↑)	↑(↑)	↓↓	↓↓	↓(↓)	↓(↓)	∅	∅	wheel- chair	wheel- chair	(YES)	nursing speech gymnastics

R=RESUME, ↑ increase, ↓ decrease , ∅ not impaired or not possible | pre–op. | post–op. |

Fig. 2. Clinical data and the effect of SCS in tetraplegics MS patients (n = 4)

Results

MS with quadraparetic spasticity (Fig. 2)

The early and also the late results depend on the neurological findings at the time of the operation and the patients' expectations. Patients with quadraparetic spasticity had the worst outcome. In one patient, who could hardly move her hands and who suffered from very intense spasticity, no change of the neurological state and life-quality were noted. Another patient who had had too high an expectation as to the method personally reported no improvement although we found a decrease in spasticity as well as an improvement in speech and in nursing. Only one patient, with a quadraparetic spasticity who however had only a slight increase in tonus and could walk with canes, noticed an improvement of standing and walking ability, of urinary dysfunction and of nursing.

Pat.	Elec.	pain		spasticity		motor function		coordination		standing		walking ability		Improved life style
S., J. ♂ 60 y.	D9 D11 RF	↑(↑)		↑↑(↑)		↓↓(↓)		(↓)		ø		wheel-chair		
S., H. ♀ 37 y.	D10/11 D11 RF	ø	ø	(↑)	ø	↓(↓)	↓(↓)	(↓)	(↓)	↑(↑)	↑↑	one cane	autonomous	YES walking bladder endurance
F., W. ♂ 38 y.	D1 ITREL 2x	ø	ø	↑↑	↑(↑)	↓	↓	ø	ø	↑↑	↑↑	autonomous	autonomous	YES walking writing bladder
G., M. ♀ 45 y.	D10 ITREL 2x	↑	↓)	↑↑	↑	↓↓(↓)	↓↓(↓)	↓(↓)	↓(↓)	ø	ø	wheel-chair	wheel-chair	YES sitting pain relief
B., W. ♂ 62 y.	D9 ITREL	ø	· ø	↑↑	↑	↓(↓)	↓(↓)	↓	↓	(↑)	↑	autonomous	walker	YES walking sitting standing
H., H. ♀ 42 y.	D10 ITREL	ø	ø	↑↑	↑↑	↓(↓)	↓(↓)	(↓)	(↓)	↑(↑)	↑(↑)	one cane	one cane	NO
S., A. ♀ 38 y.	D11 ITREL 2x	ø	ø	(↑)	ø	(↓)	(↓)	(↓)	(↓)	↑↑	↑↑	autonomous	autonomous	YES walking endurance climb stairs
P., A. ♂ 53 y.	D10 ITREL	ø	ø	↑↑	↑(↑)	↓↓	↓↓	↓(↓	↓(↓)	↑	↑	two canes	one cane	YES walking endurance

↑ Increase,   ↓ decrease,   ø not impaired or not possible | pre-op. | post-op. |

Fig. 3. Clinical data and the effect of SCS in paraplegics MS patients (n = 8)

Pat.	Elec.	Pain		Spasticity		motor function		coordination		standing		walking ability		Improved life style
W., K. ♂ 64 y.	D9 D10/11 ITREL	↑↑	ø	↑↑	(↑)	↓	↓	ø	ø	↑	↑↑	two canes wheel-chair	autonomous	YES walking standing endurance pain relief
R., C. ♀ 62 y.	D10 ITREL	↑	↑	↑(↑)	(↑)	↓↓	↓↓	(↓)	(↓)	(↑)	(↑)	two canes	two canes	NO bladder
A., E. ♂ 82 y.	D9 ITREL x2	↑	ø	↑	(↑)	↓	↓	ø	ø	(↑)	↑	wheel-chair	one cane ↓ two canes	YES walking standing bladder pain relief

↑ Increase,   ↓ decrease,   ø not impaired or not possible | pre-op. | post-op. |

Fig. 4. Clinical data and the effect of SCS in myelopathy patients (n = 3)

MS with paraparetic spasticity (Fig. 3):

We have 7 histories ranging from 3 months up to 4 years of 8 patients altogether. In 6 of 7 patients we found a slight to a distinct decrease of spasticity. Two patients with a slight increase in tonus before the operation showed a normal tonus or even hypotonus postoperatively. We found no improvement of motor function or coordination. 6 of 7 patients were satisfied with the result of the treatment, they reported advances in standing, walking, sitting up from bed into a wheel chair. 50% of the above reported an improved urinary function, all of them reported an increase in endurance which could also be observed by the physiotherapists.

Myelopathy (Fig. 4)

3 patients with a myelopathic syndrome and with a medium to high degree of spasticity mainly received treatment for pain. Two of them reported pain relief up to complete freedom from pain accompanied by a slight or distinct decrease in spasticity. Two other patients achieved improvement of standing and walking ability, one patient could walk up to 50 m without any help. In the case of the third patient we observed a decrease in spasticity and an improvement of urinary function, but she herself was displeased with the result.

In the group of patients with degenerative and vascular disorders of the spinal cord (Fig. 5) we also

pat.	Elec.	pain		spasticity		motor function		coordination		standing		walking ability		Improved life style	
K., M. ♂ 50 y.	D9 D10 RF	ø	ø	↑↑	↑↑	↓	↓(↓)	↓	↓	↑↑	↑	autonomous	one cane	NO	
E., A. ♀ 51 y.	D10 ITREL	↑↑	ø	(↑)	(↓)	ø	ø	(↓)	ø	↑↑	↑↑	autonomous	autonomous	YES	walking pain relief endurance
W., A. ♂ 19 y.	D11 ITREL	ø	ø	↑(↑)	↑	(↓)	(↓)	↓(↓)	↓	↑↑	↑↑	autonomous	autonomous	YES	endurance

↑ Increase   ↓ decrease ,   ø not impaired or not possible | pre-op. | post-op. |

Fig. 5. Clinical data and the effect of SCS in degenerative/vascular disorders of spinal cord (n = 3)

	MS in tetraplegics (n = 3)					MS in paraplegics (n = 7)					myelopathy (n = 3)					deg/vascular disorders (n = 3)					cer. palsy (n = 1)				
	++	+	±	0	-	++	+	±	0	-	++	+	±	0	-	++	+	±	0	-	++	+	±	0	-
dressing		2	1					3	4				1	2					3					1	
undressing		2	1					3	4				1	2					3					1	
eating			3						7					3					3					1	
drinking			3						7					3					3					1	
washing			3					3	4					3					3					1	
writing			3				1	1	5					3					3					1	
sitting		1	2					2	5					3					3					1	
standing		1	2				2	1	4		1	1	1						3					1	
walking		1	2			2	3	2			1	1	1				2			1	1				
climb stairs			3					2	5					3			1	1		1	1				
endurance		1	2			2	2	3				2	1				2			1	1				
speech	1		2						7					3					3					1	
bladder	1	1	1			4	3					2	1						3					1	
nursing	2	1				2	1		4			2	1						3					1	
ataxia			3					6	1			1	2			2	1							1	

++ good   + moderate   ± no change   0 not impaired   − worse function

Fig. 6. Improved or unimproved life style (n = 17). One year after SCS

observed pain relief, decrease in spasticity and in two patients an improvement of coordination. Within a follow-up to 4 years only one patient reported a deterioration in his condition, in whom preoperatively had found a syringomyelia.

In the group of patients suffering from cerebral palsy, we have a follow up of up to 1 year in one patient, who had an improvement of gait pattern and a decrease in tonus.

Fig. 6 shows the result sheet from 17 out of the 20 patients one year after spinal stimulation, with regard to distinct activities. 5 of the 17 patients reported an improvement of standing ability, 11 of the 17 an improvement of walking ability, in 6 of them accompanied by a distinct decrease in tonus. 10 patients reported an advance in endurance, 7 an improved urinary function. All patients and their relatives reported some alleviation in nursing and in physiotherapy. In comparison with patients with an RF receiver, these 14 patients who received an impulse generator did not show any additional advantage. However, a striking observation was that patients with ITREL system more often used a lower frequency.

Discussion

The therapeutic effect of stimulation of the spinal cord still remains controversial and the pathophysiological basis are not yet settled. It is still questionable whether the stimulation affects the ascending or the descending reticular formation[8] or if it has an indirect effect by means of chemical neurotransmitters or perhaps on both.

In four patients (Fig. 7) with a loss of stimulation because of battery depletion and in one further patient with a cable defect we observed an increase in spasticity and a deterioration of gait. This was confirmed by the patients themselves. After the defect had been cleared up the improvement was obvious immediately. To achieve a decrease in tonus one patient had to take 70 mg baclofen, immediately after the operation only 30 mg and 3 months later the baclofen could be taken off her medication.

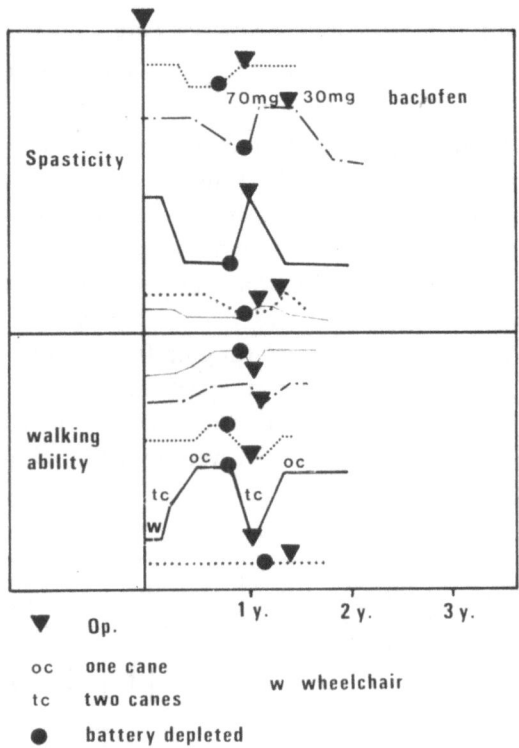

Fig. 7. Special cases with loss of stimulation (n = 5)

References

1. Cook AW, Weinstein SP (1973) Chronic dorsal column stimulation in multiple sclerosis. NY St J Med 73: 2268–2872
2. Cook AW, Taylor JK, Nidzgorski F (1979) Functional stimulation of the spinal cord in multiple sclerosis. J Med Eng Technol 3: 18–23
3. Dooley DM (1975) Percutaneous epidural electrical stimulation of the spinal cord. 43rd Annu Meet Am Ass Neurol Surgeons, Miami Beach
4. Klinger D, Kepplinger B (1982) Quantifikation of the effect of ESES in central motor disorders. Appl Neurophysiol 45: 221–224
5. Ray CD (1977) New electrical stimulation methods for therapy and rehabilitation. Orthop Rev 6: 29–39
6. Rosen JA, Barsoum AH (1979) Failure of chronic dorsal column stimulation in multiple sclerosis. Ann Neurol 6: 66–67
7. Siegfried J, Lazorthes Y, Broggi G (1981) Electrical SCS for spastic movement disorders. Appl Neurophysiol 44: 77–92
8. Waltz JM, Reynolds LO, Riklan M (1981) Multi-lead spinal cord stimulation for control of Motor disorders. Appl Neurophysiol 44: 244–257
9. Young RG, Goodman SJ (1979) Dorsal spinal cord stimulation in the treatment of multiple sclerosis. Neurosurg 5: 225–238

Acta Neurochirurgica, Suppl. 39, 117–120 (1987)

Computer-controlled 22-Channel Stimulator for Limb Movement

R. Davis*, R. Eckhouse, J. F. Patrick, and **A. Delehanty**

Veterans Administration Medical and Regional Office Center, Togus, Maine, moco Inc., Scitvate, Mass., U.S.A., and Nucleus Cochlear Pty. Ltd., Lane Cove, N.S.W., Australia

Summary

The Nucleus multichannel implantable hearing prosthesis (Nucleus Ltd., Sydney, Australia) has been modified by computer programming (MOCO, Inc., Scituate, Mass.) into a 22-channel neural stimulator for use in functional electrical stimulation (FES). Individual or multiple channels can be sequenced and adjusted for their individual pulse amplitude, width and frequency so that activation of single and multiple nerves can be achieved. Anesthetized rabbits' sciatic nerve branches (lateral and medial) were stimulated to produce single, or co-contraction at the ankle and simultaneous bilateral joint movements. The spiral (helix) electrode was also found suitable in these experiments. In planning for the stimulator's use in paraplegic subjects, information was obtained for human lower extremity peripheral nerves. In 9 patients undergoing lower extremity amputation, threshold and maximal levels of stimulation plus nerve diameters were obtained on 21 nerves. The design for the wearable ambulation unit to control the 22-channel stimulator and electrode tresses with helix electrodes has been completed.

Keywords: Neural prosthesis; limb movement.

Introduction

In 1963, Kantrowitz[9] was able to raise a paraplegic subject into a standing position using FES through surface electrodes applied to the quadriceps muscles. Kralj *et al.*[10] in 1973, expanded this concept by designing a program of isotonic exercises for muscle restrengthening through FES, so overcoming the fatigue factor before standing. Petrofsky and Philips[13] did incorporate surface stimulation in a closed-loop system which closely monitored muscle force and limb position. Problems of spastic reflexes initiated by skin surface stimulation, and the lack of specificity plus the inability to penetrate to deep muscles in paraplegic subjects have been overcome by implanting stimulating electrodes directly onto peripheral nerves[14, 6, 4] or using deeply placed percutaneous wire electrodes to muscle motor points[12]. In 1970 Willemon *et al.*[14] implanted epineural cuff electrodes onto the femoral and inferior gluteal nerves in a complete T-5 paraplegic subject who was able to stand in the erect posture. In 1972, Davis and Gesink[6] implanted a T-5 paraplegic subject with bilateral peroneal nerve stimulators fitted with exercise modules; over an 8-month period the calf muscle groups increased in circumference by 24%. In 1973, Cooper *et al.*[4] implanted electrodes onto the femoral and sciatic nerves in a T-11 paraplegic to achieve walker assisted gait up to 40 feet. Brindley *et al.*[3] in 1978 was able to produce standing with crutches and a "swing-through" gait through a series of radio receiver-stimulating systems. Thoma *et al.*[7, 8] implanted a radio linked 16-channel stimulator in 4 paraplegic patients in 1983. Following the training period the 4 subjects could walk 100 meters and stand for 20 minutes. Marsolais *et al.*[12] uses percutaneous wire electrodes to deliver precise patterns of sequencing through 32 channels of stimulation for standing, walking, and stepping for stair climbing; 6 of their 15 subjects were able to walk up to 250 meters. As a result, we are encouraged to develop an implantable 22-channel nerve stimulating system for control of the lower extremities in the paraplegic subject.

Methods

Nucleus Stimulator

From 1979 to March 1986, Nucleus Ltd., has had their multichannel implantable hearing prostheses im-

---

* Ross Davis, M.D., Veterans Administration Medical and Regional Office Center, Togus, ME 04330, U.S.A.

planted in 200 human patients for cochlear disorders with only one radiostimulating unit failure. In January of 1985, Nucleus Ltd., supplied a stimulator to us consisting of a) diagnostic and programming system consisting of the Sanyo NBC-1000 microcomputer with a special interface printed circuit board for its connection to b) speech processor interface (SPI): Z 80-A processor, programming and erasure circuit, c) speech processor (SP) connects into the SPI unit; the SP has an antenna which overlies d) the implantable receiver-stimulator unit (RSU). Nucleus Ltd., adapted a connector so that we could connect into the 22 channels. MOCO, Inc., developed new computer software programs for use with the microcomputer that connected to the external SPI-SP-RSU. This allowed single and multiple channels to be activated according to the program. The electrical characteristic of the stimulus is a biphasic current pulse adjustable in 3 percent steps from 25 µA to 2.3 mA (future units will have 3.5 mA output). The output impedance is greater than 1 Mohm. Pulse width is adjustable in 0.4 µsec steps from 20 µsec to 0.4 msec per phase. It takes approximately 0.42 msec to complete the burst envelope of instructions then a further 0.4 msec as a maximum for each phase 1 and phase 2 of the biphasic waveform. As a result a maximum time of 1.2 msec for each pulse would allow the highest repetition rate of 833 pulses per second (pps) if only one channel is used. However, this is a multiplex system which means that a second or third channel could be activated in between the repetition of the first channel. In the physiological setting, 15 to 50 pps are used, which would allow multiple use of other channels before the first channel is reused at 66 to 20 msec. A series of calibrations of the RSU was done at varying temperatures between 21.8 and 39.0 °C, with no variation occurring except for one test over a 6-month period.

Animal Experiments

Barbiturate anesthetized adult rabbits were used in 4 experiments. The first 2 experiments were conducted using the Medtronic (Minneapolis, MN.) cuff electrode similar to those used in human foot-drop investigations[11]. In the third and fourth animal experiments the electrode used was the Huntington helix electrode[1, 2, 15]. In all 4 experiments the sciatic nerve was exposed with separation of the medial and lateral branches. The threshold found for each branch was 0.1–0.2 mA at 0.2 msec at 50 pps. Maximal stimulation was achieved usually between 0.5–1.0 mA. The lateral branch when stimulated caused paw dorsiflexion whereas the medial

branch caused paw plantar flexion. Both lower extremities were exposed in one animal so that both lateral sciatic branches were stimulated through the multiplex RSU to produce simultaneous dorsiflexion of both paws. Two strain gauges were used to record the anterior and posterior muscle groups of the leg after the tendons were reattached to 2 strain gauges and a recording system. In 2 other rabbit experiments, the medial sciatic nerve was stimulated using the helix electrode. An EMG needle was inserted into the gastrocnemius muscle for recording.

Human Lower Extremity Peripheral Nerve Studies

We measured the current levels that were necessary to cause a threshold and maximal contraction of muscles in patients undergoing lower extremity amputation. There were 9 patients in varying degrees of health, 7 had severe arterial obliterative disease (4 had diabetes mellitus). The other 2 patients had trauma to the leg. Twenty-one nerves were stimulated and the diameter measured. Four nerves were in very poor condition, either due to swelling and/or hypoxia (2 cases had tourniquets used). Stimulation was carried out at 0.2 msec pulse duration with 20 pps frequency, using a portable, battery-operated, calibrated constant current unit (Cordis Corp., Miami, Fl., Model 910 A).

Nerve Sizes—Diameters

Despite a search of the literature, we have been unable to locate any human lower extremity nerve diameter measurements at various crosssections. However, we were able to locate "Color Atlas of Human Anatomy" edited by R. M. H. McMinn and R. T. Hutchings (Yearbook Medical Publishers, Inc., Chicago, IL), whose dissections were reproduced as life-size photographs allowing measurements of the diameters at different points along the nerve.

Results

In the results ansthetized rabbit experiments, 2 procedures were observed.

Ankle Movements

With the computer stimulation program set in the *single* mode, the stimulation amplitude was adjusted to 0.2 mA (0.2 msec, 50 pps). The right lateral sciatic nerve (LSN) showed a maximal contraction by dorsiflexing the right paw, following this the left LSN was stimulated with similar results to the left paw, and then the left medial sciatic nerve (MSN) was stimulated to produce plantar flexion of paw. Then the program was changed to *multi* mode. The two LSNs were stimulated simultaneously with an ensuing bilateral dorsiflexion of the ankles and paws.

Strain Gauge Recordings

With the tendons from the anterior and posterior leg groups attached to the Grass strain gauges, the left LSN when stimulated showed threshold muscle activity at 0.07 mA using 0.2 msec at 50 pps and a maximal contraction at 0.17 mA. The maximal responses were reduced by $1/3$ with a reduction of frequency from 50 pps to 20 pps. The computer program was then set into the *multi* mode using 0.2 msec pulse width at 50 pps, the LSN pulse amplitude was set at 0.1 mA and the MSN was set at 0.12 mA. Firstly, individual contractions were produced from each nerve branch as seen in Fig. 1. Then the program was adjusted so that both nerves were stimulated simultaneously to produce a combined increase in the strain gauge recording indicating that cocontraction in the anterior and posterior groups of leg muscles could be achieved simultaneously. When stimulating with the helix electrode the resulting paw movements were similar to those seen above when using the Medtronic cuff electrode.

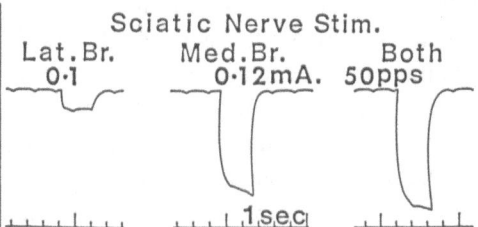

Fig. 1. Strain gauge recordings of muscle tension following stimulation of the lateral and medial sciatic nerves, individually and simultaneously

Human Lower Extremity Stimulation Parameters

Table 1 shows the human lower extremity nerve stimulation parameters and diameters. The Atlas measurements are listed and do show a relatively close correlation. The main pulse amplitudes for producing maximal stimulation and contraction in the largest of the nerves (medial sciatic) falls well within the range of the receiver-stimulating unit to be used (maximum output 3.5 mA using 0.4 msec pulse width).

Discussion

Our project has shown that the multichannel Nucleus Implantable Hearing Prosthesis could be modified and is usable for FES of peripheral nerves. The software

Table 1. *Human Lower Extremity Nerve Measurements*

Nerves	Number	Stimulation		Measured diameter (mm)	Atlas (mm)
		Amp.	(mA)		
Sciatic:					
Medial br	1	thr	0.4	7 × 6	7.3
		max	0.6		
Lateral br	2	thr	1.0	4 × 8.8	4.1
		max	2.0		
Peroneal:					
Superficial	2	thr	0.38	1.5	2.0
		max	1.15		
Deep	4	thr	0.43	2.3	—
		max	1.0		
Whole	1	thr	0.5	—	—
		max	1.0		
Tibial:					
(10 cm below popliteal fossa)	6	thr	0.7	4.9 × 3.9	5.0
		max	2.4		
Large br	4	thr	0.5	2.0	3.0
		max	2.5		
Small br	1	thr	0.1	1.0	1.0
		max	0.2		

program can be used in *single* or *multi* modes so that single nerves can be activated independently or in groups to produce cocontraction or simultaneous bilateral movements. The data of threshold and maximal currents required to produce muscle contractions in human peripheral nerves has determined that the output of the receiver-stimulating unit at its designated 3.5 mA with a maximum of 0.4 msec pulse width is adequate. Because of our selection to use the Huntington helix electrode for 11 motor nerves in each lower extremity of the paraplegic subjects, we have the nerve diameters in the lower extremity measured as in Table 1. The above results will bring our plans closer for clinical trials. The external hardware is undergoing marked modification so that the paraplegic subject will have a belt-worn microprocessor with programs that will enable individual or groups of muscles to be stimulated for exercise, standing and walking.

Acknowledgements

The study has been funded by the Division of Rehabilitation Research and Engineering, Veterans Administration, Washington, DC.

References

1. Agnew WF, McCreery DB, Bullara LA, Yuen TG, Yeh Y (1983) Development and evaluation of safe methods of intracortical and peripheral nerve stimulation, Quarterly Report, Neural Prosthesis Project, NINCDS, July 1
2. Agnew WF, McCreery DB, Bullara LA, Yuen TGH, Yeh Y-S (1985) Development and evaluation of safe methods of intracortical and peripheral nerve stimulation. Quarterly Report, Neural Prosthesis Project, NINCDS, April 1
3. Brindley S, Polkely C, Rushton D (1979) Electrical splinting of the knee in paraplegia. Paraplegia 16: 428–435
4. Cooper EB, Bunch WH, Campa JF (1973) Effects of chronic human neuromuscular stimulation. Surg Forum 24: 477–479
5. Cybulski G, Penn R, Jaeger R (1984) Lower extremity functional neuromuscular stimulation in cases of spinal cord injury. Neurosurg 15: 132–146
6. Davis R, Gesink JW (1974) Evaluation of electrical stimulation as a treatment for the reduction of spasticity. Bull Proseth Res 302–309
7. Holle J, Gruber H, Frey M, Kern H, Stohn H, Thoma H (1984) Functional electrical stimulation on paraplegics. Orthopedics 7: 145–160
8. Holle J, Thoma H, Frey M, Gruber H, Kern H, Schwanda G (1983) Epineural electrode implantation for electrically induced mobilization of paraplegics. Proc. Ist International Workshop on Functional Electrostimulation, Vienna, Abst 5.3
9. Kantrowitz A (1963) Electronic physiologic aids. A report of the Maimonides Hospital of Brooklyn, NY
10. Kralj A, Grobelnik S (1973) Functional electrical stimulation of paraplegic patients—a new hope for paraplegic patients. Bull Prosthet Res 75: 10–20
11. McNeal D, Waters R, Reswick J (1977) Experience with implanted electrodes. Neurosurg 1: 228–229
12. Marsolais B, Kobetic R (1985) Walking of paraplegic subjects with computer-controlled electrical stimulation, Motion Study Laboraty, Technical Report, MSL 3, Nov.
13. Petrofsky JS, Phillips CA, Heaton HH (1984) "Feed-back control system for walking in man". Comput Biol Med 14: 135–149
14. Willemon WK, Mooney V, McNeal D, Reswick J (1970) Surgical implanted peripheral neuroelectric stimulation. Rancho Los Amigos Hospital, Los Angeles, Internal Report
15. Yuen TGH, Agnew WF, Bullara LA (1984) Histological evaluation of dog sacral nerve after chronic electrical stimulation for micturition. Neurosurg 14: 449–455

Acta Neurochirurgica, Suppl. 39, 121–123 (1987)

Intrathecal Application of Baclofen in the Treatment of Spasticity

J. Siegfried* and **G. L. Rea**

Neurosurgical Department, University Hospital, Zürich, Switzerland

Summary

Baclofen, a derivative of g-aminobutyric acid (GABA) has been known for many years to be a useful drug in the treatment of spinal spasticity. However, when the spasticity is severe, the systemic administration has to be increased, often without therapeutic effects but frequently with central side-effects. Baclofen given intrathecally however, in microgram doses has been previously reported to be effective and safe.

A personal experience is reported of 9 severely spastic patients residing in chronic care facilities who were treated from July 1984 to March 1986 with intrathecal baclofen. The spasticity was causing significant nursing care problems, and 6 patients were reduced to a completely bedridden state. Each patient initially received a percutaneous intrathecal drug injection of 0.2–0.7 mg of baclofen to test its efficacy. A subcutaneous intrathecal system for further injections was placed in 6 patients. In 3 patients a decreased level of consciousness was observed. In the 3 cases of multiple sclerosis, intrathecal baclofen resulted in significant reduction of spasticity for 24 to 48 hours after each injection. The spasticity was improved in only one of the 2 cases of posttraumatic paraplegia. The effect was not convincing in the 2 cases of spinal cord tumour, and in the case of cerebral palsy the effect was improvement in spasticity, but also significant drowsiness.

Baclofen, in comparison with some other drugs such as morphine or midazolam, also tried intrathecally by the authors, is the most effective in reducing spasticity. Its use however warrants caution, for it can cause decreased consciousness, and there is currently no antagonist.

Keywords: Spasticity; intrathecal; GABA.

Introduction

Intrathecal application of drugs was performed at the University of Zürich beginning in July 1983. Following the first publication of Penn and Kroin in May 1984 on the use of intrathecal baclofen in humans to allivate

spinal cord spasticity[9], we have used this drug in 9 patients since July 1984. The experience with single dose injections and long-term results in patients with a subcutaneous intrathecal delivery system is reported.

Material and Methods

Nine patients in chronic care facilities who had severe spasticity that was not controlled medically were treated with intrathecal baclofen. Informed consents were obtained from all patients. The spasticity was causing significant nursing care problems in 6, and 4 of these were reduced to a completely bedridden state. The spasticity was due to multiple sclerosis in 3 cases, posttraumatic tetra-/paraplegia in 2 cases, spinal cord tumour in 2 cases, cerebral palsy in 1 case, and Laurence-Moon-Biedl syndrome in 2 case (Table 1). The 6 females and 3 males, from 18 to 63 years, were hospitalized in the neurosurgical department during the therapeutic trials. Through a lumbar puncture, baclofen (received courtesy of Ciba-Geigy Corporation, Basle, Switzerland) was administered intrathecally. Neurological examination was performed hourly and if no improvement of spasticity could be observed, a single application with a higher dosage was repeated. When the spasticity showed improvement, a drug release system (Cordis, Miami, USA) designed for repeated single applications, was placed with a reservoir in a subcutaneous pocket in the abdomen and connected to an intrathecal catheter. Spasticity was evaluated clinically. Spasms were assessed by attempting to induce them with sensory and motor stimulation. Patients were questioned about bladder function and side-effects such as drowsiness, weakness in the upper extremities, and any other changes in motor performance.

1. Multiple Sclerosis

In all 3 cases, a marked reduction of spasticity was observed up to 48 hours with single intrathecal injections of 0.2 to 0.5 mg of baclofen (Table 1). In one case there was the release of fixed appearing contractures. Although initially one patient exhibited increased weakness of the neck and upper extremities,

---

* J. Siegfried, M.D., Neurosurgical Department, University Hospital, CH-8091 Zürich, Switzerland.

Table 1. *Results on Spasticity After Intrathecal Application of Baclofen*

Age (years), sex	Diagnosis	Condition	Single dose (mg)	Delivery system	Follow-up (months)	Repeated (doses (mg)	Comments
38, female	multiple sclerosis	severe spasticity, bedridden	up to 5 (!)	yes	21	0.5/24–48 hours	spasticity markedly reduced, care facilitated
33, female	multiple sclerosis	severe spasticity, bedridden	up to 1	yes	15	0.2/48 hours	with 1 mg, severe drowsiness with 0.2 spasticity markedly reduced, care facilitated
53, female	multiple sclerosis	severe spasticity, wheelchair	up to 0.5	yes	12	0.2/48 hours	spasticity markedly reduced, care facilities
34, male	posttraumatic tetraplegia	severe spasticity, bedridden	up to 5 (!)	no	—	—	almost no effect
21, male	posttraumatic paraplegia	severe spasms, wheelchair	up to 0.5	yes	6	0.2–0.5/48 hours	spasticity markedly reduced with 0.2 up to 12 hours with 0.5 up to 24 hours
63, female	spinal cord tumour	spasticity and deafferentation pain	up to 1	yes	12	different doses, also trial with morphine, midazolam	almost no effect
29, female	spinal cord tumour	severe spasticity	up to 0.5	no	—	—	slight improvement of spasticity, slight drowsiness
18, male	cerebral palsy	severe spasms	up to 0.15	no	—	—	disappearance of spasticity, but pronounced drowsiness
18, female	Laurence-Moon-Biedl syndrom	severe spasticity, bedridden	up to 1.5	yes	5	different doses, also trial with morphine, midazolam	almost no effect

this effect was completely avoided at a lower dosage that still alleviated the spasticity. All patients spontaneously reported increased use of their upper extremities. One case had some drowsiness lasting 4 hours, but this required no intervention and was followed by complete recovery.

2. Posttraumatic Tetra/Paraplegia

In one case of C 6 tetraplegia almost no effect on spasticity was observed, even with a dose of 5 mg (!). In contrast, one case of paraplegia showed improvement in spasticity that was marked enough to warrant the implantation of a lumbar drug release system (Table 1).

3. Other Indications

In two cases of spinal cord tumors (one chordoma and one ependymoma), the decrease in spasticity was slight in one and accompanied by slight drowsiness, and in the other case there was no change in tone (Table 1). Spasticity in one case of cerebral palsy responded very well to a low baclofen dosage, but pronounced drowsiness for 4 hours discouraged further applications. The case of severe spasticity associated with Laurence-Moon-Biedl syndrome did not respond to this treatment.

Discussion

Baclofen, a 4-chlorophenyl derivative of gamma aminobutyric acid, has a low lipid solubility and is 30% protein bound in the plasma, thus limiting its movement accross the blood-brain barrier[4–6, 13]. Because of its close structural relationship to GABA, an inhibitory neurotransmitter in the spinal cord, it was originally thought to act postsynaptically on GABA receptors, but this is probably not the case[8]. The action of GABA and baclofen have been found to be quite different, with GABA depolarizing the superior cervical ganglia and dorsal root fibers, while baclofen produces depression in spinal cord preparations by hyperpolarization[1, 3, 16]. Baclofen has selective properties for the GABA-B receptor (bicuculline insensitive GABA receptor), and GABA exerts its primary influence on the GABA-A receptor (bicuculline sensitive GABA receptor)[11]. Baclofen, again unlike GABA, exerts its effects presynaptically, thus decreasing transmitter release[2]. The stereospecificity of the baclofen induced inhibition is indicated by the 20 times a greater potency of (—) baclofen as compared to (+) baclofen[12].

Although baclofen is often successful as a chronic oral agent in the treatment of spasticity[8, 17], it also has the side-effects of confusion, depression, nausea, and vomiting, even with subtherapeutic levels[4, 6]. With intrathecal delivery however, both in animals[7] and in humans[9, 10, 14, 15], it has given good relief of spasticity with extremely low dosages. However, the possible appearance of central nervous system depression, as observed by us in 3 cases, limits its application since currently there is no available antagonist.

References

1. Ault B, Evans RE (1981) The depressant action of baclofen on the isolated spinal cord of the neonatal rat. Eur J Pharmacol 71: 357–364
2. Bowery NG, Hill DR, Hudson AL et al (1980) Baclofen decreases neurotransmitter release at a novel GABA receptor. Nature 283: 92–92
3. Fukada H, Kudo Y, Oho H (1977) Effects of GABA on spinal synaptic activity. Eur J Pharmacol 44: 17–24
4. Gilman AG, Goodman LS, Gilman A (1980) The pharmacological basis of therapeutics. Macmillan Publishing, New York
5. Grahame-Smith DG, Aronson JK (1984) Oxford textbook of clinical pharmacology and drug therapy. Oxford University Press
6. Knutsson E, Lindblom U, Martensson A (1974) Plasma and cerebrospinal fluid levels of baclofen (Lioresal) at optimal therapeutic responses in spastic paresis. J Neurol Sci 23: 473–484
7. Kroin JS, Penn RD, Beissinger RL et al (1984) Reduced spinal reflexes following intrathecal baclofen in the rabbit. Exp Brain Res 54: 191–194
8. Pederson S, Arlien-Soborg P, Mai J (1974) The mode of action of the GABA derivative baclofen in human spasticity. Acta Neurol Scand 50: 665–680
9. Penn RD, Kroin JS (1984) Intrathecal baclofen alleviates spinal cord spasticity. Lancet (i): (1078)
10. Penn RD, Kroin JS (1985) Continuous intrathecal baclofen for severe spasticity. Lancet (ii): 125–127
11. Price GW, Wilkin GP, Turnbull MJ et al (1984) Are baclofen sensitive GABA-B receptors present on primary afferent terminals of the spinal cord. Nature 307: 71–74
12. Regan JW, Roeske WP, Yamamura HI (1980) The benzodiazepine receptor and its development and modulation by GABA. J Pharm Exp Ther 212: 137–143
13. Reynold JEF (1982) The extra pharmacopeia. The Pharmaceutical Press, London
14. Siegfried J, Lazorthes Y (eds) (1985) Neurochirurgie de l'infirmité motrice cérébrale. Neurochirurgie 31: 95–101
15. Siegfried J, Rea GL (1986) Intrathecal application of drugs for muscle hypertonia. Scand J Rehab Med, in press
16. Yaksh TL, Reddy SVR (1981) Studies in the primate on the analgetic effects associated with intrathecal actions of opiates, adrenerigc agonists, and baclofen. Anesthesiology 54: 451–467
17. Young RR, Delwaide PJ (1981) Spasticity, Parts I and II. NEJM 304: 28–33, 96–99

Acta Neurochirurgica, Suppl. 39, 124–125 (1987)

Stimulation of the Dentate Nuclei for Spasticity

J. R. Schvarcz*

School of Medicine, University of Buenos Aires, Buenos Aires, Argentina

Summary

It has been assumed but not been proved that cerebellar cortical stimulation activates the Purkinje cells, thereby inhibiting the deep nuclei. Furthermore, the destruction of these particular cells, consistently demonstrated in biopsy material, has not been accounted for. These has led the author to introduce chronic stimulation of the dentate nuclei for spasticity in 1978. The long-term results of stimulation in 22 such cases are reported, followed-up between 2 and 6 years.

Both the clinical results and the electrophysiological changes produced by stimulation are analyzed. They seem to indicate that it is a rational approach to relieve spasticity and to improve motor function.

Keywords: Stereotaxis; stimulation; spasticity; involuntary movements; cerebellum; dentate nucleus.

Introduction

It has been assumed but has not been proved that cerebellar cortical stimulation (CCS) activates the Purkinje cells[3, 4], thereby inhibiting the deep cerebellar nuclei. However, other more likely phenomena may also occur[2]. Furthermore, the preexistent destruction of these particular cells, consistently demonstrated in biopsy materal[10, 9], has not been accounted for. This has led the author to introduce, in 1978, chronic stimulation of the dentate nuclei (DNS) to relieve spasticity and to improve motor function[7].

The long-term results in 22 such cases are reported

Material and Methods

The technique has already been reported[7]. Briefly, patients were operated on under general anesthesia, using a modified Hitchcock

* Prof. J. R. Schvarcz, M.D., Beruti 2926, Buenos Aires 1425, Argentina.

apparatus. The ventricular system was outlined by water-soluble positive contrast, and bilateral standard DBS electrodes were stereotactically placed in the dentate nucleus, along its main axis, with their tip aimed at the fastigium of the fourth ventricle in the coronal and horizontal planes, 6 mm lateral to the midsagittal plane[1]. The electrodes were then externalized for a period of trial stimulation, and subsequently connected to a dual channel receiver. Stimulation periods of 2 hours on/3 hours off were typically used.

Twenty-two patients with spastic motor disorders underwent protracted stimulation. They all had a long-standing history of severe, disabling spasticity, which interferred with voluntary movements. It was due to cerebral palsy in 17 cases, to head trauma in 4, and to cardiac arrest in 1. Their age ranged from 9 to 47 years. They all have had extensive medical treatment without improvement. They were highly motivated to surgery, had no intellectual deficit, and had a favorable family environment. The follow-up period ranged between 2 and 6 years.

Results

It is rather difficult to assess the results objectively. Both the changes in the disability status and in the activities of daily living as well as the neurophysiological changes should be considered.

Almost all patients have had variations in their daily living activities, but these do not necessarily imply a significant change in the overall rating of their disability score. They do seem to be, nonetheless, significant to the patients themselves.

DNS consistently produced an improvement in motor function. Co-contraction and spasticity were relieved, enabling the performance of some complex voluntary movements which were not possible prior to stimulation. Posture, balance and gait were also improved. The quality of speech and respiration was enhanced.

DNS consistently produced distinctive electrophysiological changes, viz., a marked improvement in the pattern of the H reflex recovery curve, a lengthening of the silent period, an increase in the coefficient of excitability and a decrease in the T/M ratio. They were more prominent ipsilaterally but they had a bilateral distribution. These protracted changes clearly showed both a residual and a cumulative effect.

Discussion

The clinical results of CCS are still moot[6]. Its presumed rationale is the activation of Purkinje cells, with subsequent inhibition of their target nuclei[3, 4]. However, a severe loss of Purkinje cells has been consistently demonstrated in biopsy material obtained prior to electrode placement in patients undergoing cerebellar implantation[10, 9]. This absence should then minimize their stimuli-induced excitation. Indeed, other different alternative structures can be stimulated. E.g., the antidromic activation of cerebellar afferents, the synaptic activation of neurons in the dentate nucleus, and the transsynaptic activation of neurons in the reticular formation have been reported[2].

DNS has produced motor gains by relieving spasticity and improving motor control, enabling complex voluntary movements which were not performed prior to stimulation. Posture, balance and gait were improved, as well as speech and respiration.

Although electrophysiological changes do not necessarily reflect the degree of clinical improvement, they do provide objective, measurable evidence of stimuli-related modifications in the excitability of spinal motoneurons. They have already been comprehensively reported elsewhere[8]. The changes in the H reflex recovery curve conceivably indicate a decreased descending facilitation, and the effect on the silent period an increased descending inhibition, both acting onto alpha-motoneurons, whereas the reduced T/M ratio conceivably indicates a diminished facilitation acting onto gamma-motoneurons. The change in the coefficient of excitability presumably reflects an improved program of voluntary activation of alpha-motoneurons. The overall cumulative effects strongly suggest plastic modifications induced by stimulation.

Fox and Hitchcock[5] have reported significant effects on the F wave size produced by acute dentate stimulation.

Both the clinical and the electrophysiological data suggest that the clinical results of CCS are not due to prostetically induced inhibition of the dentate nuclei. Furthermore, they indicate that their stimulation is a rational approach to relieve spasticity and to improve motor control.

References

1. Afshar F, Watkins ES, Yap JC (1978) Stereotactic atlas of the human brainstem and cerebellar nuclei. A variability study. Raven Press, New York
2. Bantli H, Bloedel JR, Tolbert D (1976) Activation of neurons in the cerebellar nuclei and ascending reticular formation by stimulation of the cerebellar surface. J Neurosurg 45: 539–554
3. Cooper IS (1973) Effect of chronic stimulation of the anterior cerebellum on neurological disease. Lancet 1: 206
4. Cooper IS (1973) Effect of chronic stimulation of the posterior cerebellum on neurological disease. Lancet 1: 1321
5. Fox JE, Hitchcock ER (1982) Changes in F wave size during dentatotomy. J Neurol Neurosurg Psychiatry 45: 1165–1167
6. Penn RD (1982) Chronic cerebellar stimulation for cerebral palsy: a review. Neurosurg 10: 116–121
7. Schvarcz JR, Sica RE, Morita E (1980) Chronic self-stimulation of the dentate nucleus for the relief of spasticity. In: Gillingham FJ, Gybels J, Hitchcock ER, Rossi GF, Szikla G (eds) Advances in stereotactic and functional neurosurgery 4. Acta Neurochir (Wien), [Suppl] 30: 351–359
8. Schvarcz JR, Sica RE, Morita E, Bronstein A, Sanz O (1982) Electrophysiological changes induced by chronic stimulation of the dentate nuclei for cerebral palsy. Appl Neurophysiol 45: 55–61
9. Schvarcz JR, Caputti E (1982) unpublished data
10. Urich H, Watkins ES, Amin I, Cooper IS (1978) Neuropathologic observations on cerebellar cortical lesions in patients with epilepsy and motor disorders. In: Cooper IS (ed) Cerebellar stimulation in man. Raven Press, New York, pp 145–159

Acta Neurochirurgica, Suppl. 39, 126–128 (1987)
© by Springer-Verlag 1987

Cerebellar Stimulation for Cerebral Palsy—Double Blind Study

R. Davis*, J. Schulman, and **A. Delehanty**

Kennebec Valley Medical Center, Augusta, Maine, U.S.A., and Pacesetter Systems Incorp. Sylmar, California, U.S.A.

Summary

Twenty spastic cerebral palsy (CP) patients undergoing chronic cerebellar stimulation (CCS) for reduction of spasticity and improvement in function have participated in a double-blind study. Seven US centers involving 9 neurosurgeons (1984-6) have replaced the depleted Neurolith 601 fully implantable pulse generator (Pacesetter Systems Incorp.-Neurodyne Corp., Sylmar, CA) with new units in 19 CP patients, 1 patient entered the study following his initial implant. A magnetically controllable swith was placed in line between the Neurolith stimulator and the cerebellar lead, so allowing switching sequences for the study. Physical therapists, living in the vicinity of the patient's home, carried out two quantitative evaluations: 1. Joint angle motion measurements (passive and active). 2. Motor performance testing was done when possible and included: reaction time, hand dynamometry, grooved peg board placement, hand/foot tapping, and rotary pursuit testing. Testing was done presurgery, at 2 weeks postimplant, then the switch was activated either "on" or "off" to a schedule, with testing and reswitching at 1, 2 and 4 months, then the switch was left turned "on". Of the 20 patients, 16 finished the tests, 2 patients failed to finish and 2 had switch problems and were deleted from the study. Two of the 16 patients were "off" through the entire testing. Of the 14 that had periods of the stimulator being "on", 10 patients (72%) had quantitative improvements of over 20%, (1 pt: 50 + % improvements; 4 pts: 30–50%, 5 pts: 20–30%); while 1 patient (7%) had improvements in the 10–20% level, whereas 3 patients (21%) showed no improvement.

Keywords: Cerebellar stimulation; cerebral palsy.

Introduction

Double-blind CCS studies by Gahm *et al.*[3], Whittaker[10], Reynolds and Hardy[7] and Ratusnick *et al.*[6] did not use quantitative measurements nor were stimulation levels recorded in their series on spastic CP patients. Penn *et al.*[5] published a 10-patient double-blind study using joint compliance as the quantitative measurement as well as observation assessments of developmental reflexes and motor skills. On comparing their "on" periods to "off" periods they did not see any significant correlation. Details of their compliance test have shown that this is not an adequate quantitative test particularly when it comes to motor improvements as have been seen not only in their series[5] but in a series by Gray *et al.*[4]. Six of the 10 were able to correctly tell when the stimulator was "on" by feeling "more relaxed". Of their 14 patients undergoing a prospective study[5] from 1 to 44 months, compliance measurements improved in 9 (56%), 9 showing improvements in primitive reflexes (64%), and 11 (79%) were functioning better. The authors stated that the effects of CCS over the months may be irreversible when attempting double-blind maneuvers. The equipment in these studies[3, 5–7, 10] has been the radiofrequency linked systems which have been known to have high failure rate and unknown and inconstant current pulse deliverance[1, 2]. It is also known that there is a "window" effect for physiologic improvement, too high a current application causes a failure of results or a paradoxical effect[2, 9].

Methods

From 1979 onwards our series[2] of 160 spastic patients (146 with cerebral palsy—91%) were given the fully implantable neurolith 601 pulse generator (Pacemaker Systems Incorp.-Neurodyne Corp., Sylmar, CA), used in conjunction with the two 4-button electrode array (E-341 adult—0.45 cm^2, E-354 child—0.28 cm^2, Avery Lab., Farmingdale, NY). The electrode pads were situated bilaterally on the superior paramedian cerebellar surface and received charged densities in the range of 1.1–1.8 µCoul/cm^2 per phase. The pulse was

* R. Davis, M.D., Kennebec Valley Medical Center, 6 East Chestnut street, Augusta, ME 04330, U.S.A.

1.40 mA, 0.5 msec, at 150 pps applied for 4 minutes on and 4 minutes off. Of this group, 19 patients had depleted their battery in the neurolith 601 unit by 1984 onwards. In order to have the pulse generator replaced they have had to undergo a double-blind study whose design was approved by the US Food and Drug Administration. This required the insertion of a magnetically controlled switch placed in line between the pulse generator and the lead going to the cerebellar electrodes. This switch was placed adjacent to the pulse generator, but in one case it was placed medial to the scapula and the wire broke when the patient bent over. One patient was implanted for the first time and joined the study with the switch. The average age was 17 (range, 10–32 years), 13 males and 7 females. They ranged in disability from those thad had very little, if any, voluntary movements (2 patients) to those that had limited range of motion (4 patients) to those that were able to undergo quantitative measurements of motor skills (14 patients). Two of the 20 patients did not complete the testing, 2 had problems with the switch and were dropped from the study, 2 patients were "off" during the entire sequencing procedures.

The 20 spastic CP patients were reoperated in 7 different neurosurgical units by 9 neurosurgeons, and were tested quantitatively by physical therapists who were in the vicinity of the patient's home. Neither the patient nor the therapist knew the stimulation's status as this was monitored and switched by an engineer or EEG technician. The testing protocol involved 1. joint motion angle measurements (passive and active). 2. Motor performance testing was done when possible and included: reaction time to light/buzzer (depress and release modes), hand dynamometry, grooved peg board placement, hand/foot tapping, and rotary pursuit testing. These tests were performed: (PRE) before surgery; (test 1) approximately 2 weeks after the reimplantation with the stimulator switch off; (test 2) was carried out at the end of the next 4 weeks, switch during this period was either "on" or "off", according to a schedule for this and the following 2 periods; (test 3) at the end of another 4 weeks, (test 4) at the end of another 8 weeks. The switch was then left turned "on". The motor performance tests were purchased from the Lafayette Instrument Company, Lafayette, Ind. These tests had been used in a quantitative study made on 17 spastic CP patients of our series by Ryan *et al.*[8], but not as a double-blind study. Because of the switch problems, a more reliable switch was developed in a hermetically sealed can and has been used since March 1986. There was 1 infection in the tissues around the stimulator pocket which resulted when a faulty switch was changed halfway through the study. The whole system was removed after the patient had been treated with antibiotics. The system is scheduled to be reinserted in June 1986.

Stastical analysis was carried out on all data generated which involved particularly the mean, standard deviation, and the minimum and maximum values. Since the stimulators had run down prior to implantation the PRE implant test as well as test 1 were used as the nonstimulation basis for correlation, however, during the tests 2, 3, or 4 at certain times the stimulator was "off" in 8 of the 10 patients who completed the motor performance testing. This "off" period was also used as a basis for comparison to the data from the "on" periods. The results were examined for percentage change during the "on" and the "off" periods. Six seriously affected patients were only able to have their joint motion tests measured. Seven of the 10 more capable patients had joint motion tests assessed.

Results

Joint Motion Tests

Of the 13 patients tested, 11 had test periods which were "on" so correlations could be made. A total of 14 angles (8 in the upper extremity and 6 in the lower extremity) were measured for passive and active movements, these were done bilaterally resulting in 56 measurements for each patient. Eight (73%) showed overall increases in range of motion of 30%+ (6 had 30–50% improvements while 2 had 50%+ improvement). However, one (9%) showed only 10–20% improvement while 2 (18%) showed no improvements. Ankle dorsiflexion showed the major improvements in range of motion (6 pt: 135% passive, 196% active). Hip abduction also improved in 4 patients averaging 125% improved in active range of motion. Forearm pronation and supination showed considerable gains of pronation (4 pt: 49% active) and supination (5 pt: 75% active, 6 pt: 77% passive). Shoulder abduction improved in 5 patients with a 30% increase in active range of motion.

Motor Performance Testing

Six of the 7 tests involved performance timing, while hand dynamometry showed improvement in strength. Of the 7 tests in the 10 patients, 64 results were obtained (Table 1), six tests were unable to be done (foot tapping: 4, grooved peg board: 1 and rotary pursuit: 1). Overall, there was a 61% improvement rate and 39% showed no improvement. In 8 of the 10 patients there were test periods which were "off", while 2 of the patients were "on" throughout testing. A search was made for improvements during the "off" test periods which showed that 38% of the tests had an improvement while 62% showed no improvement. If one excluded the most difficult test which was rotary pursuit, then the improvements during the "off" testing reduced to 30% while no improvements occurred in 70%. On the other hand, when the stimulator was "on" improvements were seen at approximately 60% in these tests. The next most difficult test for the patient was reaction time and if these were also excluded with rotary pursuit from the results during the "on" and the "off" periods, 70% improvements were measured during the "on" period, while 30% were observed during the "off" period. Of the 10 patients (Table 1), 8 (80%) showed improvements of over 20% increase in performance, while 2 (20%) showed no improvements. Of the improved group, 1 patient showed over 50% improvement, 2 showed 30–50% improvement range and 4 showed 20–30% improvement ranges.

Table 1. *Double-Blind Study 10 Cerebral Palsy Pts. Stimulator: "On" Motor Performance*

Grade Imprv	Reaction		HND/DY	Tapping		GRV/PG BRD	Rotary pursuit Number	%	
	Deprs	Releas		Hand	Foot				
(50 + %)		1	2			3	3 = 9	14%	Imp
(30–50%)				1	1	3	3 = 8	13%	61%
(20–30%)		2	2	2	1		= 7	11%	
(10–20%)	3	2	1	4	3	2	= 15	23%	
No imp	7	5	5	3	1	1	3 = 25	39%	No imp ——39%
Totals	10	10	10	10	6	9	9 = 64	100%	

Discussion

The overall results of the 14 spastic CP patients who had undergone quantitative measurements during the 4 months of double-blind testing, 10 (72%) showed improvements over 20% in a combination of the joint motion and motor performance testing. One patient (7%) showed a 10–20% improvement; while 3 (21%) showed no improvements at all during the "on" testing. Of the 10 (72%) patients that showed the improvements 1 patient (7%) had over a 50% improvement in testing, 4 patients showed 30–50% improvements while 5 showed 20–30% improvements. The study indicates that CCS has caused a quantitative improvement in range of motion of joints, and particularly in the quickness and skills of doing motor performance testing. The above results also correlate well with the quantitative measurements performed on another 17 spastic CP patients studied in 1982 by Ryan *et al.* using the neurolith-avery stimulating system[8]; which showed in the 1.5–13.5 months of CCS that 8 out of 10 patients had improvements above 50% of performance increases. The patients in the present study who have had the most ability before testing showed the most improvements.

Acknowledgements

We wish to thank Drs. R. Britt and L. Shuer (Stanford, CA), M. Nanes and M. Flitter (Miami Beach, FL), M. Fleming (Atlanta, GA), P. Gildenberg (Houston, TX), R. Rydell (Tampa, FL), and M. Sukoff (Santa Ana, CA) for their participation in the neurosurgical aspects of this double-blind study. Special thanks to Mr. Thomas Ryan, Biomedical Engineer (Stanford, CA), Mr. Ken Zib and Ms. Erin Kelly (Pacesetter Systems Incorp.-Neurodyne Corp.), the physical therapists and switching managers.

References

1. Davis R, Gray EF (1980) Technical problems and advances in cerebellar-stimulating systems used for reduction of spasticity and seizures. Appl Neurophysiol 43: 230–243

2. Davis R, Kudzma J, Gray E, Engle H, Ryan T (1984) Graded clinical effects in spastic cerebral palsy groups following chronic cerebellar stimulation. In: Davis R, Bloedel JR (eds) Cerebellar stimulation for spasticity and seizures. CRC Press, pp 223–243

3. Gahm N, Russman B, Cerciello R, Fiorentino MR, McGrath DM (1981) Chronic cerebellar stimulation for cerebral palsy: a double-blind study. Neurology 31: 87–90

4. Gray E, Davis R, Cohn M (1984) Respiratory and joint compliance changes in spastic patients following chronic cerebellar stimulation. In: Davis R, Bloedel JR (eds) Cerebellar stimulation for spasticity and seizures. CRC Press, Boca Raton, FL, pp 143–167

5. Penn RD, Mylkebust BM, Gottlieb GL, Agarwal GC, Etzel ME (1980) Chronic cerebellar stimulation for cerebral palsy prospective and double-blind studies. J Neurosurg 53: 160–165

6. Ratusnik DL, Wolfe VI, Penn RD, Schewitz S (1978) Effects on speech of chronic cerebellar stimulation in cerebral palsy. J Neurosurg 48: 876–882

7. Reynolds AF, Hardy TL (1980) Cerebellar stimulation in four patients with cerebral palsy. Appl Neurophysiol 43: 114–117

8. Ryan TP, Davis R, Gray EF (1984) Quantitative study of neurological performance in spastic cerebral palsy patients with chronic cerebellar stimulation. In: Davis R, Bloedel JR (eds) Cerebellar stimulation for spasticity and seizures. CRC Press, Boca Raton, FL, pp 169–186

9. Upton A (1978) Neurophysiological aspects of spasticity and cerebellar stimulation. In: Cooper IS (ed) Cerebellar stimulation in man. Raven Press, New York, pp 101–122

10. Whittaker C (1980) Cerebellar stimulation for cerebral palsy. J Neurosurg 52: 648–653

Acta Neurochirurgica, Suppl. 39, 129–131 (1987)
© by Springer-Verlag 1987

Motor and Psychological Responses to Deep Cerebellar Stimulation in Cerebral Palsy (Correlation with Organization of Cerebellum Into Zones)

M. Galanda* and **O. Zoltán**

VÚLB Bratislava, KÚNZ Banská Bystrica, Czechoslovakia

Summary

The study includes 68 cases of cerebral palsy stereotaxically operated on from 1977. Deep cerebellar stimulation treatment was performed. The motor and psychological responses to electrical stimulation of 305 points of subcortical regions of cerebellum, mostly lobus anterior were analysed. The characteristic response—slight motor jerk immediately—followed by relaxation and feeling of pleasure, even laughing, to the electrical stimulation from selected points was always found. The level of stimulating current must be adjusted individually. The higher current increased pathological posture, muscular tonus and was conducted with the state of fear. The lower current was without detectable influence on the patient. On the trajectory of electrode, nearly perpendicular to the sagittal plane were narrow areas, which recurred as the strips, from where it was possible or not to elicit characteristic response. The most convenient target is in the region of brachia conjunctiva cerebelli. Localization of the point of stimulation in respect to organization of cerebellum into sagittally oriented zones and the parameters of stimulation seem to contribute to the diversity of responses to cerebellar stimulation.

Keywords: Cerebral palsy; cerebellar stimulation.

Introduction

Cooper[1] introduced superficial cortical stimulation of the cerebellum for treatment of cerebral palsy in the early 1970's. Since 1977 we have also used the stimulation technique for the same purpose but in a different way[2]. Multipolar electrodes were stereotaxically implanted into various subcortical regions of the cerebellum, especially into lobus anterior. Stereotaxic placement of electrodes allows accurate topographical recording of the points which were stimulated and permits clinicoanatomical correlations.

* MUDr. Mirsolav Galanda, CSc, KÚNZ, nám. gen. Svobodu 1, 975 15 Banská Bystrica, Czechoslovakia.

Material and Methods

The study includes 68 cases of cerebral palsy. Ages ranged from 3 through 25 years. A stereotaxic method (Riechert instrument) was used. Multipolar electrodes were introduced transtentorially from a unilateral burr hole in the parieto-occipital region into the subcortical regions of cerebellum[3, 4]. In the first period of our work the electrodes were externalized and neurophysiological tests were performed prior to setting up the program for therapeutic stimulation. In the second period—the system for stimulation (Tesla) was directly implanted.

Results

As we gained experience with this method, two things were found to be of great importance: Firstly the necessity of selecting a stimulating current of appropriate quality to reach the response with optimal energy demand[4]. Secondly to apply this stimulation in certain region of subcortical areas of cerebellum, from where was possible to observe the typical "characteristic" response to stimulation. The quality of this "characteristic" response depends on the amount of energy, which is applied and of the position of the electrode in the cerebellum.

Table 1 describes the response, which has two aspects—motor and psychological. The pattern of stimulation reflects the amount of energy according to the response. Too high energy of stimulation—"overstimulation"—causes an overall increase of muscular tonus, the pattern of which partially depends upon the position of the stimulating contact in the cerebellum. The pathological posture is deeply pronounced and accompanied with the psychical state of fear (Fig. 1). Just below the level of stimulation where hypertonus of muscles is reached (threshold level), is the optimal level of current for stimulation. Here

Table 1. *Immediate Characteristic Response Elicited by Stimulation of Cerebellum*

Pattern of stimulation	Motor response	Emotional expression
"overstimulation"	hypertonus	fear
optimal stimulation	relaxation	pleasure
"understimulation"	not detectable	not detectable

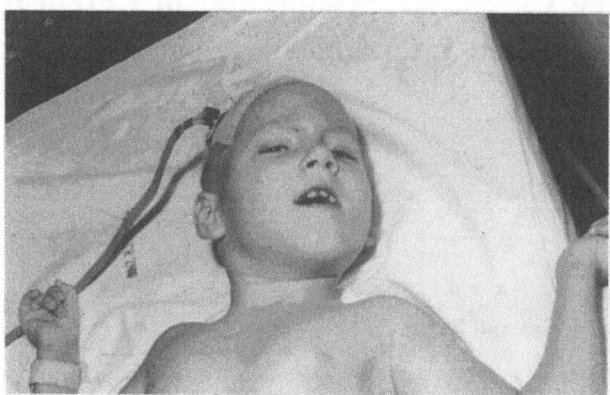

Fig. 1. "Overstimulation"—feeling of fear

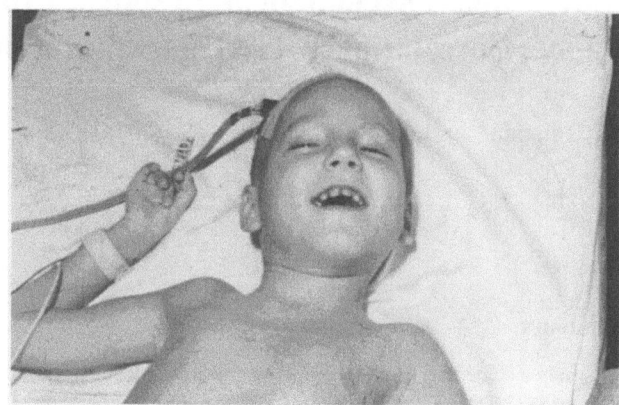

Fig. 2. Optimal stimulation—feeling of pleasure

immediately relaxation and decrease of pathological hypertonus is observed, which is more pronounced on the ispilateral side. This response is accompanied with a feeling of pleasure, and sometimes the patient smiles, even laughs (Fig. 2). Below this level of stimulation the immediate result of stimulation is not detectable neither motor nor psychological—"understimulation". Optimal stimulation forms a narrow band, usually 0.2 mA, which should be individually found. The best way to identify this level of stimulation is to apply the current at the threshold level, which elicits a slight contraction and to set stimulating current approximately 0.1 mA lower. The patient is conscious of the stimulating effect. The elicitation of this contraction is used for quick verification of the proper area in the cerebellum suitable for stimulation during stereotaxic introduction of electrode. According to our experience any point in the cerebellum, which elicits the jerk, achieves decreasing pathological hypertonus.

In the first period of stereotaxic treatment of cerebral palsy multiple points of cerebellum, mostly in lobus anterior were examined. Our transtentorial approach to the cerebellum is practically perpendicular to the sagittal plane of cerebellum. On the trajectory trial stimulation tests were taken. The positioning of the electrode must be done very precisely to elicit the "characteristic response". On this trajectory, nearly perpendicular to the sagittal plane, there are narrow areas, which recur as the strips, from where it was possible or impossible to elicit the "characteristic response". And variation in the positions from lateral to medial side of stimulating electrode also produce variation in responses.

To find out points in the cerebellum, which would be easy detectable and from where the appropriate characteristic response occurs was an important objective. Every point which was stimulated and the characteristic response elicited on monopolar stimulation was labelled with the identification number of the patient and the coordinates. The data were stored in the computer. The results of stimulation of our first 47 cases, that represents 305 points in the cerebellum, mostly in lobus anterior, were analysed. The most convenient region for therapeutic stimulation was identified in the area of brachia conjunctiva cerebelli (BCC). The target point was determined—6 mm on the line from fastigium to commissura posterior, 2 mm below this line and 5 mm laterally from midline. Two electrodes were transtentorially placed in both targets through one burr hole in the occipito-partial region and bipolar stimulation between them was made. During stimulation in this area a feeling of pleasure was always elicited.

In the second period of our cerebellar stimulation we used one bipolar electrode (Tesla). The electrode was implanted into the contralateral target in BCC after a trial stimulation for target selection. If the characteristic response was elicited, the electrode was fixed in position and a program of therapeutic stimulation begun.

Discussion

Our observations support the idea, that the relationship between the electrode position and the spatial arrangement of the cerebellar zones represents an important variable[5]. The zones in the anterior lobe as defined by Voogd[7] are narrow, rostrocaudally oriented, topographically organized areas with specificity of afferent and efferent fibers without overlapping. We agree with Haines[5], with their importance for therapeutic stimulation. The stimulating contacts of electrode should be placed precisely and the electrode must be fixed to prevent movement and changing position of stimulating areas in the cerebellum.

The level of applied current is crucial in the elicitation of motor and psychological responses of different quality, which we describe as the "characteristic response" to cerebellar stimulation. Changes were observed immediately after onset of stimulation and could be used for selection of the optimal target in the cerebellum and optimal electrical parameters for stimulation.

The feeling of pleasure expressed by patients during subcortical cerebellar stimulation was always elicited if the stimulation was within our target area for BCC. Functional connections between cerebellum and limbic system has been demonstrated[6] and correlate with our observation. The relation of locus ceruleus as the "pleasure center" to this area could be also taken into account. Activation of the patient and feeling well during stimulation is helpful during subsequent physical therapy and education.

References

1. Cooper IS, Riklan M, Amin I (1976) Chronic cerebellar stimulation in cerebral palsy. Neurology 26: 744–753
2. Galanda M, Fodor S, Nádvorník P (1978) Paleocerebelárna stimulácia v liečbe detskej mozgovej obrny. Bratisl Lek Listy 70: 99–105
3. Galanda M, Nádvorník P, Fodor S (1980) Stereotactic approach to therapeutic stimulation of cerebellum for spasticity. Acta Neurochir (Wien) [Suppl] 30: 345–349.
4. Galanda M, Nádvorník P, Fodor S (1985) Therapeutic deep stimulation of the CNS in adults with infantile cerebral palsy. Čs Neurol Neurochir 48/81: 360–366
5. Haines DE (1981) Zone in the cerebellar cortex. Their organization and potential relevance to cerebellar stimulation. J Neurosurg 55: 254–264
6. Heath RG (1972) Physiologic basis of emotional expression: evoked potential and mirror focus studies in rhesus monkeys. Biological Psychiatry 5: 15–31
7. Voogd J (1969) The importance of fiberconnections in the comparative anatomy of the mammalian cerebellum. In: Llinas R (ed) Neurobiology of cerebellar evolution and development. AMA Education and Research Foundation, Chicago, pp 493–514

Acta Neurochirurgica, Suppl. 39, 132–135 (1987)

The Treatment of Hemifacial Spasm with Percutaneous Radiofrequency Thermocoagulation of the Facial Nerve

G. Salar*, I. Iob, C. Ori[1], D. Fiore[2], M. Mattana, and C. Battaggia

Institute of Neurosurgery, University of Padova, [1] Institute of Anesthesiology, University of Padova, [2] Department of Neuroradiology, City Hospital of Padova, Padova, Italy

Summary

The authors present their first results with the treatment of hemifacial spasm by controlled percutaneous thermocoagulation of the facial nerve.

Seven patients have been treated to date with good immediate results on the movements, although a slight paresis of the homolateral facial musculature, aesthetically acceptable, persists after treatment.

A long-term follow-up at more than 10 months was achieved in only 3 patients, who showed a complete regression of the spasm with partial disappearance of the facial hemiparesis.

Introduction

Hemifacial spasm is characterized by paroxysmal involuntary clonic contractions of the muscles of one side of the face, innervated by the seventh cranial nerve. The spasmodic contraction generally arises at the level of the muscle orbicularis orbitae and is followed by the other muscular structures of the hemiface in a caudal direction[1].

The patients are usually women generally well on in years. The symptomatology may persist even during sleep and increases during emotions. It may be sometimes be bilateral and generally is painless. Nevertheless, this condition may induce important psychological problems, with a tendency of the patient to avoid social and work relations.

The etiology is largely obscure, in spite of the great number of works on this topic[3].

The most interesting and suitable hypothesis (Jannetta 1977) is that of a vascular microcompression at the level of the root entry zone of the seventh cranial nerve into the brain stem.

Various medical and surgical treatments have been proposed generally with poor results. In 1981 Hori et al.[6] using a technique proposed by Kao[8], introduced percutaneous thermocoagulation of the facial nerve for the treatment of the hemifacial spasm, under radiographic and electrophysiological control.

We report our experience, begun 15 months ago, on a group of 7 patients affected by hemifacial spasm and treated by this technique.

Material

From 1985 we examined seven patients affected by hemifacial spasm; 4 of them were males, 3 females.

The side affected was the right in 4 cases, the left in the remainder. The mean age of the patients was 58 years; the youngest subject was 36, the oldest 68 years old. A complete neuroradiological examination was carried out in all cases, including cerebral CT scan with medium contrast and vertebral angiography, resulted completely negative. Altogether the general conditions of the patients were good and neurological examination was negative, except for the muscular contractions of the hemiface. The intensity and the frequency of the latter showed a large variability, depending particularly upon the degree of relaxation of each patient.

Method

The patient is placed on a radiological chair for pneumoencephalography in an halfsitting position; the neck is moderately hyperextended to obtain skull radiograms both in lateral and in submento-

---
* G. Salar, M.D., Istituto di Neurochirurgia, Università di Padova, via Giustiniani 5, I-35100 Padova, Italy.

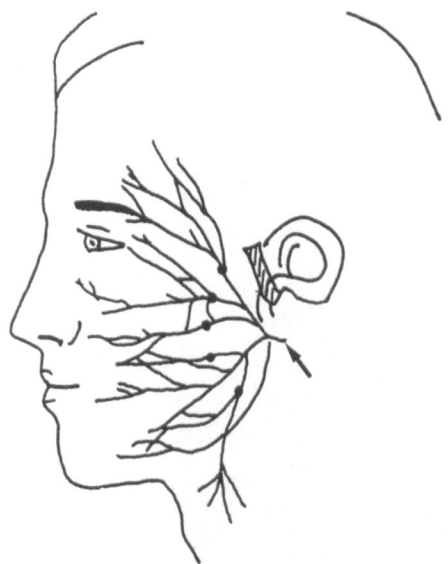

Fig. 1. The arrow indicates the needle-electrode introduced immediately under the external acoustic meatus, before the mastoid process

vertex projections, particularly for identifying of the styloid foramen which is not always easily detectable. The patient is under anesthetic surveillance, with blood pressure and heart rate monitoring.

Local anesthetic is then injected near the point of introduction of the needle-electrode without deep penetration, for the risk of a functional blockade of the facial nerve, which could interfere with the surgical procedure.

The apparatus employed for thermocoagulation is a Radionics Mod. RFGF, *i.e.*, the same normally used for trigeminal or glossopharyngeal nerves thermolesions. The uninsulated tip of the cannula is generally 5 mm long.

The needle-electrode is introduced immediately under the external acoustic meatus, closely before the mastoid process (Fig. 1); it penetrates in a medial direction with a slight anterior inclination for about two centimeters, in till reaches the styloid process.

At this point a slight movement of the needle, under fluoroscopic control, allows the final correct position of the tip, immediately under the styloid foramen. This maneuver becomes easier if the latter is radiologically well recognizable. If the styloid foramen cannot be detected, one may utilize other radiographic points of reference, the styloid process is anterior and medial with respect to the foramen and the foramen jugulare has a more medial placement (Figs. 2–3). Avoid penetration of the needle into the styloid foramen, leaving it close to the latter, for the risk of a direct damage to the facial nerve.

At this point a series of low electrical stimulations of the nerve should be carried out at about 0.1–0.2 volt. The position of the needle is correct when electrical stimulation induces rhythmic contractions of the homolateral face muscles at low frequency or, vice versa, a tonic contraction at high frequency. After repeated radiographic controls and the electrical stimulation of the nerve one may begin the lesion starting with temperatures of 55–60 °C for 15–30 seconds, repeatedly testing the function of the nerve. The surgical procedure must be interrupted immediately after the appearance of a slight facial hemiparesis; at this point a contemporary and complete disappearance of the spasm should be observed. In the case of partial results, the lesion may be repeated using the same parameters. It must be kept in mind that the damage to the seventh cranial nerve should be slight and in every case acceptable for the patient, both from an esthetic and a functional point of view.

Results

Four of the seven patients had good immediate results after only one operation. In 3 of them the lesion was performed at 60 °C for 30 seconds, in one at 55 °C for 15 seconds. In the remainder on the contrary, the correct localization of the nerve was more difficult and complex both at radiograms and at the electrical stimulation. Probably for this reason also a satisfactory clinical result was achieved only after repeated operations: two cases needed a series of two lesions at 60 °C each for 30 seconds and one patient 3 series of lesions, performed respectively at 60, 65 and 70 °C each for a time of 30 seconds.

In six patients the immediate effect on spasm was satisfactory, with complete disappearance of the spasmodic contractions of the whole hemiface. Only in one case a good control of the contractions in the orbicularis orbitae muscle was impossible. All our patients presented a variable degree of paresis of the facial nerve aesthetically and functionally acceptable, with the exception of one female patient, who considered it too severe. In six cases the facial paresis was extended to the whole hemiface and in one of them was confined to the inferior branch of the facial nerve. The good surgical results persisted in 3 patients at of 10, 13 and 14 months, while the remainder had short follow-up 6, 4, 3 and 3 months respectively. The patients of the first group showed no clear recurrence of symptomatology, except but one case, who presented rare and slight contractions of the orbicularis orbitae muscle 6 months later. Among the patients of the second group, who had a shorter period of observation, the muscular contractions recurred in one case at about 4 weeks, and the patient refused another operation. In all cases the facial paresis persisted although with a certain tendency to decrease in time. Finally, none of our patients needed tarsorrhaphy or developed keratitis, owing to the moderate deficit of the facial nerve function.

Discussion

Various surgical treatments have been proposed for facial nerve spasm. The general purpose is that of producing a less or more extended lesion to the extracranial portion of the seventh cranial nerve or to the intracranial part, near the entry zone to the brain

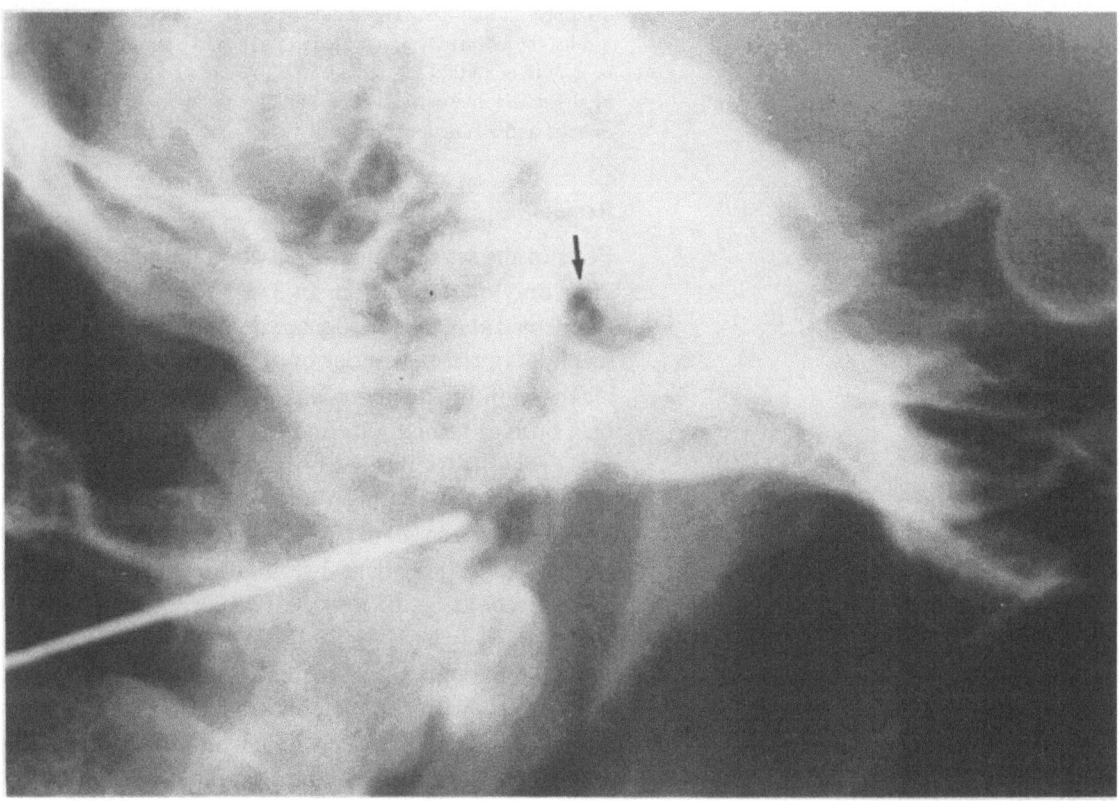

Fig. 2. Skull roentgenogram in lateral view. The needle has been introduced anteriorly to the mastoid process and under the external acoustic meatus (arrow)

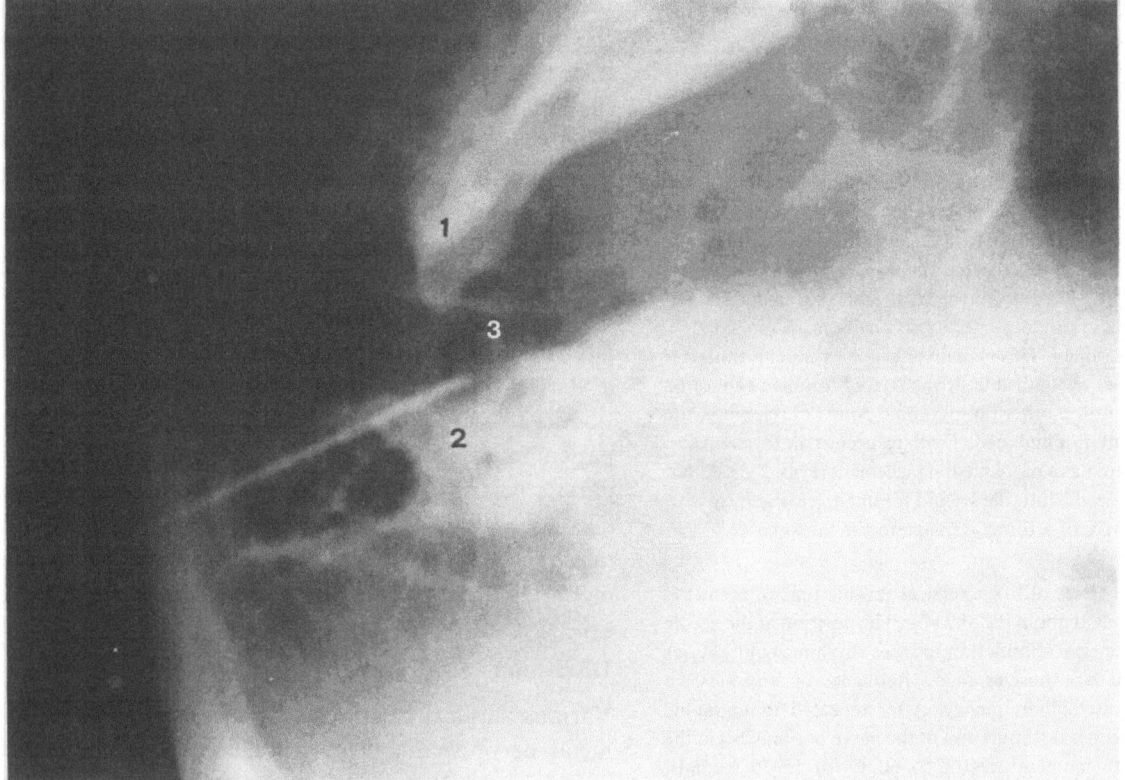

Fig. 3. Skull roentgenogram, projection for the base. The tip of the electrode is correctly positioned in proximity to the styloid foramen. *1* condilar process of the mandible, *2* mastoid, *3* jugular foramen

stem. To achieve this numerous, less or more invasive operations have been proposed. The easiest is facial nerve alcoholization, described by Harris[5] in 1932 at the level of the styloid foramen. Nevertheless this method produces a marked paresis and has a high risk of short-term recurrence[15]. In 1983, Takahashi[14] proposed a variant of this procedure more refined and accurate, consisting of adequate electrical stimulation, to define the position of the nerve, followed by alcoholization of the latter, with the advantage of a more precise localization of the side and the limits of the lesion.

Coleman[2] many years ago (1937), and more recently, Mielke[11] and Samii[12], proposed a particularly difficult method of determining a selective surgical section of the peripheral divisions of the nerve. A few years later, in 1942, German[4] and, in 1955, Scoville[13] proposed on the contrary, a partial surgical section of the main nervous trunks. Harris[5], also in 1932, suggested a cross anastomosis between the seventh and the hypoglossal nerves. More recently, in 1977, Jannetta[7] proposed a new procedure of exploration of the posterior fossa by microsurgical technique. This operation is probably the most effective in relieving spasm and the less traumatic for the facial nerve function. Its purpose is that of eliminating compressions or deformations on the intracranial part of the nerve, particularly near the dorsal root at the emergency point from the brain stem[9, 10].

Nevertheless this procedure may damage other cranial nerve in this area and needs general anesthesia with its consequent risks, particularly, in our opinion, in elderly patients or in subjects in bad general conditions. On the other hand, although in many cases this surgical approach would be preferable, particularly in subjects in good general conditions, there may be problems to convince them to accept it. On the contrary, percutaneous thermolesion under local anesthesia, such as that proposed by Hori[6], is generally easily accepted. This procedure surely represents an easier operation than that proposed by Jannetta[7], even with the limits connected to a functional damage of the

facial nerve, it is well tolerated by the patient, does not require general anesthesia, and needs a short hospitalization. In addition, the localization of the nerve is made exact by the use of radiographic and electrophysiologic controls and clinical observation. Finally, recurrencies over time are not a problem for the technical simplicity and repeatibility of this technique.

References

1. Auger RG (1979) Hemifacial spasm: clinical and electrophysiologic observations. Neurology 29: 1261–1266
2. Coleman CC (1937) Surgical treatment of facial spasm. Anh Surg 105: 647–657
3. Gardner WJ, Sava GA (1962) Hemifacial spasm: a reversible pathophysiologic state. J Neurosurg 19: 240–247
4. German WJ (1942) Surgical treatment of spasmodic facial tic. Surgery II: 912–914
5. Harris W, Wright AD (1932) Treatment of clonic facial spasm: a) by alcohol injection, b) by nerve anastomosis. Lancet I: 657–662
6. Hori T, Fukushima T, Terao H, Takakura K, Sano K (1981) Percutaneous radiofrequency facial nerve coagulation in the management of facial spasm. J Neurosurg 54: 655–658
7. Jannetta PJ, Abbasy M, Maroon JC *et al.* (1977) Etiology and definite microsurgical treatment of hemifacial spasm. Operative techniques and results in 47 patients. J Neurosurg 47: 321–328
8. Kao MC, Hung CC, Chen RC *et al.* (1978) Controlled thermodenervation of the facial nerve in the treatment of hemifacial spasm. Taiwan I. Hsuech Hui Tsa Chih 7: 226–233
9. Kondo A, Ishikawa JI, Konishi T (1981) The pathogenesis of hemifacial spasm: characteristic changes of vasculatures in vertebro-basilar artery system. In: Samii M, Jannetta PJ (eds) The cranial nerves. Springer, Berlin Heidelberg New York, pp 494–501
10. Maroon JC (1978) Hemifacial spasm. A vascular cause. Arch Neurol 35: 481–483
11. Miehlke A (1981) Management of hemifacial spasm. In: Samii M, Jannetta PJ (eds) The cranial nerves. Springer, Berlin Heidelberg New York, pp 478–483
12. Samii M (1981) Surgical treatment of hemifacial spasm. In: Samii M, Jannetta PJ (eds) The cranial nerves. Springer, Berlin Heidelberg New York, pp 502–504
13. Scoville WB (1955) Partial section of proximal seventh nerve trunk for facial spasm. Surg Gynecol Obstet 101: 495–497
14. Takahashi T, Dohi S (1983) Hemifacial spasm a new technique of facial nerve blockade. Br J nesth 50: 333–336
15. Wakasugi B (1972) Facial nerve block in the treatment of facial spasm. Arch Otolaryngol 95: 356–359

Section III

Pain and Miscellaneous

Acta Neurochirurgica, Suppl. 39, 139–141 (1987)
© by Springer-Verlag 1987

Long-Term Results of Percutaneous Gasserian Ganglion Lesions

J. Piquer*, V. Joanes, P. Roldan, J. L. Barcia-Salorio, and **G. Masbout**

Servicio de Neurocirugía, Hospital Clínico Universitario, Valencia, Spain

Summary

Percutaneous radiofrequency lesion of the Gasserian ganglion was performed between 1974 and 1984 in ninety-eight patients for the relief of trigeminal neuralgia. The average follow-up period was 4.5 years. Age, sex, and duration of illness were unrelated to outcome. Satisfactory analgesia was achieved in 68 patients. Thirty-one percent had return of pain (30 cases). Recurrent neuralgia occurred most frequently during first postoperative year (46%). Patients with marked sensory deficits had a reduced risk of recurrence. Post-operative complications included: reduced or absent corneal reflex (18 cases), corneal keratitis (3 cases) and anesthesia dolorosa (2 cases).

Keywords: Trigeminal neuralgia; radiofrequency rhizotomy.

Introduction

Percutaneous radiofrequency coagulation of the Gasserian ganglion is one of the commonly used surgical methods in the treatment of trigeminal neuralgia and short-term results are often excellent[3]. However, the long-term results of this procedure has been less often studied. The objective of this study is to present the results of percutaneous radiofrequency coagulation in 98 patients with trigeminal neuralgia after a follow-up of 2 to 10 years (average, 4.5 years).

Material and Method

The results obtained are of in 98 patients treated at the University Hospital of Valencia between 1974–78. Ages ranged from 31 to 88 years (mean, 62 years). There were 33 men and 65 women. At the time of the procedure, the duration of the disease ranged between 6 months and 11 years (mean, 4 years).

---

* J. Piquer, M.D., Servicio de Neurocirugía, Hospital Clínico Universitario, paseo Blasco Ibañez, 17, E-46010 Valencia, Spain.

Prior to the radiofrequency coagulation, some patients had been treated by other procedures: infraorbital nerve section (8 cases), alcohol nerve block (12 cases) and stereotactic radiosurgery (9 cases). In 6 cases pain affected the first division, in 21 cases the second division, in 26 cases the third division and two or three trigeminal divisions were affected in the remaining 54 patients.

The surgical technique has been reported previously[2, 3] and is similar to that described by Sweet and Wepsic[12].

To evaluate the results, recurrence was particularly looked for. Recurrences were defined and included a temporary return to medication or the need for further therapy. The possible influence of age, sex, duration of the disease, previous treatment, localization of the neuralgia and degree of sensory loss after the procedure on the results, was also evaluated. Statistical evaluation was carried out by Chi-square analysis with a significance level of p less than 0.05.

Results

Recurrences were observed in 30 patients (31%) of whom 23 underwent another radiofrequency coagulation. Four of the patients developed a second recurrence. Recurrent neuralgia occurred most frequently during the first postoperative year (Fig. 1).

The age, sex, duration of the disease and previous treatment did not significantly modify the obtained results. With respect to degree of sensory loss produced after the initial radiofrequency coagulation it was observed that in patients in whom the lesion created a substantial or moderate sensory loss, the recurrence rate was less than in patients in whom a minimal or no permanent sensory loss were observed (P ± 0,008, Table 1). Concerning the localization of pain, patients with neuralgia confined to the first trigeminal division had the least favorable outcome.

The following complications were observed: Corneal sensory loss were present in 18 patients of whom three cases developed a keratitis; two patients developed anesthesia dolorosa and one patient developed an herpes on the affected trigeminal division.

Table 1. *Relationship Between Degree of Sensory Loss and Recurrence Rate*

Degree of sensory loss	Number of cases	Number of cases with recurrence
Substantial deficit*	31	4 (13%)
Moderate deficit**	42	11 (26%)
Partial or no permanent deficit	25	15 (60%)
Total number of cases	98	30 (31%)

    * Analgesia or anesthesia in the affected division(s).
    ** Hypalgesia in the affected division(s).

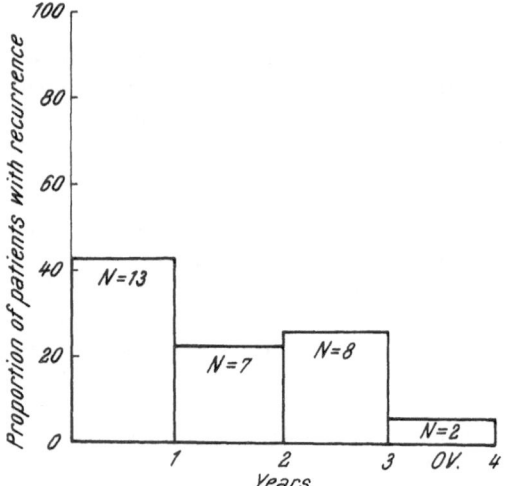

Fig. 1. Histogram illustrating lapse of time between first operation and recurrence of symptoms in trigeminal neuralgia

Discussion

Percutaneous radiofrequency coagulation of Gasserian ganglion was introduced by Kirschner in 1929[5] and then modified and perfected by Sweet and Wepsic in 1974[12]. Radiofrequency rhizotomy has recently undergone several technical modifications that have significantly reduced the complications associated with earlier techniques[14].

Long-term follow-ups including only 35% of the patients at 5 years have been reported[4, 6, 12]. In the present study, a recurrence rate of 31% after an average follow-up period of 4.5 years was seen. It has been found that the recurrence rate is higher during the first year after radiofrequency coagulation, which confirms the results reported by other authors[9, 11].

Among possible factors that might have influenced the results it was found that in patients with relatively marked sensory loss the risk of relapse was reduced. This observation has also been reported by other authors[6-9, 12].

The most frequent complication was corneal sensory loss observed both in the present series and in others[9, 13]. Corneal keratitis appears in 0 to 5%[8, 9, 11] and it occurred in 3 of our patients. Anesthesia dolorosa developed in two cases who needed stimulation of the nucleus ventralis posteromedialis thalami as further treatment[10]. The incidence of this complication ranges from 1 to 19%[1, 8, 9, 11, 12].

In conclusion, our results confirm that percutaneous radiofrequency coagulation of Gasserian ganglion is an effective method in the treatment of trigeminal neuralgia. The procedure is well tolerated and the rate of complications is low. However, there is a delicate balance between the risk of producing serious side-effects in the form anesthesia dolorosa-like conditions on a channel and of a high propability of recurrence on the other. This relationship is dependent on the degree of sensory loss which thus becomes of critical importance.

References

1. Apfelbaum RI (1977) A comparison of percutaenous radiofrequency trigeminal neurolysis and microvascular decompression of the trigeminal nerve for the treatment of tic douloureux. Neurosurg 1: 16–21
2. Barcia Salorio JL, Barberá J, Broseta J (1974) La termolesión de la raiz del trigémino en la neuralgia esencial. Rev Esp Oto-Neuro-Oftalm 32: 195–201
3. Broseta J, Gonzalez Darder J, Roldán P, Barberá J, Barcia Salorio JL (1979) Resultados de la termocoagulación por

radiofrequencia del ganglio trigeminal y raices espinales en el tratamiento del dolor cronico. Rev Cir Esp 33: 239–247

4. Burchiel KJ, Steege TD, Howe JF, Loeser JD (1981) Comparison of percutaneous radiofrequency gangliolysis and microvascular decompression for the surgical management of tic douloureux. Neurosurg 9: 111–119

5. Kirschner M (1936) Zur Behandlung der Trigeminusneuralgie. Erfahrung an 250 Fällen. Arch Klin Chir 186: 325–334

6. Latchaw JP Jr, Hardy RW Jr, Forsythe SB, Look AF (1983) Trigeminal neuralgia treated by radiofrequency coagulation. J Neurosurg 59: 479–484

7. Menzel J, Piotrowski W, Penfcholz H (1975) Long-term results of Gasserian ganglion electrocoagulation. J Neurosurg 42: 140–143

8. Nugent GR (1982) Technique and results of eight hundred percutaneous radiofrequency thermocoagulations for trigeminal neuralgia. Appl Neurophysiol 45: 504–507

9. Onofrio BM (1975) Radiofrequency percutaneous Gasserian ganglion lesions. Results in 140 patients with trigeminal pain. J Neurosurg 42: 132–139

10. Roldán P, Broseta J, Barcia Salorio JL (1982) Chronic VPM stimulation for anaesthesia dolorosa following trigeminal surgery. Appl Neurophysiol 45: 112–113

11. Siegfried J (1977) 500 percutaneous thermocoagulations of the Gasserian ganglion for trigeminal pain. Surg Neurol 8: 126–131

12. Sweet WH, Wepsic JG (1974) Controlled thermocoagulation of trigeminal ganglion and rootless for differential destruction of pain fibers. I. Trigeminal neuralgia. J Neurosurg 40: 143–156

13. Turnbull IM (1974) Percutaneous rhizotomy for trigeminal neuralgia. Surg Neurol 2: 385–389

14. Van Loveren H, Tew JM Jr, Keller JT *et al* (1982) A 10-year experience in the treatment of trigeminal neuralgia. Comparison of percutaneous stereotaxic rhizotomy and posterior fossa exploration. J Neurosurg 57: 757–764

Acta Neurochirurgica, Suppl. 39, 142–143 (1987)
© by Springer-Verlag 1987

Percutaneous Microcompression of the Gasserian Ganglion: Personal Experience *

M. Meglio **, **B. Cioni,** and **V. d'Annunzio**

Istituto di Neruochirurgia, Università Cattolica, Roma, Italy

Summary

The effectiveness of percutaneous microcompression of the gasserian ganglion, performed according to the technique proposed by Mullan and Lichtor, was evaluated in 74 patients affected by essential trigeminal neuralgia and compared to the results obtained with the RF thermocoagulation of Sweet and Wepsic.

Similar results were obtained with the two methods but adverse side-effects were less common with microcompression.

It is concluded that the microcompression should be recommended as a first choice among the different percutaneous procedures in the treatment of trigeminal neuralgia.

Keywords: Trigeminal neuralgia; RF rhizotomy; trigeminal ganglion compression.

Introduction

A new surgical procedure for the treatment of essential trigeminal neuralgia has been described by Mullan and Lichtor in 1983[1]. Following the old concept of Taarnhoj[4] and Shelden[2], they proposed a percutaneous technique to compress the gasserian ganglion.

In this paper we report the results obtained with such a technique and compare them to the results obtained with the RF lesion technique of Sweet and Wepsic[3] in another group of patients.

---

* Supported by MPI 71365 and by CNR progetti finalizzati sottoprogetto dolore.

** M. Meglio, M.D., Istituto di Neurochirurgia, Università Cattolica, Largo A. Gemelli 8, I-00168 Roma, Italy.

Material and Method

47 patients suffering from drug resistant trigeminal neuralgia had percutaneous microcompression of the gasserian ganglion. The foramen ovale was penetrated under general anesthesia and fluoroscopic control with a cannula large enough to let pass a Fogarty arterial catheter. The catheter was introduced through the cannula into the Meckel's cave and inflated with contrast media (iopamidol $0.75 \, cm^3$). The compression was maintained for periods ranging between 4 and 6 minutes.

33 different patients also affected by trigeminal neuralgia and unresponsive to medical treatment had percutaneous selective thermocoagulation of the gasserian ganglion as described by Sweet and Wepsic[3].

Only patients previously unoperated were included in this study.

Results

Percutaneous microcompression (Table 1): Immediate pain relief was achieved in all the 47 patients treated with the percutaneous microcompression of the gasserian ganglion. At 24 months of follow-up recurrence had occurred in 6 out of 14 (42.8%). The average recurrence time was 4.2 months. Marked dysesthesia requiring medical treatment occurred in 4 of the 47 patients (8.5%), while significant motor weakness occurred in two (4.25%).

Radiofrequency lesion (Table 2): Immediate pain relief was accomplished in 81.8% of our 33 patients treated with this method. At 24 months of follow-up recurrence had occurred in 15 of them (45.4%). The average recurrence time was 19.3 months.

Eight patients (24.2%) complained of permanent dysesthesia; no motor deficits were noticed.

Table 1. *Microcompression (47 patients)*

Follow-up (months)	3	6	12	24	36
Patients pain free	43	35	10	8	5
Patients with pain	4	4	6	6	6
Recurrence rate (%)	8.5	10.2	37.5	42.8	54.5

Table 2. *Radiofrequency Lesion (33 Patients)*

Follow-up (months)	3	6	12	24	36
Patients pain free	27	25	21	19	18
Patients with pain	6	8	12	14	15
Recurrence rate (%)	18.2	24.2	36.3	42.4	45.4

Conclusions

Our experience confirms the efficacy of the percutaneous microcompression of the gasserian ganglion in relieving pain of trigeminal neuralgia.

Comparing the results obtained with the RF procedure, which is presently the most common technique, shows that compression of the ganglion gives similar results as far as pain is concerned. The percentage of recurrence is similar with the two techniques. However, the one which may follow microcompression has a shorter delay. On the other hand, microcompression carries less risk of side-effects.

On the basis of the present results we suggest microcompression of the gasserian ganglion (with 4 to 6 minutes of compression) as the first choice among the different percutaneous procedures available for the treatment of trigeminal neuralgia.

References

1. Mullan S, Lichtor T (1983) Percutaneous microcompression of the trigeminal ganglion for trigeminal neuralgia. J Neurosurg 59: 1007–1012
2. Shelden CH, Pudenz RH, Freshwater DB (1955) Compression rather than decompression for trigeminal neuralgia. J Neurosurg 12: 123–126
3. Sweet WH, Wepsic JG (1974) Controlled thermocoagulation of trigeminal ganglion and rootlets for differential destruction of pain fibers. Part 1: Trigeminal neuralgia. J Neurosurg 40: 143–156
4. Taarnhoj P (1952) Decompression of the trigeminal root and the posterior part of the ganglion as treatment in trigeminal neuralgia. Preliminary communication. J Neurosurg 9: 288–290

Acta Neurochirurgica, Suppl. 39, 144–146 (1987)

Electrical Stimulation of the Gasserian Ganglion for Facial Pain: Preliminary Results

G. Broggi*, **D. Servello**, **A. Franzini**, and **C. Giorgi**

Department of Neurosurgery, Istituto Neurologico "C. Besta", Milano, Italy

Keywords: Atypical facial pain; postherpetic neuralgia; neurostimulation.

Introduction

Electrical stimulation of the Gasserian ganglion for the relief of intractable trigeminal pain is a method introduced by Shelden[6] using an apparatus inserted via the classical subtemporal approach[1]. Ten years later this technique was reintroduced in the treatment of atypical facial pain by Meyerson and Håkanson[5] who applied the modern technology of neurostimulation. A few years later Steude[7, 8] and Meglio[4] modified the technique for percutaneous application.

Patients and Methods

Of 8 patients (7 females), age 39–72 years (mean 63.7 years), with intractable facial pain, 5 had postherpetic neuralgia and 2 were diagnosed as anesthesia dolorosa following surgical treatment for tic douloureux. The eighth case presented with pain due to a intracavernous carotid aneurysm (Table 1).

Preoperatively, all patients had a test of intravenous morphine[3] and neuropsychology (Hamilton MMPI).

The surgical procedure employs the traditional approach of Härtel[2], used for RF thermorhizotomy; a small (10 mm) skin incision is made at the inferior border of the jaw where the scar is hardly noticeable. The foramen ovale is penetrated under fluoroscopic control with a 15 gauge needle. The stylet is withdrawn and a specially

designed electrode (Medtronic) is introduced within the Gasserian cystern (Fig. 1).

A trial of temporary stimulation via an external connection is performed for one month in order to avoid a placebo effect. At the end of this period, if pain relief is effective, the device is internalized. A passive stimulator is used in patients who require longlasting, continuous stimulation, while an ITREL system (Medtronic) is utilized when a cycling type of stimulation is preferable. Parameters of stimulation vary a great deal between one patient and another (see Table 2).

Results

During the temporary test, 7 patients had satisfactory but incomplete pain relief but in spite of this only partial effect, they had internalization.

The postoperative follow-up range between 7 and 20 months.

In 3 patients (1 case of anesthesia dolorosa, 1 postherpetic and 1 symptomatic neuralgia) the pain was significantly reduced: during the temporary test period the analgesic effect permitted withdrawal of all drugs, and at follow-up the therapeutic outcome remained unchanged, through minor changes of the stimulation parameters were required.

In 2 patients the pain was reduced to about 50% of preoperative level and during follow-up the efficacy of the stimulation was largely retained.

Finally, in the last 2 patients only slight, discontinuous pain relief was achieved after the inter-

* G. Broggi, M.D., Department of Neurosurgery, Istituto Neurologico "C. Besta", I-20133 Milano, Italy.

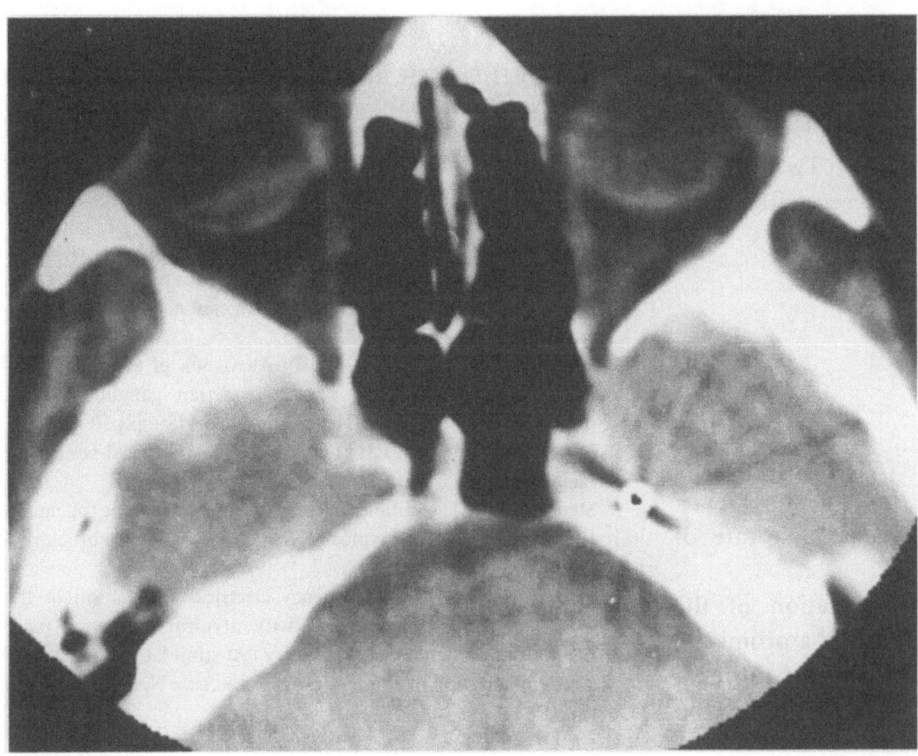

Fig. 1. CT scan image showing the electrode within the right Gasserian cistern

Table 1

Name	Age, sex	Etiology	Division involved	Temporary implant	Definitive implant	Results	Follow-up (months)
F. E.	69, female	postherpetic neuralgia	I–II L	yes	yes	poor	(removed after 6 months)
B. M.	70, female	postherpetic nueralgia	II–III L	yes	yes	poor	18
G. D.	68, female	postherpetic neuralgia	I–II–III L	yes	yes	good	15
T. S.	66, male	postherpetic neuralgia	I–II R	yes	yes	fair	12
V. V.	72, female	postherpetic neuralgia	II–III L	yes	no	poor	
C. G.	56, female	anesthesia dolorosa	II–III R	yes	yes	good	7
M. P.	39, female	anesthesia dolorosa	II–III L	yes	yes	fair	7
B. C.	72, female	atypical facial pain	I–II R	yes	yes	good	5

Table 2. *Stimulation Parameters*

Name	Amplitude (volt)	Rate (PPS)	Width (msec)	Cycle-on (sec)	Cycle-on (min)
T. S.	0.50	85	210	15	15
G. D.	0.25	85	210	64	15
B. C.	3.25	85	210	64	8
				period of stimulation	
F. E.	3.00	100	210	5 min × 4/day	
B. M.	1.50	120	210	5 min × 6/day	
V. V.	2.50	85	210	5 min × 4/day	
C. G.	4.00	25	210	continuous	
M. P.	2.50	100	210	5 min × 4/day	

nalization. At six months' follow-up one of them had the stimulator removed as there was no more pain relief.

Side-effects consisted of disconnection between the electrode and the stimulator in 4 patients. Two patients suffered from wound infection and necessitated removal of the entire system, which was later reimplanted.

Conclusions

The results obtained in this small series of patients have been quite unsatisfactory. In fact, no patient achieved complete pain relief and there was a high rate of complications and technical failures.

Nevertheless, we think that stimulation of the gasserian ganglion can still be considered a promising therapeutic tool for the treatment of facial pain resistant to conventional medical and surgical therapy.

References

1. Frazier CH, Lewy SH, Rowe SN (1937) Origin and mechanisms of paroxysmal neuralgia pain and surgical treatment of central pain. Brain 60: 44–45
2. Hartel F (1914) Über die intracranielle Injektionsbehandlung der Trigeminusneuralgie. Med Klinik 10: 582–584
3. Hosobuchi Y (1982) Personal communication
4. Meglio M (1984) Percutaneously implantable chronic electrode for radiofrequency stimulation of the Gasserian ganglion. A new prospective in the management of trigeminal pain. Acta Neurochir (Wien) [Suppl] 33: 521–525
5. Meyerson BA, Håkanson S (1980) Alleviation of atypical trigeminal pain by stimulation of the gasserian ganglion via an implanted electrode. Acta Neurochir (Wien) [Suppl] 30: 303–309
6. Shelden CM, Pudenz RH, Doyle J (1967) Electrical control of facial pain. Am J Surg 114: 209–212
7. Steude U (1978) Percutaneous electrostimulation of the trigeminal nerve in patients with atypical trigeminal neuralgia. Neurochirurgia 21: 66–69
8. Steude U (1984) Radiofrequency electrical stimulation of the gasserian ganglion in patients with atypical trigeminal pain: methods of percutaneous temporary test-stimulation and permanent implantation of stimulation devices. Acta Neurochir (Wien) [Suppl] 33: 481–486

Acta Neurochirurgica, Suppl. 39, 147–150 (1987)
© by Springer-Verlag 1987

Tractotomy and Partial Vertical Nucleotomy—for Treatment of Special Forms of Trigeminal Neuralgia and Cancer Pain of Face and Neck

C. A. Plangger*, J. Fischer, V. Grunert, and **I. Mohsenipour**

Department of Neurosurgery, School of Medicine, University of Innsbruck, Innsbruck, Austria

Summary

The therapy of face and neck pain has often been elusive. We attempted to improve the condition of these patients and tried to influence 1. pain of trigeminal neuralgia, where other forms of therapy had failed, 2. pain due to tumours in the distribution of the Vth, IXth and Xth nerve, when all other methods had proved to be unsuccessful, 3. pain due to a traumatic lesion of the Vth nerve after severe injury of the face and 4. pain in the first division of the Vth nerve after herpes zoster infection, when other forms of therapy had failed. After tractotomy the subnucleus caudalis n.V. is partially destroyed. Aim of the partial vertical nucleotomy is the interruption between the first and second neuron of the Vth nerve conveying pain and thermal sensibility, but also of the IXth and Xth nerve, which end in the subnucleus caudalis n.V. as well. Tactile and some thermal sensibility in the face is so retained, and anesthesia dolorosa or keratitis neuroparalytica avoided. Medially of and vertically to the tractotomy a 4–6 mm long incision both cranially and caudally of the tractotomy was made. For the first division of the Vth nerve the nucleotomy is performed on the lateral end of the tractotomy incision. In the patients with cancer of the face and neck a rhizotomy C 1/2 was added. 7 of the 12 patients with trigeminal neuralgia and 3 of the 6 patients with tumors of the face and neck were pain-free. The rest also showed a marked improvement. The patient with post-traumatic neuralgia of the Vth nerve and the herpes zoster neuralgia V_1 patient were freed from their complaints. The success rate in trigeminal neuralgia appears to depend on the duration of the disease.

Keywords: Trigeminal neuralgia; facial pain; neck pain; cancer pain; tractotomy; vertical nucleotomy; rhizotomy.

Introduction

Tractotomy and partial vertical nucleotomy are not indicated as the first procedure for tic douloureux. But in the therapy resistant idiopathic trigeminal neuralgia, symptomatic trigeminal neuralgia or pain in the distribution of the VIIth, IXth and Xth nerve it may be an useful adjunct to neurosurgical therapy. The mandibular fibers have been identified most dorsally in the spinal tract of n.V., the ophthalmic fibers most ventrally (and laterally), the maxillary fibers (V_2) between those of the other branches of the trigeminal nerve. Stimulation studies and postoperative observations of Kunc *et al.*[17–21] locate the somatic sensory fibers of cranial nerves VII, IX and X in a zone between the mandibular fibers of the trigeminal nerve and the most lateral fibers of fasciculus cuneatus (column of Burdach[17]). This places the fibers adjacent to the mandibular branch of V, which supplies the bordering cutaneous areas of the face and the mucous membranes of the tongue and oral cavity.

Dejerine[1] described five concentric zones, which end in various parts of the subnucleus caudalis n.V., the "onion-skin" pattern. This pattern has been confirmed by trigeminal tractotomies[21, 23]. McKenzie[23] noted that if the spinal tract is not sectioned high enough to give analgesia of all areas supplied by the trigeminal nerve, the upper lip is most commonly spared. The fibers of the trigeminal nerve overlap with those of the upper cervical roots, and this explains why a tractotomy may not suffice and we have to add a partial vertical nucleotomy to cope with the pain.

* Dr. C. A. Plangger, Department of Neurosurgery, University of Innsbruck, Anichstrasse 35, A-6020 Innsbruck, Austria.

The selective tractotomy varies according to the division of the Vth or VIIth, IXth of Xth nerve involved in the neuralgia. The level of the tractotomy, depends on the onion-skin pattern of the pain. The more caudally the tract is incised, the further the margin of analgesia extends from the midline of the face, whereas in more rostral incisions, analgesia appears in the center of the face.

Material and Methods

Four groups of patients were in treated by tractotomy and partial vertical nucleotomy after other methods of pain treatment had failed: 1. Pain of trigeminal neuralgia (12 patients); 2. pain due to tumors in the distribution of the Vth, VIIth, IXth and Xth nerve, when all other methods had proved to be unsuccessful (6 patients; mean age 60 years (44–72 years) 4 men, 2 women); 3. pain due to a traumatic lesion of the Vth nerve after severe injury of the face (1 patient, 45 years, male); 4. pain in the first division of the Vth nerve after herpes zoster infection, when other forms of therapy had failed (1 patient, 72 years, male).

The mean age of the patients with idiopathic trigeminal neuralgia was 68 years (between 54 and 84 years). Tractotomy and partial vertical nucleotomy was done in the mean 8 years after initial treatment. 67% of the patients were women, 33% men. Tractotomy and partial vertical nucleotomy was a last resort in cases of pain recurrences after more conventional forms of drug and surgical therapy had failed (Table 1). Almost all patients in this group were treated with carbamazepine, 10 of the 12 patients had up to three alcohol blocks of the gasserian ganglion and 8 of 12 patients up to three thermocoagulations of the gasserian ganglion. The patients treated with tractotomy are therefore a very selected group of problem patients.

The preferred level for tractotomy is 14 mm above the first filament of the second sensory cervical root within a range of 2 to 4 mm. The tractotomy should not be too rostral as incisions placed more cranially increase the possibility of postoperative ataxia. After tractotomy the subnucleus caudalis n.V. is partially destroyed. Medially and vertically to the tractotomy a 4–6 mm long incision is made above and below the tractotomy. For the first division of the Vth nerve the nucleotomy is performed on the lateral end of the tractotomy incision. The partial vertical nucleotomy interrupts not only fibers leading to the secondary ascending tract of the trigeminal nerve but spinothalamic fibers coming from the sensory cervical roots as well. In patients with cancer of the face and neck a rhizotomy C 1/2 was added.

Results

There is some correlation between the duration of pain and postoperative results in the patients with tic douloureux (Table 1). On average patients had been treated for 8 years before tractotomy and partial vertical nucleotomy. 7 of the 12 patients of this group were almost pain-free. The success rate of tractotomy

and nucleotomy seems to decrease with duration of the disease. Unfortunately the postoperative follow-up of patients is only 6 years.

The second group of patients are those with cancer pain, traumatic lesion of the Vth nerve or postherpetic pain. Pain in all patients with cancer could be alleviated, 3 patients have remained pain-free for the rest of their lives (Table 2). All cancer patients were poor general condition and in the final stages of their primary disease and their follow-up period is naturally short. The patient with postherpetic neuralgia responded well to tractotomy and partial vertical nucleotomy.

Discussion

Pain and temperature sensation of the Vth, VIIth, IXth and Xth nerve descend in the spinal tract of V. ending in the subnucleus caudalis n.V.[2, 6–8, 26]. Knowledge of this anatomic distribution of the fibers has led to the surgical operation of tractotomy and selective tractotomy[22]. Because the sponal tract of the trigeminal nerve and the trigeminal nuclear complex are both found in the upper two to three cervical cord segments, there is an overlap of spinal nerves C 1–C 3 and the spinal tract of V. at these spinal cord levels[10–14, 19, 21, 25, 27]. This overlap is most marked at the level of C 2, since the sensory root of C 1 may be small or absent[10, 15, 21] and only a small group of fibers of the spinal tract of V. enter C 3. There is evidence experimentally to show that both spinal and trigeminal nerve fibers may converge on the same neuron at these levels[14, 16]. There is also clinical evidence that this region of overlap may transmit painful sensations to higher centers over both trigeminothalamic and spinothalamic fiber systems (ventral secondary ascending tract of V. and lateral spinothalamic tract, respectively). Partial vertical nucleotomy also aims to destroy pain fibers coming from the cervical roots. In patients with cancer of the head and neck and pain in the Vth, VIIth, IXth and Xth nerve tractotomy and partial vertical nucleotomy is the procedure of choice and usually combined with cervical rhizotomy.

The method has many advantages. The sacrifice of sensory functions in order to suppress the pain is slight. After cutting the spinal trigeminal tract, only analgesia and thermanesthesia occur. This is very important because to a large number of patients the complete loss of sensation in the face and mouth is just as unbearable

Table 1. *Tractotomy and Partial Vertical Nucleotomy in Trigeminal Neuralgia*

Number	Sex	Age at nucleotomy	Previous treatment					Pain distribution			Duration of pain years	Results pain-free	Partial improvement
			Drugs	Block	Exeresis	Coagulation	Tractotomy	V_1	V_2	V_3			
1	female	76	+	2	−	3	1	−	+	+	5	−	+
2	female	61	+	2	−	−	−	−	+	+	4	+	−
3	female	62	intolerance	1	−	3	−	−	+	+	8	+	−
4	female	57	+	3	−	−	−	−	+	−	6	+	−
5	female	84	+	2	−	2	−	−	+	−	16	−	+
6	female	66	+	−	−	2	1	−	+	+	6	+	−
7	female	67	+	3	−	−	−	−	+	−	8	+	−
8	female	72	+	2	1	3	−	−	+	+	12	−	+
9	male	76	+	−	−	1	1	+	+	−	12	−	+
10	male	70	+	2	−	−	−	−	+	+	7	+	−
11	male	54	intolerance	2	−	3	1	−	+	−	8	−	+
12	male	57	+	2	−	1	1	+	+	−	5	+	−

Table 2. *Tractotomy and Partial Vertical Nucleotomy in Cancer Pain, Traumatic Lesion of the VIth Nerve and After Herpes Zoster Infection*

Number	Sex	Age at operation	Primary disease	Pain distribution						Operation			Results pain-free	Partial improvement
				V_1	V_2	V_3	VII	IX	X	Rhizotomy	Tractotomy	Nucleotomy		
1	male	67	squamous cell carcinoma	−	+	+	−	+	+	C_1, C_2	+	+	−	+
2	male	44	epipharynx carcinoma	−	+	+	+	+	+	C_1, C_2	+	+	−	+
3	female	62	malignant hemangioma	−	+	+	+	+	+	C_1, C_2	+	+	+	−
4	female	72	squamous cell carcinoma	−	+	+	+	+	+	C_1, C_2	+	+	−	+
5	male	63	larynx carcinoma	−	+	+	+	+	+	C_1, C_2	+	+	+	−
6	male	53	larynx carcinoma	−	+	+	+	+	+	C_1, C_2, C_3	+	+	+	−
1	male	45	posttraumatic neuralgia	+	+	+	−	−	−	−	+	+	+	−
1	male	72	herpes zoster neuralgia	+	−	−	−	−	−	−	+	+	+	−

as the neuralgic pains. Tactile sensation after tractotomy and partial vertical nucleotomy is adaequate for normal function[14, 21, 28]. The afferent impulses necessary for tongue, facial and other reflexes participating in mastication and in swallowing are intact and the location of food particles on the lips, tongue or gingiva is recognized. The corneal reflex, although sometimes markedly diminished, is not entirely lost and particles in the eye are felt and keratitis is rare[3–5, 9, 19, 21]. We have never encountered complications such as keratitis neuroparalytica, paresis of masticatory muscles, facial palsy, hemiparesis or analgesia dolorosa. The procedure can be done bilaterally. Pain in the distribution of the VIIth, IXth and Xth nerve is also easily accessible to surgical therapy. Tractotomy and partial vertical nucleotomy may relieve facial pain when other methods have failed.

References

1. Dejerine J (1914) Sémiologie des affections du système nerveux. Masson, Paris
2. Fischer J (1985) Partielle vertikale Nukleotomie. Ein neurochirurgischer Eingriff bei einigen Formen des therapieresistenten Gesichtsschmerzes. Zbl Neurochirurgie 46: 195–217
3. Grant FC (1941) Experiences with intramedullary tractotomy. IV. Surgery of the brainstem and its operative complications. Surg Gynecol Obstet 72: 747–754
4. Grant FC, Weinberger LM (1941) Experiences with intramedullary tractotomy; relief of facial pain and summary of operative results. Arch Surg 42: 681–692
5. Groff RA, Lewy FH (1939) Experiences with section of the descending spinal root of the fifth cranial nerve. Trans Am Neurol Assoc 65: 162–168
6. Grunert V (1971) Die chirurgische Behandlung der kombinierten Trigeminus- und Glossopharyngeusneuralgie. Wien Klin Wschr 51: 925–926
7. Grunert V (1972) Beitrag zur chirurgischen Behandlung der rezidivierenden Trigeminusneuralgie. Wien med Wschr 10: 134–135
8. Grunert V, Witzmann A, Grunert P, Partielle vertikale Nukleotomie als Modifikation der Tractotomie des Nervus trigeminus. In press
9. Guidetti B (1956) Tractotomy for the relief of trigeminal neuralgia; observation in 124 cases. J Neurosurg 7: 499–508
10. Humphrey T (1952) The spinal tract of the trigeminal nerve in human embryos between 7 1/2 and 8 1/2 weeks of menstrual age and its relation to early fetal behaviour. J Comp Neurol 97: 143–209
11. Humphrey T (1954) The trigeminal nerve in relation to early human fetal activity. Res Publ Assoc Res Nerv Ment Dis 33: 127–154
12. Humphrey T (1982) The central relations of the trigeminal nerve. In: Schneider RC *et al* (eds) Correlative neurosurgery, 3rd ed. Ch C Thomas, Springfield, Ill, pp 1518–1532
13. Kerr FWL (1961) Structural relation of the trigeminal spinal tract to upper cervical roots and the solitary nucleus in the cat. Exp Neurol 4: 134–148
14. Kerr FWL (1961) Atypical facial neuralgias; their mechanisms as inferred from anatomic and physiologic data. Mayo Clin Proc 36: 254–260
15. Kerr FWL (1963) Mechanisms, diagnosis and management of some cranial and facial pain syndromes. Surg Clin N Am 43: 951–961
16. Kerr FWL, Olafson RA (1961) Trigeminal and cervical volleys, convergence on single units in the spinal gray at C-1 and C-2. Arch Neurol 5: 171–178
17. Kunc Z (1957) Le traitement des douleurs intraitables du glossopharyngien et du vague. Presented at the Congress Soc Internat Chir, Mexico
18. Kunc Z (1960) La localisation des trajets de la douleur des nerv. IX, X et VII dans la moelle allongée et la possibilité de leur tractotomie sélective. Acta Neurochir (Wien) 8: 327–334
19. Kunc Z (1964) Tractus spinalis nervi trigemini. Fresh anatomic data and their significance for surgery. Rozpr Českoslovenki Akad Ved 79: 3–98
20. Kunc Z (1965) Treatment of essential neuralgia of the 9th nerve by selective tractotomy. J Neurosurg 23: 494–500
21. Kunc Z (1966) Significance of fresh anatomic data on spinal trigeminal tract for possibility of selective tractotomies. In: Knighton RS, Dumke PR (eds) Pain. Little, Boston, pp 351–363
22. Kunc Z (1970) Significant factors pertaining to the results of trigeminal tractotomy. In: Hassler R, Walker AE (eds) Trigeminal neuralgia. Thieme, Stuttgart, pp 90–100
23. McKenzie KG (1952) Trigeminal tractotomy. Presented at the Fourth Annual Max M. Peet Lecture, Ann Arbor, Michigan, October 31
24. Taren JA (1964) The position of the cutaneous components of the facial, glossopharyngeal and vagal nerves in the spinal tract of V. J Comp Neurol 122: 389–398
25. Taren JA, Kahn EA (1962) Anatomic pathways related to pain in face and neck. J Neurosurg 19: 116–121
26. Taren JA, Kahn EA (1982) Trigeminal neuralgia. In: Schneider RC *et al.* (eds) Correlative neurosurgery, 3rd ed. Ch C Thomas, Springfield, Ill, pp 1532–1545
27. Torvik A (1956) Afferent connections to the sensory trigeminal nuclei, the nucleus of the solitary tract and adjacent structures; an experimental study in the cat. J Comp Neurol 106: 51–141
28. Weinberger LM, Grant FC (1942) Experiences with intramedullary tractotomy: III. Studies in sensation. Arch Neurol Psychiatry 48: 355–381

Acta Neurochirurgica, Suppl. 39, 151–154 (1987)
© by Springer-Verlag 1987

Investigation of Trigeminal SEP with Scalp and Deep Electrodes

A. Sólyom*, Sz. Tóth, I. Holczinger, J. Vajda, and **Z. Tóth**

National Institute of Neurosurgery, Budapest, Hungary

Summary

A special technique for stimulating and recording trigeminal somatosensory evoked potential (SEP) is described. Eleven healthy subjects and seven patients with chronically implanted deep electrodes have been investigated. Characteristic polyphasic waves were repeatedly observed. The recording of somatosensory evoked potential following trigeminal stimulation is a more difficult technique. This paper describes data in control subjects with scalp electrodes and data on patients with electrodes in the nucleus ventralis posteromedialis and cortical white matter. Depth recording may provide useful information about the origin, nature, and properties of trigeminal SEP.

Keywords: Trigeminal evoked potential.

Introduction

Trigeminal SEPs have been investigated in other areas where evoked potential techniques can give useful clinical data. The different techniques used have given results that are not easily understood. Recording a trigeminal SEP (TSEP) has been difficult, because of large amplitude stimulus artefacts produced by the close proximity of the stimulating and recording electrodes and the simultaneous production of muscle potentials by transcutanous stimulation. The paucity of reports on TSEP may have been due to these technical problems. Research on afferent impulses travelling through the fifth cranial nerve would be of great interest, as this innervates most of the face and may be involved in a number of intracranial pathologies, of known and unknown origin. However, little is known about trigeminal evoked potentials, and even less about their early waves and related far fields[4]. Drechsler et al.[1, 2] demonstrated negative peaks at 5, 14, and 34 msec, positive peaks at 9 and 23 msec. N 5 and N 9 were missing after thermocoagulation of the gasserian ganglion, and a negativity with a peak at 5 msec was recorded from the ganglion during surgery, suggesting that structure as the origin of N 5. Later waves were postulated as being cortical in origin[1, 2]. Salar et al.[7] stimulated at sensory threshold via a coaxial subcutaneous needle placed near the point of emergence of the branch of the trigeminal nerve and Singh et al. registered N 3, P 9, N 14, P 23 and N 34 stimulating by needles[8]. Findler and Feinsod found lower lip latencies slightly longer than upper lip latencies and polarities were reversed[3]. There are very few reports of recording from the trigeminal system in man[5, 6]. Somatosensory evoked potentials following stimulation of the trigeminal nerve have been studied in normal subjects and in patients with trigeminal neuralgia and multiple sclerosis but we have no evidence about investigations on patients with chronically implanted deep electrodes. The aim of this paper is to provide further information for the origin and properties of TSEP in man.

Methods and Material

The infraorbital nerve was stimulated by bipolar steel needles inserted into the infraorbital foramen. Rectangular wave pulses were delivered by a MEDICOR M 440 stimulator at about a rate of 3 Hz. The pulse duration was very short (0,05–0,1 msec) in order to minimize the artefact. The stimulus intensity was 3 times the sensory threshold. It ranged between 0,5–2 mA. The fine uninsulated subcutaneous needle recording electrodes were placed according to the 10–20 system on the C 5–C 6 points with Fz reference. The settlement of deep electrodes was published in detail elsewhere[10, 11]. Recording

* Dr. A. Sólyom, National Institute of Neurosurgery, Amerikai út 57, H-1145 Budapest, Hungary.

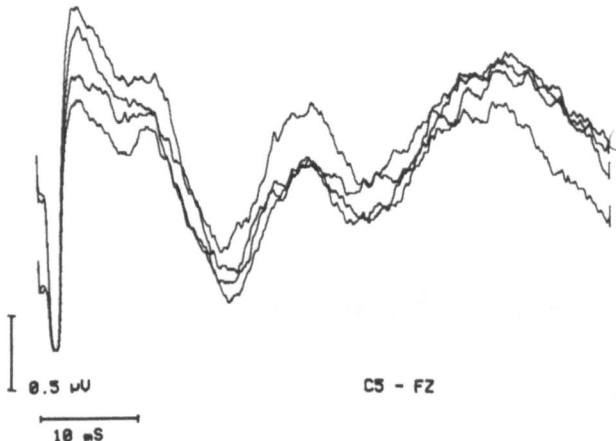

Fig. 1. Characteristical scalp recorded TSEP following contralateral stimulation of the second branch of the trigeminal nerve. Superimposed averaged evoked potentials. Components of the first ten msec are uncertain

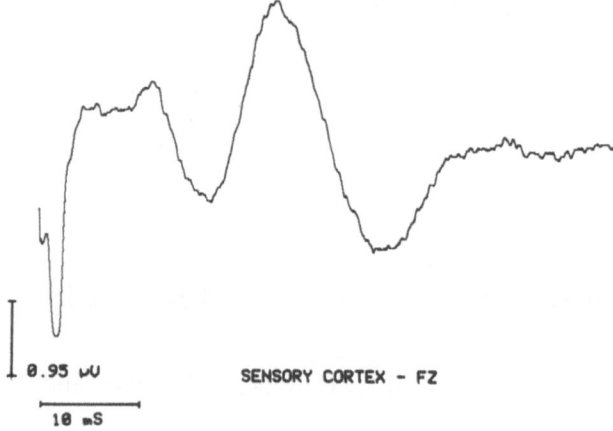

Fig. 2. TSEP recorded from the cortical white matter electrode

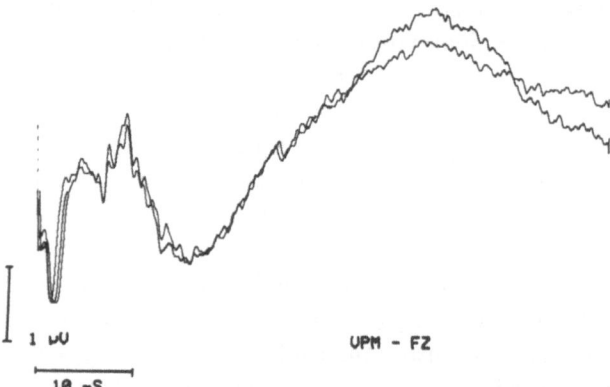

Fig. 3. Common feature of the VPM is the sharp positive deflection at about 6 msec. The earlier components are more definite. The first, second and third records were derived from a patient suffering from intractable pain

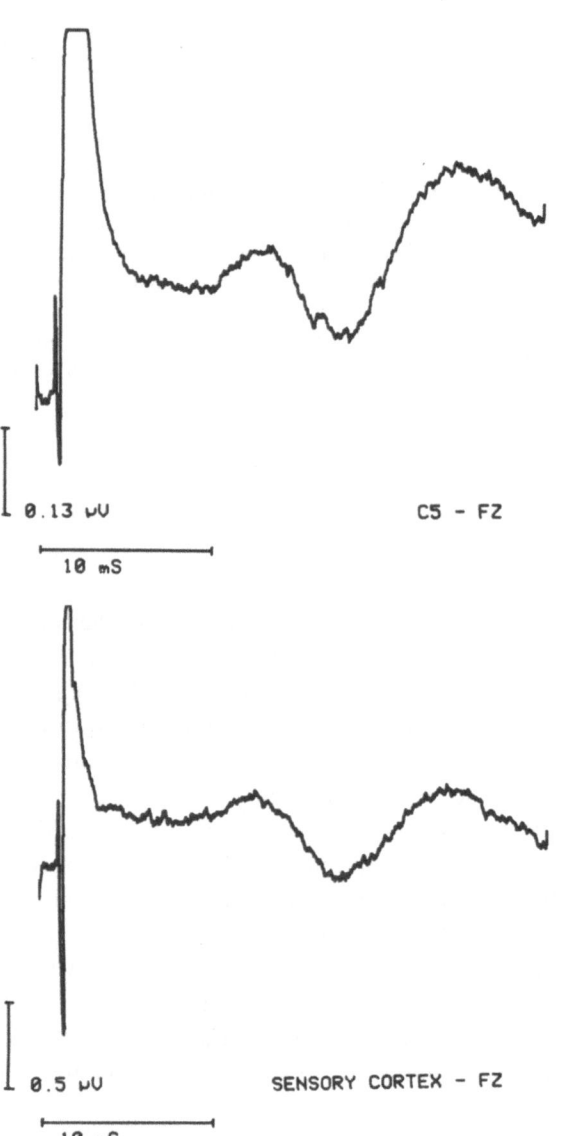

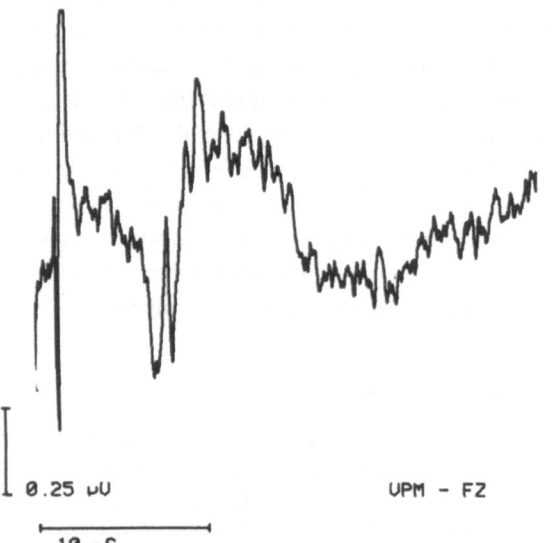

Fig. 6. The major positive deflection at about 6 msec is characteristic of the nucleus ventralis posteromedialis. The Figs. 4–6 originate from the same patient

Figs. 4 and 5. Relatively consistent triphasic complex in a patient suffering from parkinsonism. Early potentials are absent

from muscles surrounding the infraorbital foramen was performed using EMG concentric needle electrodes to assess any muscular reflex activity. Input from the recording electrodes was led to a MEDICOR M 440 differential amplifier and the output was summated by MYOPROC V 3 averager. An analysis time of 30, 60 or 100 msec was used with all patients at 1 msec delay. The horizontal resolution was 20 or 40 µsec per channel. Routinely, 256–512 responses were averaged using 10 Hz low and 10 kHz high filter. Negativity showed an upward deflection. The number, shape and polarity of the recorded waves were analyzed visually. Latencies were measured, the mean and standard deviation were calculated for all the waves. TSEPs were recorded in 11 volunteers and 4 patients suffering from parkinsonism, 1 patient from depression, 1 patient from atactic tremor and another one from intractable pain with chronically implanted deep electrodes. Also the deep electrodes were coupled in a bipolar fashion.

Results

Latencies of SEP in normal subjects (N = 11)

Left	Right
$P_1 = 9.1 \pm 0.79$	9.73 ± 0.69
$N_1 = 11.14 \pm 0.52$	11.0 ± 0.35
$P_2 = 17.89 \pm 1.58$	17.55 ± 1.74
$N_2 = 26.69 \pm 2.59$	27.84 ± 3.61
$P_3 = 5.63$	38.13 ± 7.3
$N_3 = 47.8 \pm 4.64$	51.84 ± 8.17
$P_4 = 56.77 \pm 4.57$	55.64 ± 3.58

Latencies of VPM TSEP (N = 7)

$P_1 = 2.53 \pm 0.31$ (N = 3)
$N_1 = 4.09 \pm 0.62$ (N = 6)
$P_2 = 6.28 \pm 0.22$ (N = 6)
$N_2 = 8.71 \pm 0.48$ (N = 7)
$P_3 = 14.89 \pm 2.03$ (N = 6)
$N_3 = 27.8 \pm 3.89$ (N = 3)

Latencies of cortical white matter TSEP (N = 7)

$N_1 = 5.55 \pm 0.71$ (N = 4)
$P_1 = 7.67 \pm 0.48$ (N = 5)
$N_2 = 10.55 \pm 0.95$ (N = 4)
$P_2 = 16.91 \pm 0.93$ (N = 6)
$N_3 = 23.67 \pm 0.8$ (N = 6)
$P_3 = 34.63 \pm 1.84$ (N = 4)
$N_4 = 43.87 \pm 2.66$ (N = 4)

Latencies of scalp TSEP (N = 7)

$P_1 = 8.28 \pm 0.97$ (N = 5)
$N_1 = 11.26 \pm 1.05$ (N = 6)
$P_2 = 18.07 \pm 0.9$ (N = 6)
$N_2 = 25.24 \pm 3.28$ (N = 6)
$P_3 = 34.13 \pm 1.4$ (N = 6)
$N_3 = 47.17 \pm 7.4$ (N = 4)
$P_4 = 50.85 \pm 1.48$ (N = 2)

Discussion

Our patterns of TSEP elicited by infraorbital stimulation are uniform and well reproducible. Peak polarities and latencies may vary between studies as a function of recording placement of stimulating electrodes. The authors agree that the first peak, labeled N 5 is likely to be generated in the gasserian ganglion. Leandri *et al.* suspected reflection of muscular activity[4]. N 11 is probably of primary sensory cortex origin and the later components have been attributed to secondary cortical sensory areas. There is a first positivity followed by a slower negative potential in the nucleus ventralis posterolateralis following stimulation of the contralateral median nerve. This is very similar to the first positive deflection followed by a slower negative potential in the VPM when the contralateral infraorbital nerve is stimulated. The difference of latencies may derive from the length of the two nerves. Our bipolar montage in deep structure exclude the muscular contamination. We suppose that the thalamic (VPM) first positive deflection with stimulation of the infraorbital nerve might have been of medial lemniscal origin and the following negative wave may be attributable to synaptic potentials of thalamic relay neurons[9]. Our thalamic (VPM) evoked potential evoked by infraorbital nerve stimulation is a new result in this field.

References

1. Drechsler F (1980) Short and long latency cortical potentials following trigeminal nerve stimulation in man. In: Barber C (ed) Evoked potentials. University Park Press, Baltimore, pp 415–422
2. Drechsler F, Wickboldt I, Neuhauser B, Miltner F (1977) Somatosensory trigeminal evoked potentials in normal subjects and in patients with trigeminal neuralgia before and after thermocoagulation of the ganglion Gasseri. Electroencephalogr Clin Neurophysiol 43: 496

3. Findler G, Feinsod M (1982) Sensory evoked response to electrical stimulation of the trigeminal nerve in humans. J Neurosurg 56: 545–549

4. Leandri M, Parodi CI, Favale E (1985) Electroencephalogr Clin Neurophysiol 62: 99–107

5. Leandri M, Campbell JA (1986) Origin of early waves evoked by infraorbital nerve stimulation in man. Electroencephalogr Clin Neurophysiol 65: 13–19

6. Ridderheim P-A, Essen C von, Blom S, Zetterlund B (1985) Intracranially recorded compound action potentials from the human trigeminal nerve. Electroencephalogr Clin Neurophysiol 61: 138–140

7. Salar G, Iob I, Mingrino S (1982) Somatosensory evoked potentials before and after percutaneous thermocoagulation of the gasserian ganglion for trigeminal neuralgia. In: Courjon I, Manguiere F, Revol M (eds) Clinical applications of evoked potentials in neurology. Raven Press, New York, pp 359–365

8. Singh N, Sachdev KK, Brisman R (1982) Trigeminal nerve stimulation: short latency somatosensory evoked potentials. Neurology 32: 97–101

9. Sólyom A, Tóth Sz, Holczinger T, Vajda J, Tóth Z, Kálmánchey R The spread of SSEP within the nervous system (in press)

10. Tóth Sz, Vajda J (1980) Multitarget technique in Parkinson surgery. Appl Neurophysiol 43: 109–113

11. Tóth Sz, Vajda J, Zaránd P, Sólyom A (1979) Recent developments of neurobiology in Hungary VIII. 73–91, Academia Budapest

Acta Neurochirurgica, Suppl. 39, 155–158 (1987)

Dorsal Root Entry Zone Lesion versus Spinal Cord Stimulation in the Management of Pain from Brachial Plexus Avulsion

G. Garcia-March, M. J. Sánchez-Ledesma, P. Diaz, L. Yagüe, J. Anaya, J. Gonçalves, and J. Broseta*

Departmento de Neurocirugía, Hospital Clínico Universitario, Salamanca, Spain

Summary

In six patients with total or partial brachial plexus avulsion, spinal cord stimulation was tried as pain treatment. Two patients had had amputation of the arm and suffered from phantom limb and stump pain. After a mean follow-up of 14 months two patients were pain-free, one had partial relief and required analgesics and in three patients there was no effect. In eleven patients, including the three patients in whom spinal cord stimulation had failed to produce a long-lasting pain relief, dorsal root entry zone (DREZ) lesions were performed. At early follow-ups all these patients reported substantial pain relief, but after a mean follow-up of 17 months the results were less favorable: Three patients were pain-free, three had a marked improvement and five had recurrence of the original pain. Neither of the two methods of treatment produced any serious side-effects or permanent sequelae.

Keywords: Brachial plexus avulsion; dorsal root entry zone; pain; spinal cord stimulation.

Introduction

Pain from brachial plexus avulsion has traditionally been considered as resistent to all forms of therapy. Amputation, local surgical approaches to the nerve lesion and different medullary, mesencephalic and thalamic ablative procedures have proved to be ineffective[3, 5, 9–12]. Spinal cord and thalamic stimulation, introduced in the early 1970s, have been tried with promising results[1, 4]. Since dorsal root entry zone (DREZ) lesions has been claimed to be particularly suitable for dealing with this type of pain[2, 6–8], the indication for stimulation methods now seems somewhat controversial. Nevertheless, according to our experience, spinal cord stimulation (SCS) still remains the first choice of treatment due to its non-destructive nature. This study compares the long-term results obtained with SCS and with DREZ lesions in a group of 14 patients presenting with chronic pain due to brachial plexus avulsion.

Clinical Material and Methods

Table 1 summarizes the neurological condition, duration of pain, and type and outcome of treatment in 14 patients with partial or total brachial plexus avulsion. In 11 patients the cause of the injury was traffic accidents, in most cases with motorcycles, two patients had had industrial accidents and one a high fall. All patients suffered major neurological deficits. Only one patient had normal motility of the arm, and two patients had amputations. Severe disturbance of sensation was present in all patients. Horner's syndrome appeared in three of four patients with injury to the inferior part of the brachial plexus. All patients suffered chronic pain, generally with diffuse distribution in the arm and hand, and the two patients with amputation had stump and phantom limb pain. Pain and sensory loss were generally not limited to the dermatomes corresponding to the avulsed roots. The pain had started 21 to 30 days after the injury and the mean duration of pain history was 13 months. Analgesics, including narcotic drugs, had been ineffective and transcutaneous nerve stimulation and various types of nerve blocks had failed to produce long-lasting pain relief. Two patients had had sympathectomy and in one patient nerve grafting of a partial plexus lesion had been performed. Neither of these procedures had been successful. Myelography disclosed the presence of meningocoele at the place of the avulsed roots in 13 of the patients. With electrophysiological methods preganglionic lesions could be demonstrated in 11 patients and in three there were signs of postganglionic injury.

* J. Broseta, M.D., Departamento de Neurocirugía, Hospital Clínico Universitario, E-37007 Salamanca, Spain.

Table 1. *Clinical Data, Treatments and Results*

Case	Age, sex	Clinical condition	Duration of pain	Previous treatments	Current treatment	Follow-up	Results	
							Early	Late
1	44, male	diffuse weakness, sensory loss and pain in left arm	48 months	drugs, sympathectomy	SCS	58 months	excellent	excellent
2	48, female	dysesthesia and pain in right arm, flail limb	10 months	brachial pl. surgery, TNS	SCS	43 months	good	good
3	30, male	stump and phantom phantom pain in left upper limb, Horner's syndrome	16 months	amputation, TNS, drugs	SCS, DREZ	19 months 46 months	fair excellent	poor good
4	31, male	weakness, sensory loss and pain in right forearm and hand	22 months	nerve blocks, TNS	SCS, DREZ	22 months 41 months	good excellent	poor good
5	51, male	stump and phantom pain in left upper limb	15 months	sympathectomy, amputation, stump revision	SCS, DREZ	14 months 40 months	fair excellent	poor excellent
6	37, female	diffuse weakness, sensory loss and pain in right arm	10 months	drugs, TNS	SCS	12 months	excellent	good
7	28, male	flail limb and left arm pain, Horner's syndrome	7 months	nerve blocks, TNS	DREZ	39 months	excellent	fair
8	45, male	dysesthesia and pain in right arm	10 months	nerve blocks	DREZ	34 months	excellent	excellent
9	52, male	flail limb and right arm pain, Horner syndrome	12 months	nerve blocks, TNS	DREZ	30 months	excellent	poor
10	57, male	proximal palsy, sensory loss and pain in right extremity	9 months	nerve blocks, TNS, drugs	DREZ	18 months	excellent	fair
11	48, male	phantom pain in left extremity	11 months	amputation, stellate block, amputation	DREZ	13 months	excellent	fair
12	46, female	dysesthesia, sensory loss and pain in arm, flail limb	15 months	TNS	DREZ	9 months	excellent	poor
13	29, male	distal weakness, sensory loss and pain in right forearm and hand, Horner's syndrome	8 months	Drugs, TNS	DREZ	8 months	excellent	excellent
14	47, female	dysesthesia, sensory loss and pain in right arm, flail limb	7 months	nerve blocks, TNS	DREZ	8 months	excellent	good

TNS: transcutaneous stimulation; SCS: spinal cord stimulation; DREZ: dorsal root entry zone lesion.

In the first six patients, SCS was tried to alleviate the pain. In all cases bipolar stimulation was used. Two stimulating electrodes were percutaneously introduced in the epidural space and, under fluoroscopic control, advanced to a cervical level were stimulation provoked a tingling sensation in the painful region. Percutaneous extension wires were connected to the proximal end of the electrodes in order to permit trial stimulation which was performed during a period of about two weeks. The parameters of the stimulation were 0.5 msec pulse width and 80–120 Hz. The amplitude was kept at a level just suprathreshold to evoke paresthesiae. Stimulation was performed in periods of 30 minutes, 1–6 times daily. All six patients claimed pain relief during the period of trail stimulation and the system was therefore permanently implanted.

In the next eight patients, and in three who had failed to respond to long-term SCS, DREZ lesions were performed. A surgical technique according to the originators of this method was employed. Using general anesthesia and with the patient in sitting position wide laminectomies of the vertebra corresponding to the injured spinal cord segments were performed. After opening the dura the posterolateral sulcus was identified. A cordotomy electrode with a bare tip of 2 mm was introduced into the dorsal root entry zone at a 25 degrees angle and thermolesions made with 25–30 mA at every second millimeter along the sulcus from the upper to the lower limits of the injured roots. The electrode was maintained in a fixed position until a white coagulated area appeared around its tip. The maximal lesion time was 15 seconds.

Results

The results were classified in four categories: *Excellent*, implying 100% pain relief, no analgesics required and return to work or social life; *good*, 75–100% pain relief, light analgesics occasionally required and return to work or social life; *fair*, 25–75% pain relief, analgesics required, unable to return to work or social life; *poor*, less than 25% pain relief, narcotics required and other invasive neurosurgical treatment considered.

All six patients with SCS initially enjoyed pain relief, and two were judged to have excellent result, two good and two fair. However, after a mean follow-up of 28 months the stimulation became less effective. Only one patient remained pain-free and two had only fair relief. In the other three patients the treatment had become ineffective and they later had DREZ lesions.

The DREZ thermocoagulation produced excellent results in the early postoperative period, but after one year there was a progressive incidence of pain recurrence. After a mean follow-up of 27 months only three patients were classified as having excellent results and three had good, three fair and two poor results.

There were no side effects associated with the SCS treatment. Following DREZ lesions one patient suffered ispilateral leg paralysis which subsided after three months.

Discussion

It is postulated that deafferentation following brachial plexus avulsion leads to a local hyperactivity in dorsal horn neurons transmitting and modulating pain. This pathophysiological mechanism may account for the relative ineffectiveness of SCS in reducing this pain, and *viceversa* explain the efficacy of DREZ lesions. However, our experience as reported in this study does not support these ideas since most of the patient benefited from SCS although the effect faded after one year. The outcome of DREZ lesions was excellent during the first postoperative period but with this treatment also the pain tended to recur although at a somewhat later stage than with SCS.

In conclusion, both SCS and even more so DREZ lesions have proved to have an initially beneficial effect on plexus avulstion pain. However, both methods failed to provide long-term satisfactory pain relief in several of the patients who had benefited earlier. Nevertheless, due to its nondestructive nature SCS should still be considered a first choice for treating pain in partial brachial plexus avulsion. On the other hand, in cases with more extensive avulsion or with phantom and/or stump pain, DREZ lesions could be recommended.

References

1. Broseta J, Roldan P, Gonzalez-Darder J, Bordes V, Barcia-Salorio JL (1982) Chronic epidural dorsal column stimulation in the treatment of causalgic pain. Appl Neurophysiol 45: 190–194
2. Dieckmann G, Veras G (1984) High frequency coagulation of dorsal root entry zone in patients with deafferentation pain. Acta Neurochir (Wien) [Suppl] 33: 445–450
3. Falconer MA (1953) Surgical treatment of intractable phantom-limb pain. Br Med J 1: 299–304
4. Hood TW, Siegfried J (1984) Epidural versus thalamic stimulation for the management of brachial plexus lesion pain. Acta Neurochir (Wien) [Suppl] 33: 451–457
5. Narakas AO (1981) The effects of pain of reconstructive neurosurgery in 160 patients with traction and/or crush injury to the brachial plexus. In: Siegfried J, Zimmermann M (eds) Phantom and stump pain. Springer, Berlin Heidelberg New York, pp 126–147
6. Nashold BS, Ostdhal RH (1979) Dorsal root entry zone lesions for pain relief. J Neurosurg 51: 59–69
7. Nashold BS, Bullitt E, Ostdhal RH (1981) The role of dorsal root entry zone lesions in relief of central pain. In: Dietz J *et al.* (eds) Neurological surgery. Thieme, Stuttgart New York, pp 67
8. Samii M (1981) Thermocoagulation of the substantia gelatinosa for pain relief (preliminary report). In: Siegfried J, Zimmermann M (eds) Phantom and stump pain. Springer, Berlin Heidelberg New York, pp 156–159

9. Siegfried J, Zumstein H (1973) La douleur fantôme et son traitment neurochirurgical. Med Hyg 31: 867–868

10. Sugita K (1972) Results of stereotaxic thalamotomy for pain. Confin Neurol 34: 265–274

11. White JC, Swett WH (1969) Pain and the neurosurgeon. A forty-year experience. ChC Thomas, Springfield, Ill

12. Zorub DS, Nashold BS, Cook WA (1974) Avulsion of the brachial plexus. I. A review with implications on the therapy of intractable pain. Surg Neurol 2: 347–353

Acta Neurochirurgica, Suppl. 39, 159–162 (1987)

Radiosurgery of Central Pain

J. L. Barcia-Salorio*, P. Roldan, and **L. Lopez-Gomez**

Servicio de Neurocirugía, Hospital Clínico Universitario, Valencia, Spain

Summary

Based on experimental and clinical evidence of central pain produced by hyperactivity of deafferented neurones and associated with irritative foci at thalamic or cortical levels, stereotactic low-dose (10 Gy) radiosurgery has been performed in 3 patients with central pain syndromes. SEEG findings and results of stereotactic radiosurgery on painful conditions are presented and mechanisms of action discussed.

Keywords: Central pain; radiosurgery.

Introduction

After the initial report by Nashold[7] on the epileptic nature of thalamic pain, Toth[8] has recently reported that in central pain the spontaneous and evoked electrical activity in the specific and nonspecific thalamic nuclei are characteristically paroxysmal. Moreover we have reported the beneficial effect of focal irradiation of the epileptic focus in epileptic patients[2] and experimentally in cats[3]. We report the electrographic SEEG findings in a group of patients with central pain, as well as the effect on the pain of focal stereotactic radiosurgery of irritative foci.

Methods and Material

Case 1: A 56-year-old female suffered a stroke with right hemiparesis and painful anesthesia. CT scan showed an hemorrhagic lesion in the right temporal region. The pain was disabling, with continuous painful sensation, and several episodes of intense, burning pain, beginning in the right foot and extending up to the arm and face, resembling an epileptiform "jacksonian march". Preoperative EEG showed slow basal activity and bilateral delta waves particularly prominent in the left temporal region.

Case 2: A 60-year-old female suffered a right, mandibular postherpetic trigeminal neuralgia. A trigeminal selective radiofrequency thermolesion, produced facial anesthesia dolorosa including the first trigeminal division. The right eye had been enucleated. A stimulating device was implanted in the contralateral VPM thalamic nucleus with a good initial result, but a poor long-term effect requiring its withdrawal. Preoperative EEG showed a generalized, fast irritative theta activity.

Case 3: A female patient aged 61 had a stroke with residual left hemiparesis and anesthesia dolorosa. The pain started in the upper extremity and extended down to the leg, a "jacksonian march"-like pain attack, as in case 1. CT scan showed no abnormal findings. EEG showed slow irritative activity.

After an extensive conventional EEG study, a flexible electrode with 6 terminals for deep brain recording was stereotactically introduced to the VPL thalamic nucleus, contralateral to the pain, and electrical activity recorded for a period of about one week. The VPL coordinates were 7 mm behind the AC-PC line midpoint, 5 mm above this line, and 14 mm lateral to the midline. The target coordinates were calculated directly from stereo CT (Fig. 1).

Once the electrode was withdrawn, and the stereotactic coordinates of the epileptic-like irritative focus were obtained, stereotactic radiosurgery was performed with 1.2 MeV photons from a conventional cobalt unit coupled with a stereoguide, with the cross-fire technique and a 10 mm diameter colimeter. A total dose of 10 Gy was administered.

After radiosurgery, changes in EEG activity, pain levels and medication were assessed.

Results

Deep brain recordings of SEEG showed marked delta activity in the thalamus (terminals 1, 2) (Fig. 2) in case

---

* Prof. Dr. J. L. Barcia-Salorio, Servicio de Neurocirugía, Hospital Clínico Universitario, paseo Blasco Ibañez, 17, E-46010 Valencia, Spain.

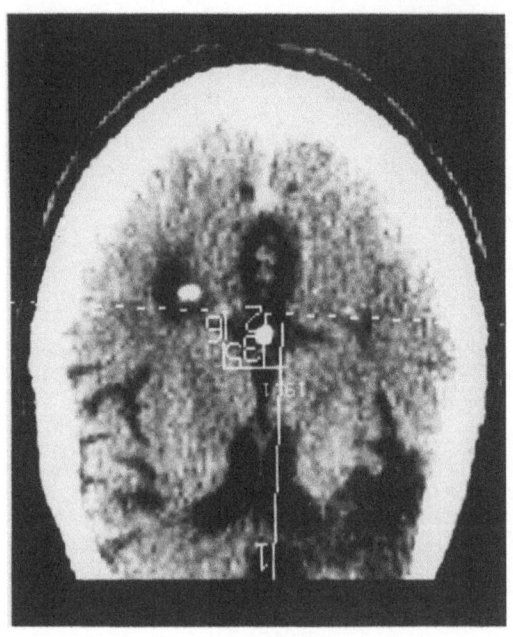

Fig. 1. CT used for the determination of thalamic target coordinates

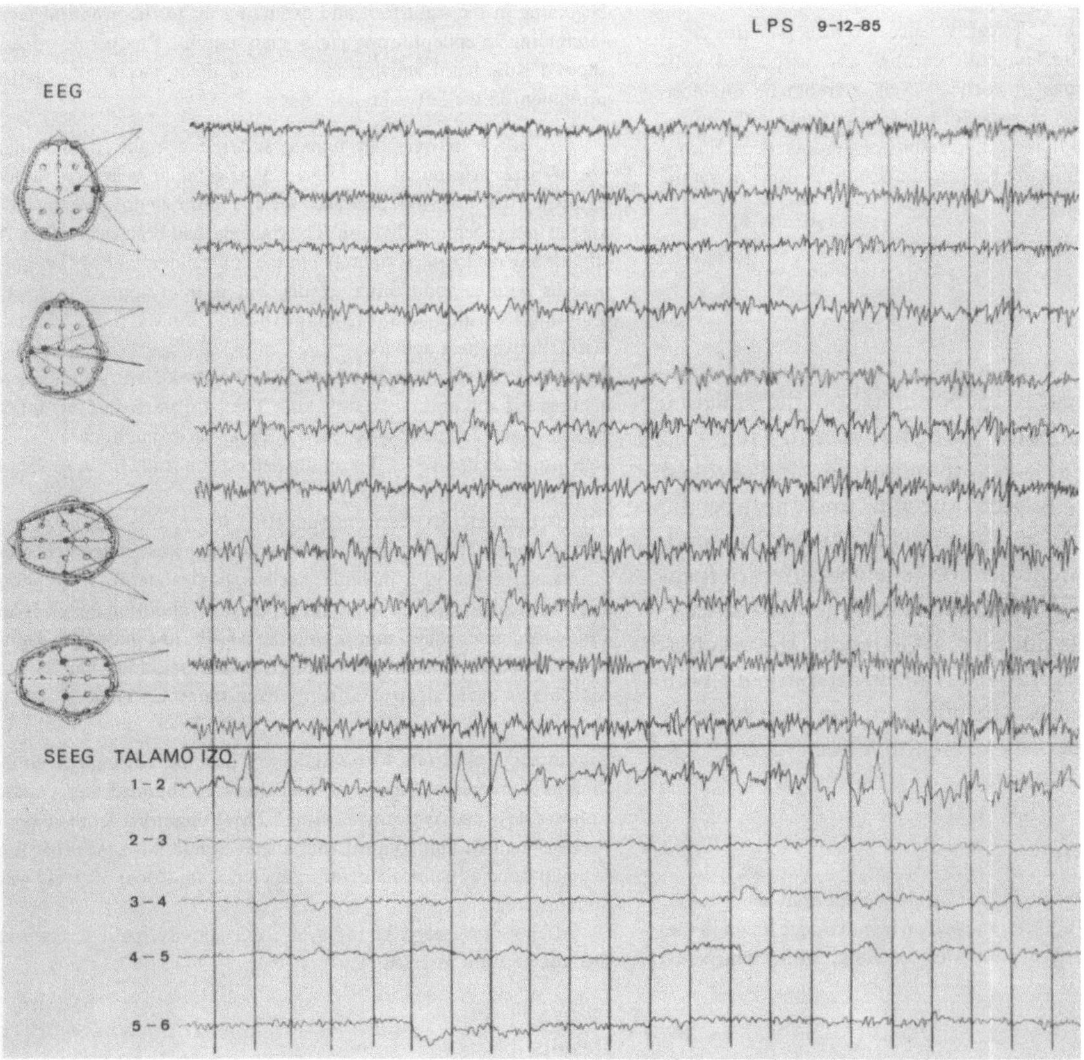

Fig. 2. Case 1: Delta activity in the thalamus (electrode terminals 1, 2)

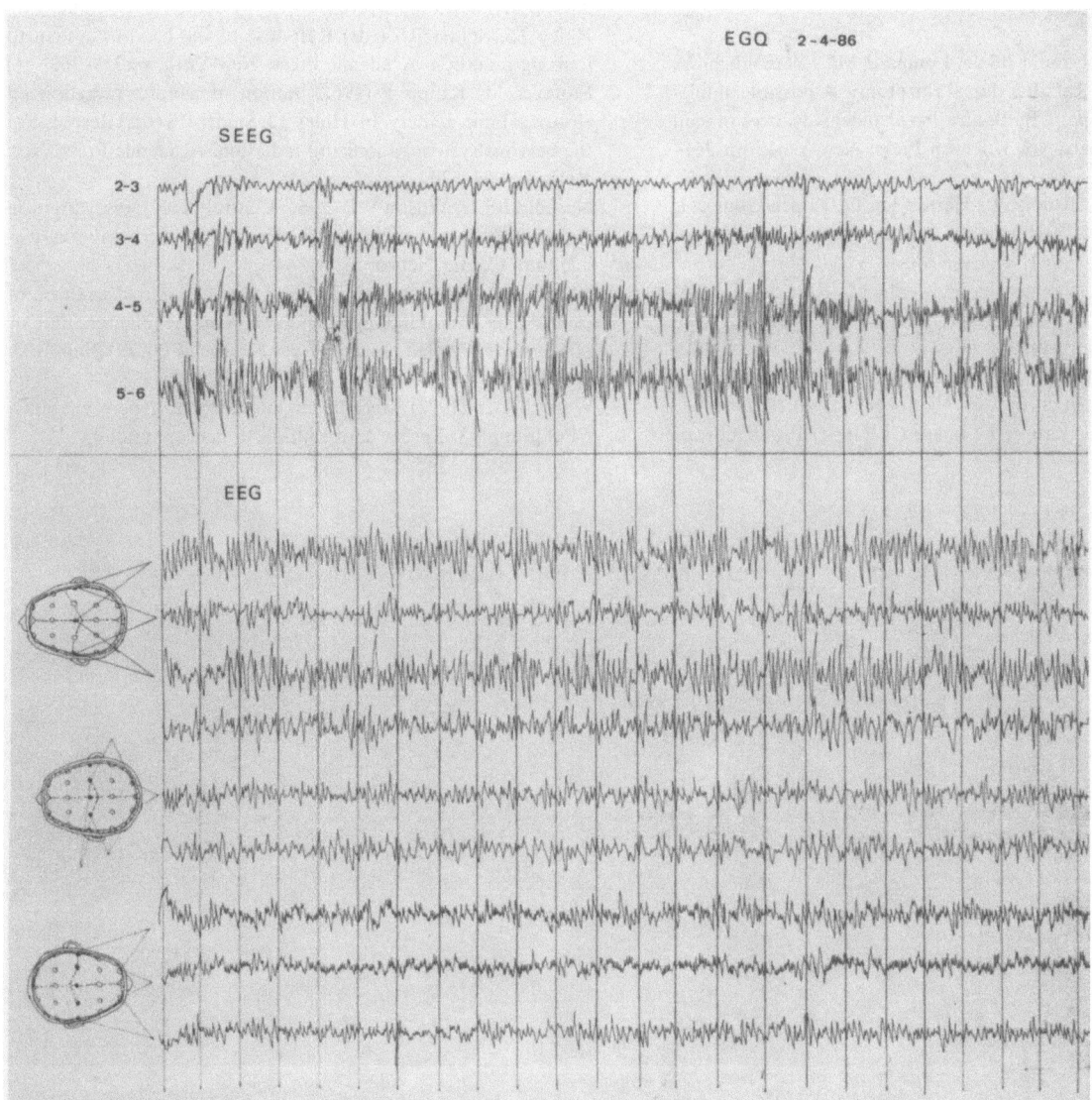

EGQ 2-4-86

Fig. 3. Case 3: Cortical focus (electrode terminals 5, 6)

1, and a cortical focus in case 3 (terminals 5, 6) (Fig. 3), but failed to show any abnormal finding in case 2.

After a mean follow-up of 6 months the EEG was unchanged. However, the painful episodes had disappeared and the background pain had diminished and was better controlled with medication.

Discussion

Barcia-Salorio has reported on the beneficial effects of focal irradiation of the epileptic focus in epileptic patients[2] and experimentally in cats[3]. Evidence for the "epileptic" or irritative nature of central deafferentation pain is still controversial. Nashold and Wilson in 1966[7] reported spontaneous epileptic activity in the mesencephalon in cases of central pain and Toth[8] reported on the paroxysmal or epileptiform character of spontaneous and evoked electrical activity in the specific and nonspecific thalamic nuclei in patients with deafferentation pain. Experimental findings of thalamic discharges after dorsal rhizotomy in the rat[1], in the dorsal horn after brachial plexus avulsion in the cat[4], and similar phenomena in the trigeminal system[9], further support this concept.

The effect of focal irradiation on deafferented neurons in pain syndromes depends on synaptic neoformation[5] or reduction of cortical excitability[6].

Our results with stereotactic radiosurgery in central pain are promising, but the follow-up is too short to allow definite conclusions. Thus, more experimental and clinical work in this field is essential.

References

1. Albe-Fessard D, Nashold BS Jr, Lombard MC, Yamaguchi Y, Boureau F (1979) Rat after dorsal rhizotomy. A possible animal model for chronic pain. In: Bonica J *et al.* (eds) Advances in pain research and therapy, vol 3. Raven Press, New York, pp 761–766

2. Barcia-Salorio JL, Roldán P, Hernández G, Lopez-Gomez L (1985) Radiosurgical treatment of epilepsy. Appl Neurophysiol 48: in press

3. Barcia-Salorio JL, Vanaclocha V, Cerdá M, Roldán P (1985) Focus irradiation in epilepsy. Experimental study in the cat. Appl Neurophysiol 48: in press

4. Loeser JD, Ward AA Jr (1967) Some effects of deafferentation on neurons of cat spinal cord. Arch Neurol (Chicago) 17: 629

5. Malis LI, Rose JE, Kruger L, Baker CP (1962) Production of laminar lesions in the cerebral cortex by deuteron irradiation. In: Haley TS, Snider RS (eds) Response of the nervous system to ionizing radiation. Academic Press, New York, pp 359–368

6. Monnier M, Krupp P (1962) Action of gamma radiation on electrical brain activity. In: Haley TJ, Snider RS (eds) Response of the nervous system to ionizing radiation. Academic Press, New York, pp 607–620

7. Nashold BS Jr, Wilson WP (1966) Central pain: Observations in man with chronic implanted electrodes in the midbrain tegmentum. Confin Neurol 27: 30–44

8. Tóth SA, Sólyom Z, Tóth Z (1984) One possible mechanism of central pain. Autokindling phenomenon on the phantom limb or sensory loss oriented patients. Acta Neurochir (Wien) [Suppl] 33: 459–469

9. Ward AA Jr (1972) Mechanisms of neuronal hyperexcitability. EEG [Suppl] 3. Recent contribution to neurophysiology

Acta Neurochirurgica, Suppl. 39, 163–165 (1987)
© by Springer-Verlag 1987

Continuous Intraventricular Morphine- or Peptide-Infusion for Intractable Cancer Pain

K. Weigl*, F. Mundinger[1], and **J. Chrubasik**[2]

[1] Department of Stereotaxy, University Hospital, Freiburg i. Br., Federal Republic of Germany, [2] Department of Anesthesiology, University Hospital, Düsseldorf, Federal Republic of Germany

Summary

The continuous intraventricular administration of small daily doses of morphine by means of an implantable pump is an effective method of obtaining considerable pain reduction for patients suffering from otherwise untractable carcinoma pain. We consider this method of treatment to be an excellent alternative to the epidural and intrathecal application. Particularly in cases with obstruction of the spinal canal or in cases suffering from untractable pain in the face, neck or upper thoracic area. During the period of treatment, none of the patients involved in the study developed tolerance to morphine or specific opiate side effects. The programmable pump allows precise dosage which is adjusted to the requirements of the individual patient. The high cost of a pump is a justifyable investment in patients in good general condition with a life expectancy longer than 3 months. In most cases the patient may be cared for at home, making further hospitalization unnecessary.

Keywords: Cancer pain; intraventricular morphine.

Introduction

The epidural administration of morphine has frequently proved to be successful in the treatment of untractable cancer pain[3]. Morphine affects the neurons of the posterior horn of the spinal cord[12], but with epidural administration only a small percentage of the drug penetrates the dura mater[1]. Therefore, with intrathecal morphine administration, the same analgesic effect could be obtained with decreased daily doses[11]. The preferential method is a daily bolus injection of morphine through a subcutaneously implanted reservoir. This allows substantial pain reduction over long periods of time.

However, some cases of morphine tolerance following intrathecal administration have recently been reported. This seams to occur also with a continuous administration of very small quantities of morphine[6, 9]. It was reported in 1982 that with intraventricular administration even smaller doses of morphine could produce the same analgesic effect in a chronic cancer pain[9]. Using this method, the morphine can exert its central inhibitory effect by occupying the receptors in the brain or spinal cord, allowing a further reduction of the daily dosage[10]. In addition, the positive analgesic effect of peptides, especially of somatostatin, has been described[1, 2].

We present the results of the treatment of 8 patients who received continuous intraventricular morphine from an implanted pump. One patient was also given somatostatin intraventricularly for a brief period of time.

Method

To obtain a constant high-level pain reduction, intraventricular morphine was administered via an implantable pump (INFUSAID 100) to 5 patients suffering from multiple bone metastases and to 2 patients with perforated larynx carcinoma. In another case presenting with uncontrollable pain caused by a malignant schwannoma in the cervical region, a telemetrically programmable pump produced by MEDTRONIC was implanted.

Before implantation of the pump system, the daily morphine dose required for sufficient analalgesia was checked by the use of an external pump. In all cases, the pumps were subcutaneously implanted below the clavicle. The pump-catheter was then connected to a ventricular catheter. In case 8, a somatostatin solution was given over a 5-day period via an external pump. Pain scores were assessed by means of a visual pain analogue scale (0 = no pain, 10 = severe pain).

---

* Dr. K. Weigel, Department of Neurosurgery, Hospital Steglitz FU Berlin, Hindenburgdamm 30, D-1000 Berlin 45.

Table 1

Patient number	Age (years)	Sex	Tumor histology	Metastases	Previous treatment
1	69	female	colon	bone	irradiation, chemotherapy, system. analgesics
2	56	male	prostata	bone	irradiation, chemotherapy, system. analgesics
3	56	male	prostata	bone	irradiation, chemotherapy, peridural analgesics
4	38	male	prostata	bone	irradiation, chemotherapy, system. analgesics
5	37	male	maligne schwannoma	none	resection, irradiation, system. analgesics
6	53	male	tongue	local (neck)	resection, isotop. implant., system. analgesics
7	61	male	larynx	local (neck)	resection, irradiation with radionecrosis, system. analgesics
8	42	female	Pancoast	local	resection, irradiation, system analgesics

Table 2

Case number	1	2	3	4	5	6	7	8
Beginning of treatment	August 1984	August 1984	February 1885	April 1985	June 1985	June 1985	September 1985	December 1985
Type of pump	400*	100	100	100	100	8600**	100	100
Flowrate (ml/day)	3.1	1.8	2.5	2.7	1.5	var.	1.5	1.5
Refilling-cycle (days)	14	25	18	16	30	var.	30	30

Months	Morphine-consumption for constant analgesia (mg)							
1	15	21	15	21	12	10.5	15	21
2	15	15	15	15	12	10.5	15	15
3	15	15	15	15	15	10.5	15	15
4	15	15	+	12	12	+	12	15
5	15	15		12	12		21	15
6	15	15		12	21		21	
7	15	15		15	+		21	
8	15	12		+			21	
9	15	12						
10	15	12						
11	15	12						
12	15	12						
13	15	+						
22	15							

+ Died. var: programmable pump.

* Manufactured by INFUSAID CORPORATION, Norwood, MA 02062.

** Programmable pump, manufactured by MEDTRONIC INC., Minneapolis, MN 55440.

Results

Starting with an hourly dose of between 0.02 and 0.013 mg of morphine, a constant high level of pain reduction was obtained in all cases. For two of the patients it was eventually possible to reduce the dose. The period of treatment lasted between 3 and 21 months. The 5 patients suffering from bone metastases died within 3 and 12 months after the implantation of the pump (Table 1). In one case, at the beginning of the therapy, a slight urinary retention occurred, but this symptom disappeared with a reduction of the daily dose of morphine (case 4). Additional side effects, in particular respiratory disorders were not observed in any of the cases.

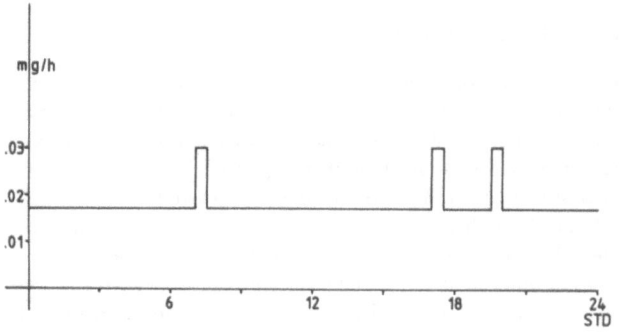

Fig. 1. Daily base dose of morphine with three additional boli (0.007 mg)

In a patient illustrated in Fig. 1, a morphine bolus of 0.007 mg was given in addition to the normal dose. It should be noted that in this patient the required total daily dose was exceptionally low. Additional analgesics were not necessary in any of the cases; however, in 4 cases antidepressive drugs were prescribed. With a daily dose of 120 µg somatostatin, a considerable pain reduction was obtained without obvious side effects during the 5-day period of treatment. Further details may be seen in Table 2.

Discussion

In cases of untractable cancer pain, continuous intraventricular administration of morphine can be used to obtain satisfactory and long lasting pain reduction. In contrast to systemic or epidural applications an analgesic effect can be obtained with a much smaller dose of morphine, and the characteristic opiate side effects are rarely observed[4, 7, 8, 11]. With the application of a hyperbaric morphine solution, the risk of a respiratory depression is minor[4].

With the aid of implanted pumps the complication rate is reduced, and the quality of life for the patients is largely improved. This improvement is due to the absence of opiate specific side effects combined with the excellent analgesic effect. The use of a programmable pump seems to offer considerable advantages. In addition, the administration of peptides appears to be promising not only on the basis of animal experiments[2] but also in view of occasional clinical observations.

References

1. Chrubasik J (1985) Spinal infusion of opiates and somatostatin. Fresenius Foundation, FRG
2. Chrubasik J, Volk B, Meynadier J, Berg G, Wuensch E (1986) Observations in dogs receiving chronic spinal somatostatin and calcitonin. Schmerz-Pain-Douleur 1: 10–12
3. Coombs DW, Saunders RL, Pageau MG (1982) Continuous intraspinal narcotic analgesia. Technical aspects of an implantable infusion pump. Reg Anesth 7: 110–113
4. Crawford ME *et al* (1983) Pain treatment on outpatient basis utilizing extradural opiates. A danish multicentre study comprising 105 patients. Pain 16: 41
5. Friedrich G, Chrubasik J, Scholler KL, Andreas P, Rupp HP, Weigel K, Roth H (1985) Peridurale Morphinapplikation: Zum Risiko der Atemdepression. Schmerz 6: 10
6. Greenberg SH, Taren J, Ensminger WD, Doan K (1982) Benefit from and tolerance to continuous intrathecal infusion of morphine for intractable cancer pain. Neurosurg 57: 360
7. Leavens ME, Hills CS Jr, Cech DH, Weyland JB, Westen JS (1982) Intrathecal and intraventricular morphine for pain in cancer patients; initial study. J Neurosurg 56: 241
8. Lobato RD, Madrid JC, Fatela LV (1983) Intraventricular morphine for control pain in terminal cancer patients. J Neurosurg 59: 627
9. Milne B, Cervenko F, Jhamandas K, Loomis Chr, Sutak M (1985) Analgesia and tolerance to intrathecal morphine and norepinephrine infusion via implanted mini-osmotic pumps in the rat. Pain 22: 165–172
10. Snyder SH, Childers SR (1979) Opiate receptors and opioid peptides. Ann Rev Neurosci 2: 35
11. Wang JK, Nauss LA, Thomas JE (1979) Pain relief by intrathecally applied morphine in man. Anesthesiology 50: 149
12. Yaksh TL (1981) Special opiate analgesia: characteristics and principles of action. Pain 11: 293

Acta Neurochirurgica, Suppl. 39, 166–169 (1987)

Hemodynamic Changes from Spinal Cord Stimulation for Vascular Pain

P. Roldan*, **V. Joanes**, **J. Santamaria**, **J. L. Barcia-Salorio**, **I. Casans**, **C. Carbonell**, and **E. Tejerina**

Servicio de Neurocirugía, Hospital Clínico Universitario, Valencia, Spain

Summary

The hemodynamic changes induced by spinal cord stimulation (SCS) have been studied in a group of 20 patients with peripheral vascular pain. The surgical technique consisted of the introduction of 1 or 2 electrodes in the subarachnoid space up to the level of the painful area, for mono or bipolar SCS.

Several techniques have been used for evaluation of hemodynamic changes induced by SCS. Peripheral blood flow speed was measured by means of ultrasound Doppler, showing a raised maximum speed during stimulation, and a tendency of the pulse wave to return to normal. Thermography showed a marked increase of temperature in the painful area. Preoperatively, plethysmography showed an absence of the typical flow waveform, whilst postoperative recordings showed a small wave of progressively increasing amplitude. Scintigraphy with ^{201}Tl showed an increase in muscular blood flow in previously hypovascularized areas. The mechanism of action of SCS on peripheral blood flow and vascular pain is discussed.

Keywords: Spinal cord stimulation; vascular pain; peripheral blood flow.

Introduction

Based on clinical and experimental observations, spinal cord stimulation (SCS) has been used to treat ischemic pain of vascular origin, for its beneficial effect upon peripheral circulation, as measured by Doppler, thermography, reography and plethysmography[1, 2, 5]. Nevertheless the mechanism of action of SCS on peripheral blood vessels remains obscure and experimental work is lacking. The aim of this study is to present the hemodynamic changes induced by SCS in a group of 20 vasculopathic patients as measured by Doppler, thermography, plethysmography and thallium scintigraphy.

---

* Prof. Dr. P. Roldan, Servicio de Neurocirugía, Hospital Clínico Universitario, paseo Blasco Ibañez, 17, E-46010 Valencia, Spain.

Methods and Material

A group of 20 vasculopathic patients were tested with SCS. The etiology of the clinical syndrome was arteriosclerosis (12 cases), Buerger's disease (3 cases), diabetic angiopathy (4 cases) and ergotamine intoxication (1 case). The patients presented in stage 2 (intermittent claudication) (2 cases), stage 3 (pain at rest) (7 cases), and stage 4 (trophic lesions) (11 cases), and previous surgery included sympathectomy (9 cases) and direct arterial surgery (10 cases).

Following standard techniques, one or two electrodes were introduced into the epidural space under fluoroscopic control, up to the level at which stimulation produced paresthesias in the painful area, usually between T 10–L 1. During a trial stimulation period of about 1 week (1 hour *on*/1 hour *off*) the effect of SCS on pain and hemodynamic parameters was measured. Initially Doppler and thermography were done (14 cases) but currently plethysmography (6 cases) and thallium scintigraphy (4 cases) are used for evaluation of hemodynamic changes.

Plethysmography was done with a model SPG-1b (Medasonic) of Indio-Galio. Patients were placed in the supine position and rings were placed in the first finger of both feet. The system was calibrated and measures were taken before and 5, 15 and 30 minutes after beginning of stimulation.

For thallium scintigraphy, an intravenous dose of 74 MBq of ^{201}Tl was administered while the patient was resting. Five minutes after injection, 10 minutes images of both lower extremities were obtained with gammacamera and computer. Quantification was achieved by counting over the midthigh, knee, midcalf and ankle, and then intraextremity (thigh/calf, thigh/knee and calf/ankle) and inter-extremity (thigh/thigh, calf/calf) perfusion ratios were obtained, according to Siegel[6]. One week later the same exploration was repeated but the ^{201}Tl injection was administered after 30 minutes stimulation, and the results of both explorations were compared.

Results

The effect of SCS has been fairly good in general terms, since the relief of pain is the main indication for implantation of the system. Ultrasound Doppler recordings showed in 7 out of 9 patients an increase in the maximum speed during stimulation and a tendency to normalization of the form of the pulse wave. These effects appear 10 minutes after stimulation and stand for the whole stimulation period (Figs. 1 and 2). Telethermography showed a marked warming of about 2 °C of the painful area, increasing as stimulation proceeds.(Fig. 3)

Preoperatively, plethysmography showed an absence of the typical flow waveform, whilst postoperatively a small wave, progressively increasing in amplitude, was observed, reaching its maximum at 30–40 minutes after the beginning of stimulation (Figs. 4 and 5).

Thallium scintigraphy has been of value in 2 out of 4 cases, showing a preoperative decrease in the affected limb and an increase of muscular perfusion after stimulation (Fig. 6). The small number of patients does not allow definite conclusions.

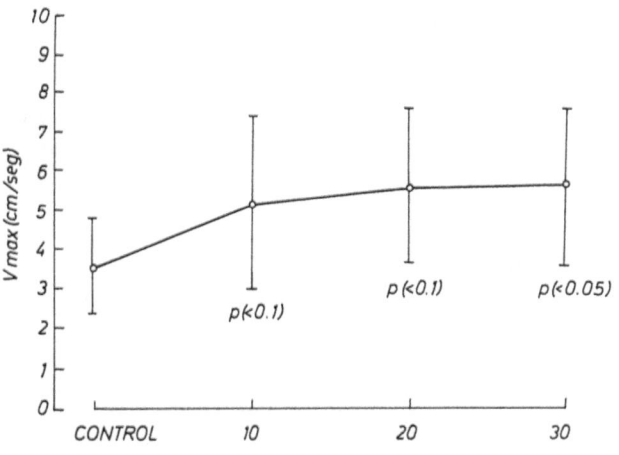

Fig. 2. Mean variation in amplitude of Doppler recordings before and 10, 20 and 30 minutes after stimulation (7 patients)

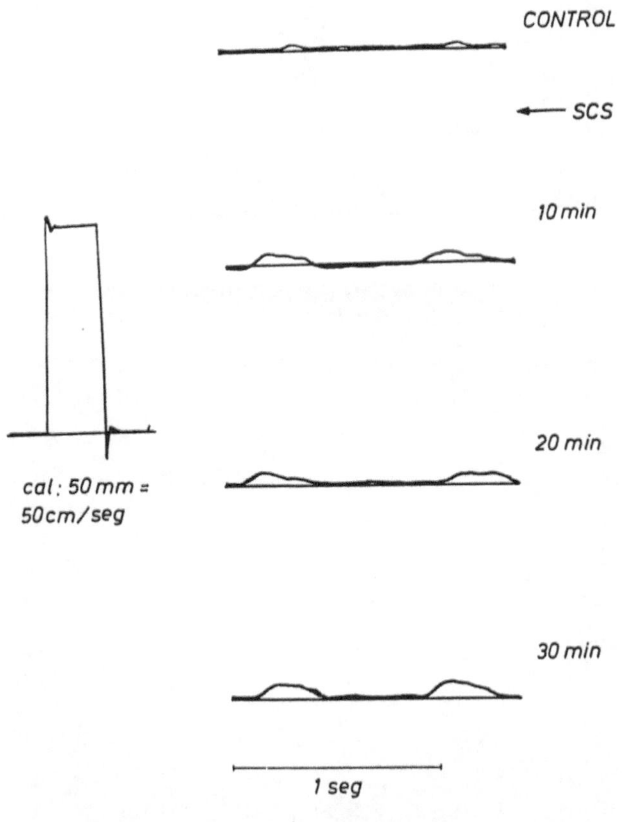

TERMOMETRIA

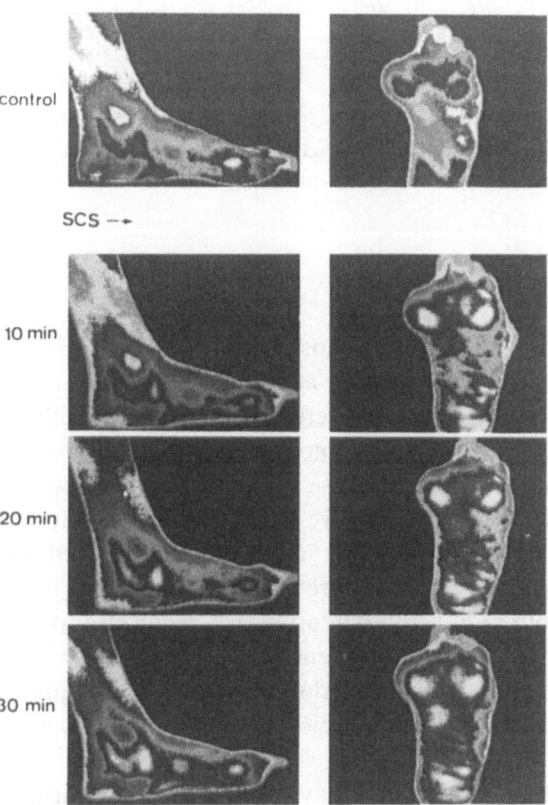

Fig. 1. Doppler recording from dorsal artery of the foot before and 10, 20 and 30 minutes after stimulation. Raise in the maximum speed and normalization of wave form

Fig. 3. Telethermography before and 10, 20 and 30 minutes after stimulation

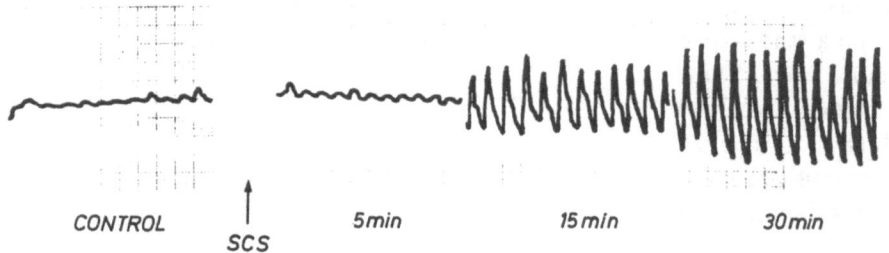

CONTROL   SCS   5min   15min   30min

Fig. 4. Plethysmography: Preoperative control shows an absence of the typical flow waveform, whilst postoperatively a small wave, progressively increasing in amplitude, is recorded

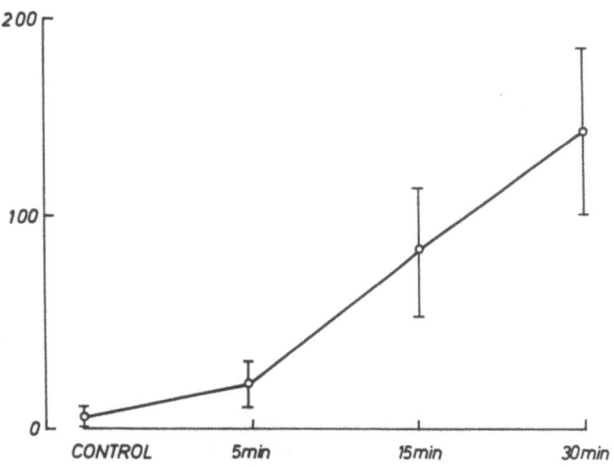

Fig. 5. Mean variation in amplitud of plethysmographic flow wave before and 5, 15 and 30 minutes after stimulation (6 cases)

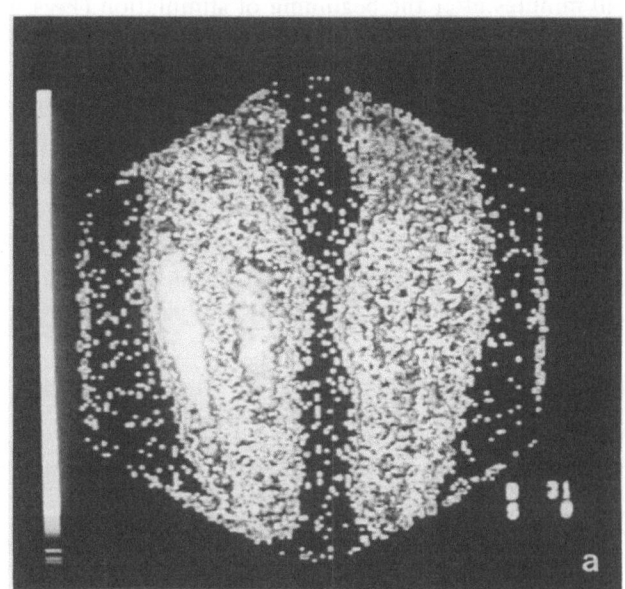

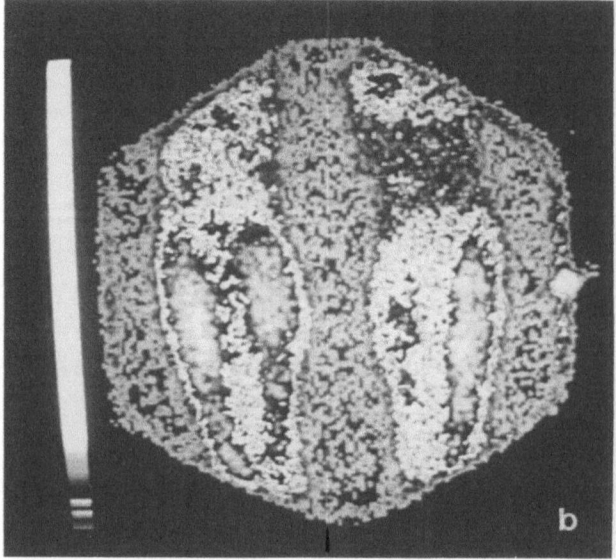

Fig. 6. Thallium scintigraphy before (a) and after (b) stimulation, showing a marked activity in previously hypoactive limb

Discussion

Our results agree with those of other authors[1, 2, 5] in respect to pain reduction and hemodynamic changes measured by Doppler, telethermography and plethysmography. Only plethysmography is now used as a routine technique, because of the poor reliability of Doppler and thermography. Thallium scintigraphy has been employed as an investigational procedure. Nevertheless other more accurate methods are available: radioactive microspheres[4], xenon clearance[8], transcutaneous oxygen tension[3], capillaroscopy and laser-Doppler. These methods may be more appropriate for evaluation of microcirculation changes as a result of SCS.

Various mechanisms have been proposed in hemodynamic changes induced by SCS[1, 8]: pain relief, autonomic nervous system activation, activation of C fibers in dorsal roots, local vascular control, VIP, prostaglandins and prostacyclins. We think that the mechanism of action of SCS is different from sympathectomy, and that it could reflect a specific action on microcirculation, operating at the level of the precapillar sphincter.

References

1. Augustinsson LE, Holm J, Carlsson CA, Jivegard L (1985) Epidural electrical stimulation in severe limb ischemia. Evidences of pain relief, increased blood flow and a possible limb-saving effect. Ann Surg 202: 104–111

2. Broseta J, Barbera J, Vera JA de, Barcia-Salorio JL, Garcia-March G, Gonzalez-Darder J, Rovaina F, Joanes V (1986) Spinal cord stimulation in peripheral arterial disease. A cooperative study. J Neurosurg 64: 71–80

3. Burgess EM, Matsen FA, Wyss CR, Simmons CW (1982) Segmental transcutaneous measurements of pO2 in patients requiring below-the-knee amputation for peripheral vascular insufficiency. J Bone Joint Surg 64 A, 3: 378–382

4. Fee HJ, Friedman BH, Siegel ME (1977) The selection of an amputation level with radioactive microspheres. Surg Gynecol Obstet 144: 89–90

5. Meglio M, Cioni B, Sandric S (1981) Spinal cord stimulation and peripheral blood flow. In: Hosobuchi Y, Corbin T (eds) Indications for spinal cord stimulation. Excerpta Medica, Amsterdam, pp 60–75

6. Siegel ME, Siemsen JK (1978) A new non invasive approach to peripheral vascular disease: Thallium-201 leg scans. Am J Roentgenol 131: 827–830

7. Silberstein EB, Thomas S, Cline J, Kempczinski R, Gottesman L (1983) Predictive value of intracutaneous Xenon clearance for healing of amputation and cutaneous ulcer sites. Radiology 147: 227–229

8. Tallis RC, Illis LS, Sedgwick EM, Hardwidge C, Garfield JS (1983) Spinal cord stimulation in peripheral vascular disease. J Neurol Neurosurg Psychiatry 46: 478–484

Acta Neurochirurgica, Suppl. 39, 170–173 (1987)
© by Springer-Verlag 1987

Short Latency Somatosensory-Evoked Potentials—Direct Recording from the Human Midbrain and Thalamus

T. Taira*, K. Amano, H. Kawamura, T. Tanikawa, H. Kawabatake, M. Notani, H. Iseki, T. Shiwaku, T. Nagao, Y. Iwata, Y. Umezawa, T. Shimizu, and K. Kitamura

Department of Neurosurgery, Neurological Institute, Tokyo Women's Medical College, Tokyo, Japan

Summary

Short latency somatosensory-evoked potentials were recorded from the human thalamus and the midbrain during stereotactic operations. Several subcomponents were recognized on the peak of N 18. These were recorded with maximal amplitude at the border between the caudal portion of the thalamus and the rostral midbrain. Two positive-negative responses, not previously shown, were observed between P 14 and N 18. These responses were prominent in the rostral midbrain. These findings indicat that the ascending phase of N 18, and the N 18 itself, are the compound potential generated in the mesodiencephalic junction.

Keywords: Human thalamus; human midbrain; evoked potentials; somatosensory function.

Introduction

The neural sources of short latency somatosensory-evoked potentials (SEP) have been extensively investigated using the method of recording during stereotactic operation[2, 6] as well as on the basis of findings in patients with lesions in the somatosensory pathway[1, 5, 7, 8]. P 14, the first intracranial component, originates in the medial lemniscus and N 20 is the earliest cortical response[3, 5]. There is, however, no unanimous conclusion on the origin of N 18 which is recognized between P 14 and N 20. This component has been considered to be of subcortical origin because of its wide distribution over the scalp.[5] In this study short latency SEP was recorded from the human midbrain and the thalamus during stereotactic operation with the aim to identify the generator of N 18.

 * T. Taira, M.D., Department of Neurosurgery, Neurological Institute, Tokyo Women's Medical College, 8-1 Kawada-cho, Shinjuku-ku, Tokyo, 162 Japan.

Materials and Methods

SEP was recorded in three patients who underwent rostral mesencephalic reticulotomy for pain relief, Vim thalamotomy for tremor, and zona incerta-tomy for spasmodic torticollis. Fifteen waveforms recorded from the thalamus and the midbrain were evaluated together with the scalp recording. SEP was elicited by stimulation of the median nerve contralateral to the operated side. Square pulses of 4 Hz with 0.1 msec pulse width were delivered to produce minimal thumb twitches. Signals were amplified using linked earlobe reference. Filters were set at 5–3,000 Hz. Five hundred responses within 20 msec after the stimulation were averaged. Scalp recording was named as A.

Results

Case 1 (Fig. 1): A large negative wave which corresponded to N 18 of the scalp recording was obtained from points C, D, E and F. Several wavelets were observed on the peak of N 18. These were recorded with maximal amplitude at point E. N 18 was preceded by four positive-negative complexes which were largest at point E. The first two complexes corresponded to P 13 and P 14 respectively. The latter two complexes have not been previously documented.

Case 2 (Fig. 2): Negative deflection of N 18 progressively increased their amplitude towards points F and G. Wavelets on the peak of N 18 and a positive-negative complex between P 14 and N 18 were largest at point G.

Case 3: The wavelets on the peak of N 18 were most prominent at a level where the medial lemiscus enters the thalamus. Two positive-negative complexes were observed at the rostral midbrain.

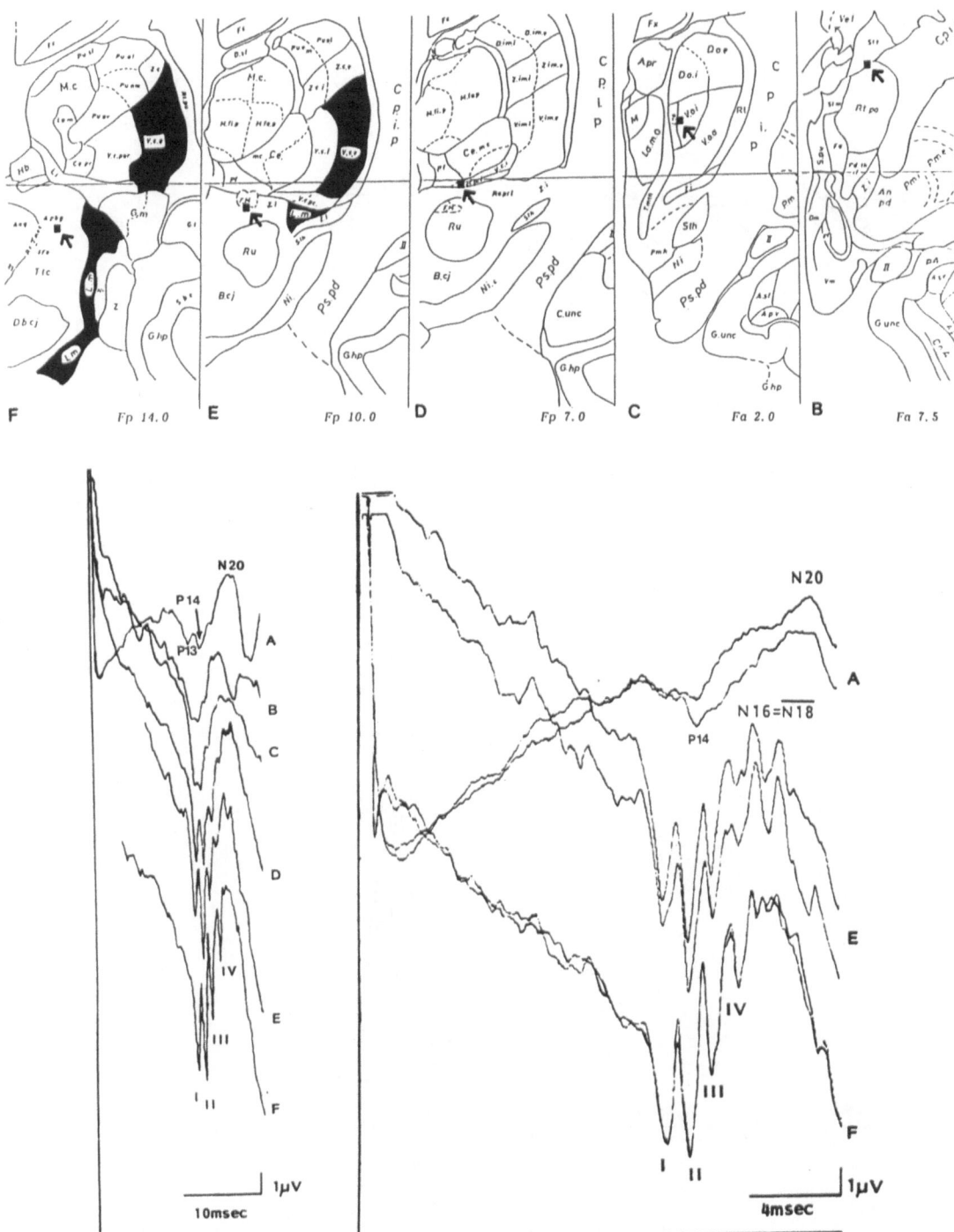

Fig. 1. Recording points and waveform of SEP in case 1

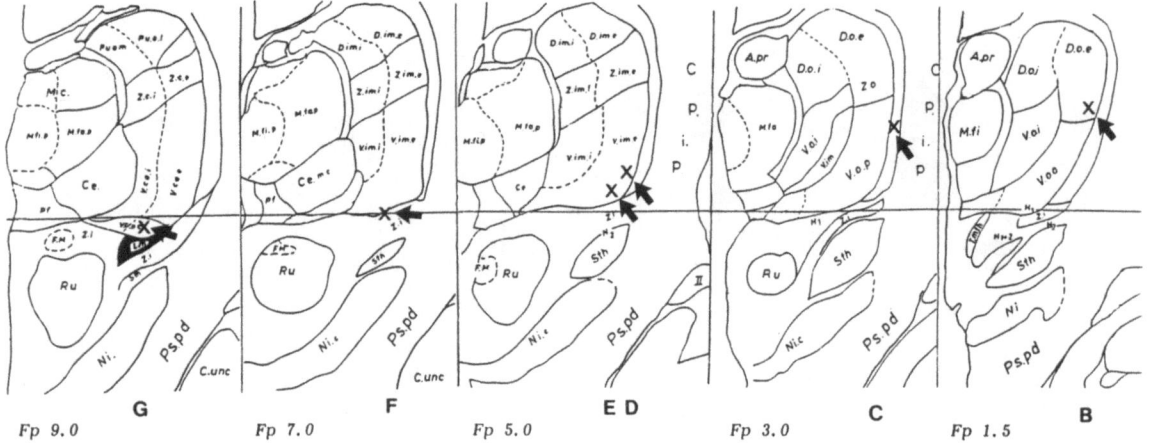

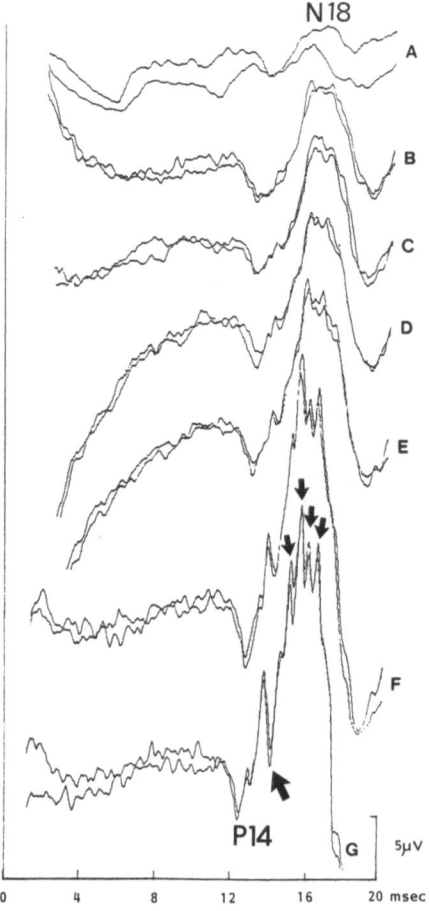

Fig. 2. Recording points and waveform of SEP in case 2

Summary of the results: N 18 has several subcomponents on its summit. They are observed with a maximal amplitude at the border between the caudal portion of the thalamus and the zona incerta, where the medial lemniscus enters the thalamus. Two positive-negative complexes followed by N 18 are prominent below this level.

Discussion

There are two hypotheses on the origin of N 18. One is that N 18 originates in the thalamus because it can be recorded from within the thalamus[6], and because of its absence in patients with thalamic syndrome[8]. The other is that it is generated subthalamically because it may be present also in cases with thalamic lesions[7]. Thus, no unanimous conclusion is yet reached on the origin of N 18. There have been many reports concerning direct recording of SEP from the human thalamus[2, 6]. Recording below the thalamus, however, has been rarely documented[4]. This is one of the reasons why direct recording has failed to identify the origin of N 18. The present study showed that N 18 has several subcomponents on its peak and that two positive-negative complexes are present between P 14 and N 18. These findings indicate that N 18 does not originate from a single generator and that the potential gradient from P 14 to N 18 is generated from the rostral midbrain to the thalamus.

It is assumed that even if some later subcomponents of N 18 disappear in patients with thalamic lesion, N 18

of scalp recording may remain unchanged because of persistence of other earlier subcomponents. Even if N 18 is completely lost, the two positive-negative complexes which are followed by N 18 may well be possible to record over the scalp as a small deflection similar to N 18. These hypotheses explain the reason why the changes of N 18 in patients with thalamic lesion show different patterns from author to author.

References

1. Anziska B, Cracco RQ (1980) Short latency somatosensory evoked potentials. Electroencephalogr Clin Neurophysiol 49: 227–239
2. Celesia G (1979) Somatosensory evoked potentials recorded directly from human thalamus and SMI cortical area. Arch Neurol 36: 399–405
3. Desmedt JE, Cheron G (1981) Non-cephalic reference recording of early somatosensory evoked potentials to finger stimulation in adult or aging normal man. Electroencephalogr Clin Neurophysiol 52: 553–570
4. Liberson WT, Voris HC, Uematsu S (1970) Recording of somatosensory evoked potentials during mesenceophalotomy for intractable pain. Confin Neurol 32: 185–194
5. Mauguiere F, Desmedt JE, Courjon J (1983) Neural generators of N 18 and P 14 far field somatosensory evoked potentials studied in patients with lesion of thalamus or thalamocortical radiations. Electroencephalogr Clin Neurophysiol 56: 283–292
6. Suzuki I, Mayanagi Y (1984) Intracranial recording of short latency somatosensory evoked potentials in man. Electroencephalogr Clin Neurophysiol 59: 286–296
7. Urasaki E, Matsukado Y, Wada S, Nagahiro S, Yadomi C, Fukumura A (1985) Origin of component N 16 in short latency somatosensory evoked potentials to median nerve stimulation. Brain Nerve (Tokyo) 37: 393–402
8. Yamada T, Graff-Rafford NR, Kimura J, Dickins QS, Adams HP Jr (1985) Topographic analysis of somatosensory evoked potentials in patients with well-localized thalamic lesions. J Neurol Sci 68: 31–46

Acta Neurochirurgica, Suppl. 39, 174–176 (1987)

Electrical Impedance Recording for Localization in Functional Neurosurgery of the Spinal Cord and Lower Brain Stem

C. Shieff*, J. F. S. Vieira, B. S. Nashold Jr., and J. Ovelmen-Levitt

Duke University Medical Center, Department of Surgery, Division of Neurosurgery, Durham, North Carolina, U.S.A.

Summary

Functional neurosurgery aims to modify or abolish neural messages. Established techniques use confirmatory electrical stimulation prior to ablation and require local anesthesia. Recently developed procedures take place under general anesthesia increasing the chance of damage to adjacent neural structures with postoperative morbidity. We describe a laboratory study correlating changes in measured electrical impedance with transition from white to grey matter in the brain and spinal cord of two mammalian species; this has not previously been easy to undertake[2, 4, 9, 11] nor felt reliable[3]. Impedances can now be measured simply and reliably. This study confirms our operating theater experience. We recommend that when stimulation cannot be used, impedance can and should be utilized to indicate the need for an electrode to be resited.

Introduction

Our investigation evolved from the knowledge that the dorsal spinocerebellar pathway lies immediately anterolateral to the substantia gelatinosa and is therefore at risk during the formation of dorsal root entry zone lesions (the DREZ operation[5, 6]) in the treatment of various chronic pain syndromes.

The electrical impedance of a tissue indicates its potential ability to resist the passage of a high frequency alternating current applied to an electrode in the electrically grounded tissue. There is no net electron flow between the active and ground electrodes. The tissue impedance is derived by fairly complex mathematical formulae from the characteristics of both the current and the electrode being used[7, 8]. The facility for its measurement is now inbuilt in the Radionics® radiofrequency lesion generator which is used in most neurosurgical operating theatres.

Materials and Methods

The study was started with the brains of seven male Sprague-Dawley rats whose weights ranged from 280 to 2,470 grams (mean 450 g). All underwent tracheostomy under general anesthesia, using intraperitoneal Ketamine® (9 mg/kg body weight) and, in several rats, Nembutal® (40 mg/kg). Spontaneous respiration was allowed in all these experiments unless depressed respiration required assisted ventilation. Biparietal craniectomy was then performed and the animals placed in a Stellar® stereotactic frame (model 51400). The anteroposterior coordinate for electrode insertion was the midpoint between the coronal and lambdoid sutures[1]. A blunted cordotomy electrode (exposed length and diameter 0.5 mm) was used and the inbuilt impedance meter of the Radionics® lesion generator (model RFG3B) provided the measurements. Electrical grounding was via a separately inserted subcutaneous needle. In rats 2, 4, 5 and 6, electrode passages were made on both sides at 2 and 4 mm from the midline. In rat 1, 3 and 5 mm were used on the left. Rat 3 died before observations could be undertaken on the second side. In rat 7, the tracks were 2.5 mm on both sides. Recordings were made of impedance values at dural contact (this then being incised) and at cortical contact. Electrode insertion was by manually operated microdrive, readings being repeated at 0.5 mm intervals during insertion until resistance indicated that the base had been reached. Recordings were repeated during electrode withdrawal. The animals were sacrificed and the whole brains removed for fixation.

---

* C. Shieff, F.R.C.S., Department of Neurosurgery, Midland Centre for Neurosurgery and Neurology, Holly Lane, Smethwick, West Midlands, U.K.

Two female cats, both weighing 3.2 kg were then studied. Anesthesia was with 40 mg/kg Nembutal® and tracheostomy was performed, spontaneous respiration being allowed. The spinal cords of both were investigated at cervical and lumbar levels, together with the brain of the first. Bilateral parietal trephines and cervical (C 1– C 3) and lumbar (L 2–L 7) laminectomies were performed. The Trent Wells® animal stereotactic frame was used together with the same electrode and lesion generator.

In the cerebral study, electrode insertion was 7.5 mm anterior to the interaural line and 7 mm from the midline[10]. Recordings were made every 0.25 mm during inward and outward passage. RF lesions were made at 15 mm (30 volt/25 mA/15 sec). In the cervical cord at C 2 and in the lumbar cord at L 5, penetration was at the medialmost insertion of the dorsal roots in order to reach the dorsal root entry zone. These animals were then sacrificed and large blocks of brain and cord removed for fixation.

250 micron coronal sections were then made from each specimen to include the electrode tracks and these were stained with cresyl violet and luxol fast blue. Graphical plots were then made of impedance value against distance and superimposed on the histological slides to allow comparison.

Results

The following values were found for the electrical impedances of the structures traversed and at boundary zones (range ± 200 ohms):

RAT cerebrum:

grey mater	2,000 ohms
grey/white transition	2,200 ohms
fascia dentata	2,500 ohms
nucleus lateralis thalami	2,600 ohms
VP	2,900 ohms
VL	2,100 ohms
zona incerta	2,900 ohms
posterior hypothalamic complex	3,100 ohms
fornix	2,900 ohms
amygdala	2,750 ohms

CAT cerebrum:

grey matter	1,800 ohms (range ± 300)
grey/white transition	2,200 ohms
white matter	2,550 ohms
nucleus caudatus	2,800 ohms
nucleus corporis	3,100 ohms
formatio hippocampalis ventralis	3,100 ohms

C 2:

dorsal colums	2,100 ohms (range ± 200)
central grey	1,900 ohms
anterior white	2,250 ohms

L 5:

DREZ	2,900 ohms
grey/white transition	2,600 ohms
dorsal horn	2,200 ohms
central grey	1,900 ohms
anterior horn	1,600 ohms

These results enable us to make the following definitive observations:

1. Impedance in white matter is higher than in grey matter.

2. Rat thalamic nuclei have impedances varying from 2,600 to 3,100 ohms while the zona incerta gives values between 2,800 and 3,100 ohms.

3. In cat, brain impedance values become higher with depth while spinal cord values are higher in the lumbar region than in the cervical region. The differential between white and grey matter is retained.

Discussion

A major importance of impedance measurement is its ability to distinguish between the white and grey areas of the central nervous system and to demonstrate the regional differences between adjacent thalamic nuclei. Its commonest use is in functional stereotaxy, allowing correlation of instrument position with anatomical location. The highest values are obtained in white matter and in peritumoral edema. Lower values are found in grey matter and blood vessels while the lowest values are obtained from cerebrospinal fluid (a current "sink"). We routinely perform impedance studies during each DREZ operation and have observed a normal range in this region from 900 to 1,200 ohms. In spinal cords which have previously undergone root avulsion or trauma resulting in painful paraplegia, impedances are low even where the macroscopic appearance is normal (below 1,000, even as low as 300 ohms), which we take to indicate scarring or gliosis. As the anatomical level approaches normally functioning tissue, the impedance values return to the normal range.

References

1. Albe-Fessard D, Stutinsky F, Libouban S (1974) Atlas stéréotaxique du diencéphale du rat blanc. Centre National de la Recherche Scientifique, Paris
2. Fugita S, Cooper IS (1976) Impedance and spontaneous electrical activity as a localizing method in percutaneous spinal surgery. Acta Neurol Scand 53: 201–208
3. Laitinen L, Johansson GG, Sipponen P (1966) Impedance and phase angle as a locating method in human stereotactic surgery. J Neurosurg 25: 628–633

4. Mori K, Iwayama K, Ito M, Shimabukuro H, Handa H (1977) Electrical impedance as a locating method in human stereotactic surgery. Appl Neurophysiol 39: 216–221

5. Nashold BS, Ostdahl RG (1979) Dorsal root entry zone lesions for pain relief. J Neurosurg 51: 59–69

6. Nashold BS, Lopes H, Chodakiewitz J, Bronec P (1986) Trigeminal DREZ for craniofacial pain. In: Gildenberg P, Samii M (eds) Surgery in and around the brain stem. Springer, Berlin Heidelberg New York Tokyo, pp 53–58

7. Robillard PN, Poussart Y (1977) Specific-impedance measurements of brain tissues. Med Biol Eng Comput 15: 438–445

8. Robinson BW (1962) Localization of intracerebral electrodes. Exp Neurol 6: 201–223

9. Robinson BW, Bryan JS, Rosvold HE (1965) Locating brain structures. Arch Neurol 13: 477–486

10. Snider R, Niemer J (1961) Atlas stereotactic of the cat brain. University of Chicago Press, Chicago

11. Taren JA, Davis R, Crosby EC (1969) Target physiological corroboration in stereotaxic cervical cordotomy. J Neurosurg 30: 569–582

Acta Neurochirurgica, Suppl. 39, 177–180 (1987)

Cryoanalgesia. Ultrastructural Study on Cryolytic Lesion of Sciatic Nerve in Rat and Rabbit

V. A. Fasano*, [1], S. M. Peirone[2], S. Zeme[1], M. Filippi[3], G. Broggi[1], M. de Mattei[4], and A. Sguazzi[1]

[1] Institute of Neurosurgery, University of Turin, Italy, [2] Institute of Veterinary Anatomy, University of Turin, Italy, [3] Energy Department, Politecnico of Turin, Italy, [4] I. Chair of Neurology, University of Turin, Italy

Summary

The sciatic nerve was exposed to cryoinjury at different freezing patterns in albino rats and rabbits and the frozen nerves were serially examined with electron microscopy from the time of cryolitic lesion (−60 °C for 3 minutes) for up to 28 days. The cryolesion was characterized by a total degeneration of the myelin fibers, while non-myelin fibers and vessels seemed less affected. Regeneration began 8 days after cryolysis. A peculiar pattern was the absence of Schwann cells, while the basal membrane around regenerating axons remained intact. The hypothesis that the basal membrane might play a role is discussed.

Keywords: Cryolysis; peripheral nerve; ultrastructure of cryolesion.

Introduction

The aim of this ultrastructural study is to analyze the immediate and late effects and the post-lesional regeneration after in vivo cooling of the peripheral nerve.

Already in the beginning of this century the effects of freezing peripheral nerves had been studied. In 1916 Trendelenburg[20] stated that freezing was the most gentle method to interrupt nerve function. He underlined the absence of neuroma and scar tissue formation at the site of freezing and the very good regeneration. Bielschowsky and Valentin[4] in 1922 described in more detail the histopathology of nerves at varying intervals after freezing. They found that after freezing a sciatic nerve of the dog for 5 minutes all the fibers degenerated and that regeneration was complete after 91 days. Denny-Brown et al.[7] in 1945 published the important study in which they described the degeneration and regeneration of the fibers of a peripheral nerve after different degrees of cooling and freezing. They pointed out that the myelin and axis cylinders are selectively damaged by exposure to cold, the largest fibers being the most sensitive and the smallest fibers being most resistant. Regeneration was rapid and complete in all degrees of injury except after complete necrosis.

Local application of cold to produce a reversible or irreversible block of neural transmission in peripheral or cranial nerves, or in the brain, has been used on many indications. Thus, it has been applied in the management of chronic and acute postoperative pain, trigeminal neuralgia, movement disorders[1, 3, 6, 9–13, 16].

The present ultrastructural study was carried out after in vivo freezing of the exposed sciatic nerve in rats and rabbits. The method of freezing was very similar to that used clinically.

Methods and Material

Sixty male albino rats (Wistar strain, average weight 200 g) and twenty albino rabbits (average weight 1800 g) were used. Under general anesthesia the sciatic nerve was exposed for a length of 2–4 cm at the thigh. The nerve was gently isolated from the surrounding tissues and marked with a 5.0 silk suture and frozen just proximal to the mark by a nitrous oxide probe with a silver tip. The tip contained a 3 mm notch in which the nerve could be placed with two thermocouples for temperature recording. A defrosting device allowed complete thawing of the frozen nerve within 20 seconds after the termination of the freezing ("Cryotom Elettronico" apparatus supplied by A.S.M.O.T. of Turin). The animals were divided in

---

* Prof. V. A. Fasano, Institute of Neurosurgery, via Cherasco 15, I-10126 Torino, Italy.

subgroups for which different temperatures (from — 5 to — 90 °C) for different periods of freezing (from 30 seconds to 3 minutes) were employed. Five animals were used as a control group. Immediately after surgery and 6 hours, 1, 8, 14, 21 and 28 days later the frozen portion of the nerve was resected and prepared for electronic microscopic examination.

We describe our observations on nerve segments frozen at — 60 °C for 3 minutes.

Results

The cryolytic lesions caused immediate local disruption of the myelin sheaths which appeared very deformed and wrinkled. There was a marked splitting of the myelin lamellae throughout the thickness of the sheath, often associated with a coartation of the axon (Fig. 1).

During the first week there was a progressive degeneration with disorganization of the axon neurofilaments, followed by the appearance of vacuoles in the axoplasm and finally by a total destruction of the nerve fibers. It is important to point out that large fibers suffered before small ones. Initially, the cytoplasm of Schwann cells was normal. However, following the first

6 hours it became progressively more wrinkled and usually finally dissolved. After 8 days a great number of degenerating myelin sheaths were surrounded by small normal axons wrapped by the same basal membrane (Fig. 2). Among the degenerated nerve fibers macrophagic-like cells appeared with dark cytoplasm and dense bodies. Neutrophilic granulocytes were tightly adherent to the degenerating myelin sheaths near the vessels, both being surrounded by the same basal membrane. The collagen fibers showed a less distinct pattern and among them a pool of amorphous and floccular material appeared. The second week after freezing was characterized by disapearance of the pattern characterized by nude axons encircling degenerating myelin sheaths. At this time there were an increasing number of axons related to cellular elements. We think that these represent new Schwann cells.

At the third week, besides some residual myelin degenerated sheaths, a great number of nerve fibers were encircled by new myelin sheaths of varying thickness. There were also groups of normal non-myelinated axons (Fig. 3). After the first month a large number of normal myelinated axons were present the collagen fibers were once again normally organized and the floccular material had decreased.

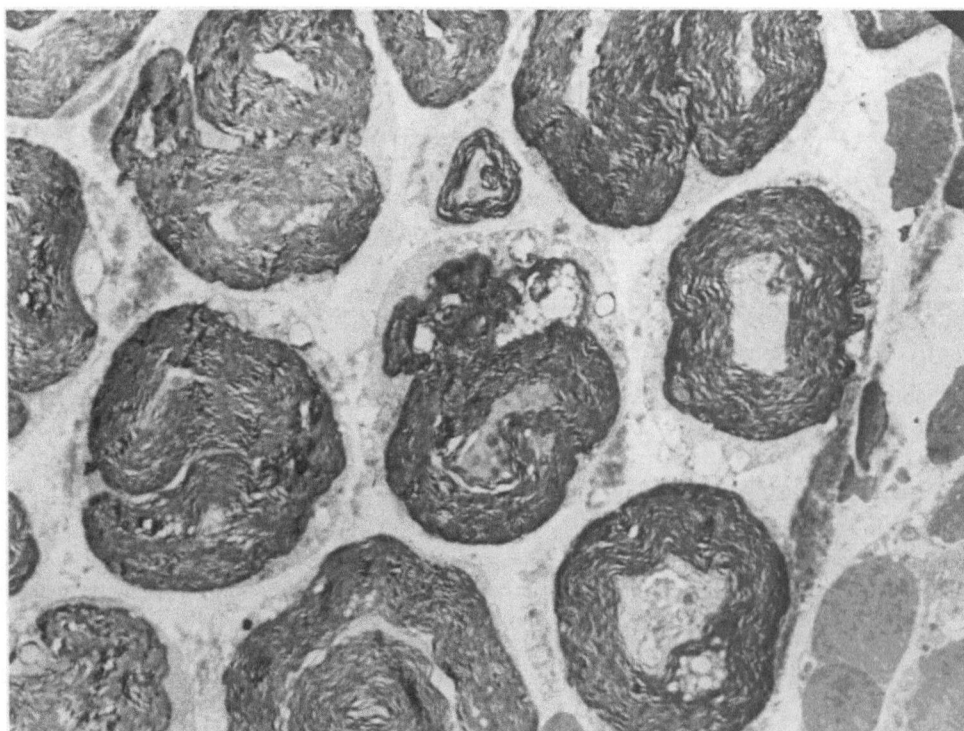

Fig. 1. Sciatic nerve immediately after a cryolysis. Degeneration of myelin sheaths and axons (x7,800)

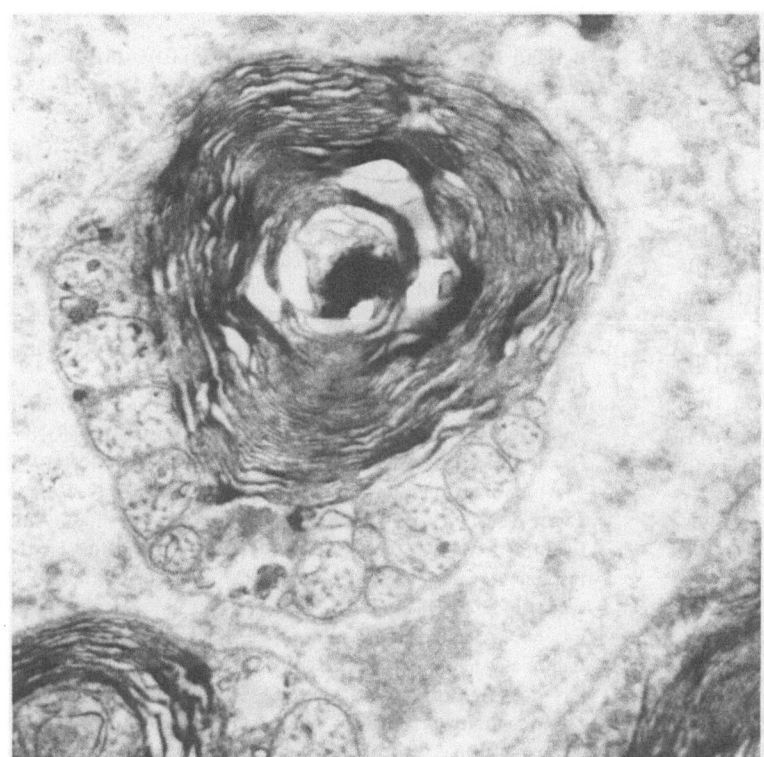

Fig. 2. Sciatic nerve 8 days after a cryolysis. An intact basal membrane wraps several small axons regenerating around a residual myelin body (x37,500)

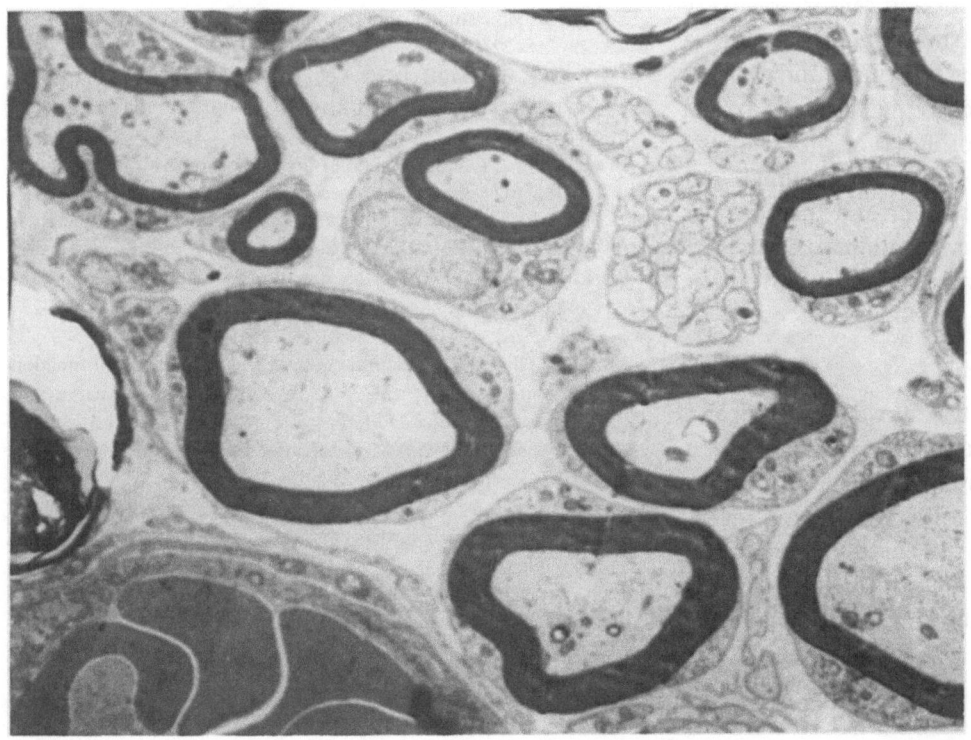

Fig. 3. Sciatic nerve 21 days after a cryolysis. Advanced regeneration: the axons are wrapped by new myelin sheaths. Groups of normal nonmyelinated axons can be seen (x7,800)

Discussion

Cryolysis of a peripheral nerve at — 60 °C for 3 minutes causes a rapid and total necrosis of myelinated fibers. The nonmyelinated fibers and vascular structures appear to be less affected. The typical pattern of the cryolytic lesion is the immediate damage to the axons and myelin sheaths[7, 21]. This indicates that cold causes direct physical disruption of these structures due to the formation of ice crystals[17, 19]. As osmotic effect during freezing and thawing has also been claimed[15, 18]. More questionable is the hypothesis that damage might be due primarily to ischemia[5, 7, 14]. Our observation that axons and myelin are damaged early, while vascular structures are relatively resistant, indicates that the primary cause of nerve damage is freezing, even though ischemic changes may become important later.

Another peculiar feature is that in most fibers the basal membrane is intact in the first phase of the myelin and axonal distruction as well as in the next phase of early regeneration (second week), when the Schwann cells are absent. These data suggest that, at least in the beginning, the basal membrane, which is the only structure in contact with the central stump, may play an essential role in guiding the bundles of axons in the reconstruction of the nerve fiber; Schwann cells appear later and are responsible for the continuation and the completion of the regenerative process. Animal and human clinical studies of motor and sensory recovery after peripheral nerve freezing show that in most cases a complete recovery is possible. The time course of the recovery is primarily related to the rate of axonal regrowth and to the distance of the cryolesion from the end organ[1, 2, 7, 8, 12, 13]. It is generally claimed that neuroma formation does not occur and that fibrosis and scarring at the site of injury is minimal. Therefore, there is little risk that a deafferentation syndrome is produced.

In conclusion, our ultrastructural study indicates that freezing of a peripheral nerve can produce a predictable lesion with a subsequent prolonged interruption of conduction. Our observations on regeneration of the initially destroyed nerve fibers indicate that this lesion is to a great extent reversible and that normal morphology of the nerve often reappears.

References

1. Barnard JDW, Lloyd JW, Glynn CJ (1978) Cryosurgery in the management of intractable facial pain. Br J Oral Surg 16: 135–143
2. Beazley RM, Bagley DH, Ketcham A (1974) The effect of cryosurgery on peripheral nerves. J Surg Res 16: 231–236
3. Bernstein EF (1963) Cryogenic surgery. Jed J Aust 5352: 267–275
4. Bielschowsky M, Valentin B (1922) Die histologischen Veränderungen in durchgefrorenen Nervenstrecken. J Psychol Neurol 29: 133–145
5. Carter DC, Lee PWR, Gill W, Johnston RJ (1972) The effect of cryosurgery on peripheral nerve function. Ann R Coll Surg Edinburgh 17: 25–36
6. Cooper IS (1962) A cryogenic method for physiologic inhibition and production of lesions in the brain. J Neurosurg 19: 853–862
7. Denny-Brown D, Adams RD, Brenner C, Doherty MM (1945) The pathology on injury to nerve induced by cold. J Neuropath Exper Neurol 4: 305–323
8. Evans PJD, Lloyd JW, Green CJ (1981) Cryoanalgesia: the response to alterations in freeze cycle and temperature. Br J Anaesth 53: 1121–1127
9. Fasano VA, Broggi G, Nunno T de, Baggiore P (1964) Cryotherapie et neurochirurgie. Neuro-Chirurgie 6: 172–179
10. Fasano VA, Broggi G, Schiffer D, Urciuoli R (1965) Lésions expérimentales provoquées par le refroidissement localisé du cerveau. Étude morphologique et histo-chimique. Neuro-Chirurgie 11: 519–528
11. Fasano VA, Broggi G, Zeme S (1982) L'utilizzazione del freddo nel trattamento analgesico: criolisi del ganglio di Gasser. 1° Simposio interdisciplinare di Algologia, Torino. Atti 79–88
12. Glynn CJ, Lloyd JW, Barnard JDW (1980) Cryoanalgesia in the management of pain after thoracotomy. Thorax 35: 325–331
13. Hannington-Kiff JG (1980) Cryoanalgesia for postoperative pain. Lancet i: 829–834
14. Harkin JS, Skinner MS (1970) Experimental and electron microscopic studies of nerve regeneration. Ann Otol Rhinol Lar 79: 218–224
15. Joy RT, Finean JB (1963) A comparison of the effects of freezing and of treatment with hypertonic solutions on the structure of nerve myelin. J Ultrastruct Res 8: 264–271
16. Le Beau J, Dondey M (1964) Premières observations humaines de repérage de structures cérébrales profondes par refroidissement localisé et réversible au cours des interventions stéréotaxiques. Neurochirurgia 7: 18–25
17. Lovelock JL (1957) The denaturation of lipid protein complexes as a cause of damage by freezing. Proc Roy Soc (biol) 147: 427–433
18. Menz LJ (1971) Structural changes and impairment of function associates with freezing and thawing in muscle, nerve and leucocytes. Cryobiol 8: 1–7
19. Meryman HT (1956) Mechanism of freezing in living cells and tissue. Science 124: 515–523
20. Trendelenburg W (1916–1917) Über langdauernde Nervenausschaltung mit sicherer Regenerationsfähigkeit. Z ges exp Med 5: 371–378
21. Whittaker DK (1974) Degeneration and regeneration of nerves following cryosurgery. Br J exp Path 55: 595–602

Acta Neurochirurgica, Suppl. 39, 181–185 (1987)
© by Springer-Verlag 1987

Shape-Factor Intensity Analyses of Brain Slices in Surgery for Epilepsy*

S. H. M. Nyström**, **P. H. Eskelinen, E. R. Heikkinen,** and **M. T. Weckström**

Department of Neurosurgery, Oulu University Central Hospital and the Department of Physiology of University of Oulu, Oulu, Finland

Summary

A special shape-factor intensity (SFI) quantification method for analysing of macro- and microrecordings of brain activity during surgery for epilepsy is illustrated. Comparative analyses show that epileptic activity may be characterized as different from normal brain activity by both macro-and microtechniques. The possible advantages of brain slice technique in comparison with direct peroperative microrecording are discussed; absence of artifacts from respiratory and pulsatory movements of the brain is stressed. The slice technique cannot be advocated for routine use in epilepsy surgery as it requires advanced neurophysiological knowledge.

Keywords: Surgery of epilepsy; cerebral recording; brain slice.

Introduction

Surgery for medically refractory epilepsy involves the application of elaborate neurophysiological techniques. Modern advanced developments in neurophysiology, and especially in computer techniques, provide promising tools for comprehensive analysis of epileptic brain activity and may be used as guidance for surgical procedures and studies on fundamental aspects of the functioning of the nervous system.

Epileptic spikes have been elucidated by various methods with such features as rise and fall time, peak amplitude and angle, second time derivative[5, 11], dura-

tion, and various combinations of these[3, 4, 6, 9]. Direct or template matching with a predetermined event in spike analysis has been successfully employed by Salzberg et al.[11], and continuous inverse filtering through estimation of the pertinent correlation coefficient model of the original EEG, as implemented by Lopez da Silva et al.[8], has provided a potential means for indicating abnormal events emerging from stationary EEG signals. A similar detection algorithm was devised by Barlow[1] by means of the Fourier transform.

A new method for analyzing the EEG in detail and reporting short temporal deviations and long-term structural alterations is utilized in the present paper. This program, called the shape-factor intensity (SFI) method, views the EEG signals as series of positive and negative processes defined by two sets of parameters, *i.e.*, the shape-factor (S) and intensity (I). The shape factor is dependent on the speed of change, duration and routes of the signals, while intensity is determined by the signal level. The signals are mapped as such onto positive and negative shape-factor intensity planes, where their characteristic structure as two sets of two-dimensional points will allow visual interpretation and further quantification. The quantitative differences between local positive and negative processes are simultaneously visualized on maps of columnar histograms and the results of numerical differences are calculated on differential histogram maps. The method has been applied to both macrorecordings and microrecordings by altering the parameters to allowing very fast high-voltage processes to be analyzed as well. The theoretical background to the method has been published elsewhere[2].

---

 * This work was supported by grant no 7455/304 from the Academy of Finland. The examinations were carried out in accordance with the Declaration of Helsinki as a research project combined with clinical treatment.

 ** S. H. M. Nyström, M.D., Department of Neurosurgery, Oulu University Central Hospital, Kajaanintie 50, SF-90220 Oulu 22, Finland.

Material and Methods

The material consists of the following types of brain recordings: surface and depth recordings in 20 temporal epilepsy patients, surface recordings in nonepileptic patients, microrecordings from epileptic brain slices and from normal human brain slices removed for surgical access to deep pathological processes, recordings from slices from normal brains of decapitated rats and slices from rat brain rendered epileptic with strychnine. Seven animals were used for the experiments. The animal work was included in order to check the methodology.

The 20 patients with drug resistant temporal epilepsy comprised 13 men and 7 women with a mean age of 34 years. In the peroperative macrorecordings the usual techniques for corticography and depth recordings in epilepsy surgery were employed[10]. Specimens from the temporal cortex or hippocampus were taken for microrecordings during lobectomy and placed in a beaker containing an oxygenated glucose-bicarbonate medium at 37 °C (NaCl (124 mM), NaHCO$_3$ (26 mM), KCl (5 mM), KH$_2$PO$_4$ (1.24 mM), CaCl$_2$ (0,75 mM), MgSO$_4$ (1,3 mM), glucose (10 mM) and 1% inulin). A gas mixture of O$_2$ + CO$_2$ (95 : 5) was bubbled through the solution 10 minutes prior to its use and throughout the time of recording.

The neurons of temporal lobe of rat brain were rendered epileptic by means of 15 µm/ml solution of strychnine. Thin slices of 350–600 µ were gently cut with razor blades[7]. Glass capillary microelectrodes (microelectrode drawer produced by T. Zaschka, Munich) were filled with a 3 M KCl solution and moved to the target with a Takahashi micromanipulator. A grass P 16 or a high impedance preamplifier by P. H. Eskelinen and a Tektronix 5113 dual beam storage oscilloscope were used to amplify the signals, which were then recorded with a HP 3960 magnetic tape recorder for analysis by a DEC PDP 11/23 computer with a 128 KB memory and floating point unit. The results were displayed on a videomonitor, a Texas SD 422 hard copy unit or a HP 7470 A digital plotter. A 16-channel Elema-Schönander EEG machine was used for analogue recordings in the operation room.

Results

All the epileptic patients had typical temporal analogue EEG abnormalities. The results of the analogue recordings were compared with those of the quantitive analyses with respect to epileptic activity, and the surgical resections were performed according to the informations obtained from all the neurophysiological studies. None of the patients had a normal SFI map preoperatively. The following examples illustrate basic transformations of normal and abnormal EEG processes onto the SFI planes and histograms. A surface macrorecording of normal adult cortical brain activity is seen in Fig. 1,I A, B, and C. Fig. 1,I A shows the results of 90-second epochs recorded from the right midtemporal region. The S axis extends from 0 to 1 and the intensity scale from 0 to 100 µV. There is a clear symmetry in the clustered positive and negative processes, as seen also on the histogram map I B. Fig. 1,I C nevertheless shows a slight asymmetry of somewhat higher S values on the negative side of the differential maps within the range of low intensity values. Fig. 1,II A shows a corresponding SFI map of a surface macrorecording of the right midtemporal area of a 27-year-old female patient with epilepsy. The dots are distributed in a different manner. The histogram maps in Fig. 1,II B show a broadening out of the "pyramid" with more processes in squares of higher intensity and higher shape-factor values than observed in the normal recordings. Fast lower-intensity processes are also present. The differential histogram maps (Fig. 1,II C) show a predominance of positive processes in the range of somewhat lower shape-factor values and of negative processes in the somewhat higher shape-factor values. Fig. 1,III A, B, C show the maps of a deep macrorecording from the right amygdala of the same subject, and Fig. 1,IV A, B, C the maps of surface recordings of the same patient after temporal lobe resection. There is a clear tendency towards normalization of the maps as compared with those in Fig. 1,I A, B, C and Fig. 1,II A, B, C. Fig. 2,I A, B, C shows the SFI maps of microrecording of a slice from the apparently healthy temporal cortex from another adult subject and Fig. 2,II A, B, C one from the temporal cortex of the above-mentioned patient with epilepsy. Background noise is filtered out in both series of maps. There is a clear tendency for an accumulation of faster and higher intensity processes in the epileptic recordings. Fig. 2,III A, B, C reveals a similar situation in the recordings from a hippocampal slice obtained from the patient with epilepsy. Fig. 2,IV A, B, C shows a microrecording of a slice of normal temporal rat cortex and Fig. 2,V A, B, C one of a slice of rat cortex with strychnine-induced epileptic neurons. The normal rat cortex gives fairly symmetrical results in comparison with the maps of processes in the slice of epileptic cortex, which depicts an accumulation of slow high-voltage negative spikes.

Discussion

As far as macroelectrode recording is concerned, the present analyses of normal and epileptic brain activity by the new SFI method yields promising results. It seems utilizable for guiding of surgical procedures. In fact, results of the operative procedures corresponded to common international standards. To be noticed is, that such an analysis saved a lot of the electroencephalographers time in the operation room. The additional contribution of microtechniques is not yet clear.

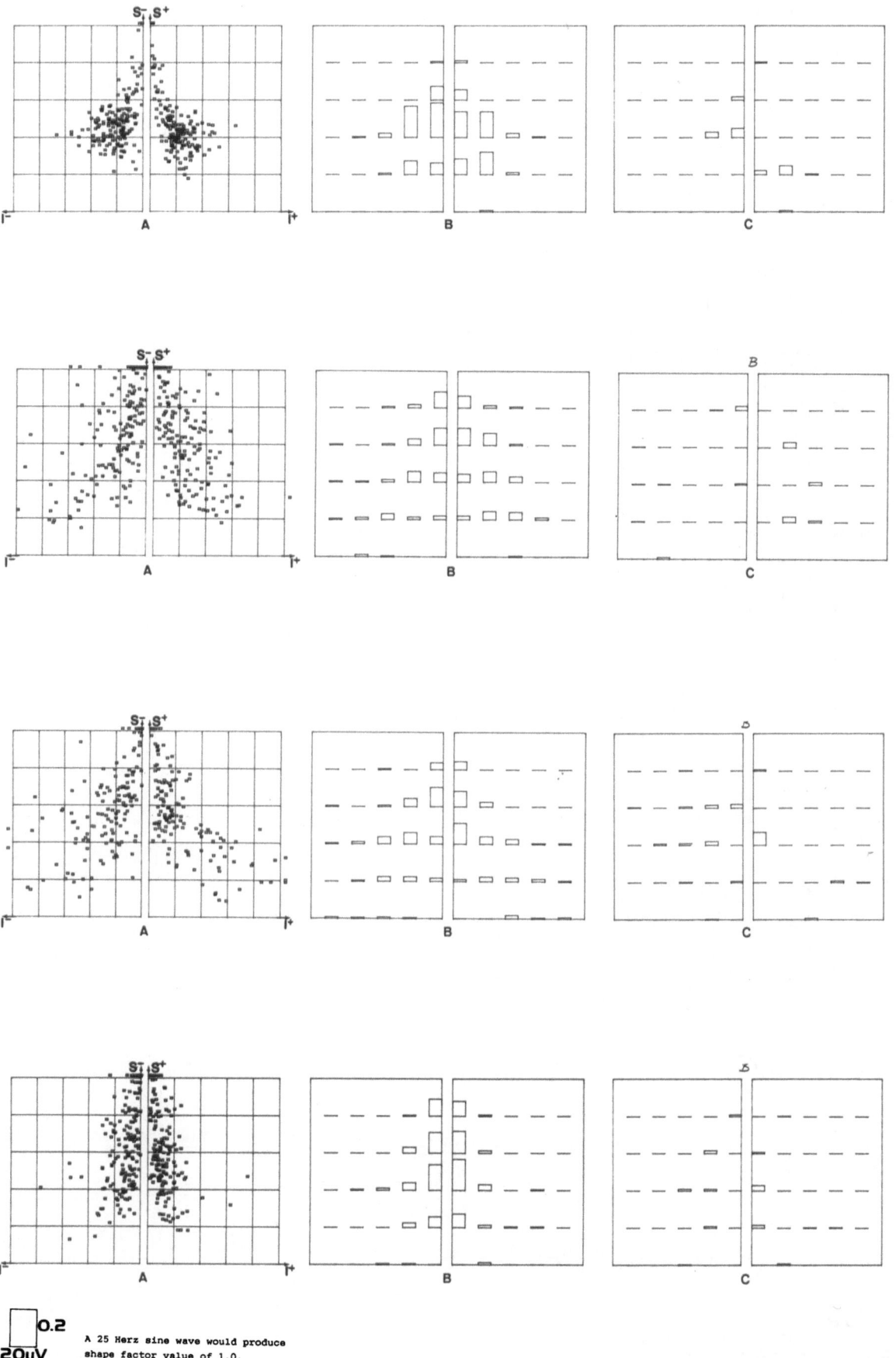

0.2
20μV

A 25 Herz sine wave would produce
shape factor value of 1.0.

Fig. 1, I–IV See text

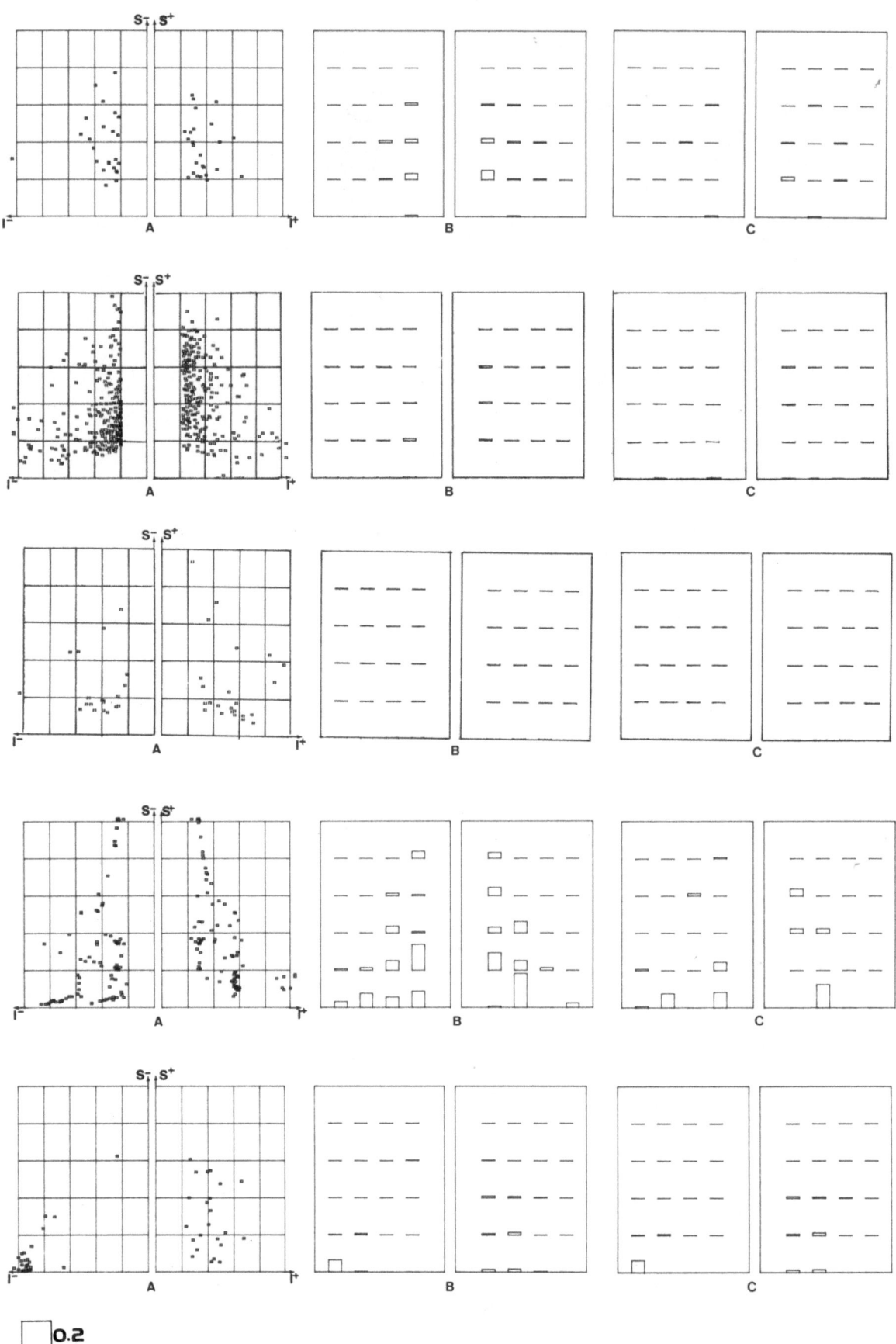

A 25 Herz sine wave would produce
shape factor value of 1.0.

Fig. 2, I–V See text

The microtechnique method requires elaborate neurophysiological experience and is sensitive to artifacts; it is therefore difficult to adopt it as a routine method in clinical work. At the same time it needs practice and animal testing of equipment and personnel at regular intervals. The specimens to be sliced have to be removed without crushing them, and the slices have to be thin enough to avoid tissue. necroses. Special oxygen bubble-protected beaker constructions must be used. Adequate oxygenation is important, and fresh solutions must always be used. The advantage of this technique in comparison with direct microrecording in situ is that there are no respiratory or pulsatory movement artifacts. The field needs to be explored further in order to assess its real value for guiding microsurgical removal of epileptic tissues.

References

1. Barlow JS (1975) Some programs for the processing of EEG data on a small general-purpose digital computer. In: Dolce G, Kunkel H (eds) Cean-computerized EEG analysis. Fisher, Stuttgart, pp 172–179

2. Eskelinen PH, Nyström SHM (1985) Shape-factor intensity analysis: a new method of EEG signal processing. IRCS Med Sci 13: 180–181

3. Gevins AS, Yeager CL, Diamond SL, Spire J-P, Zeitiin GM, Gevins AH (1975) Automated analysis of the electrical activity of the human brain (EEG): A progress report. Proc IEEE 63: 1382–1399

4. Gotman J, Gloor P (1976) Automatic recognition and quantification of interictal epileptic activity in the human scalp EEG. EEG Clin Neurophysiol 41: 513–529

5. Kooi KA (1966) Voltage-time characteristics of spikes and other rapid electroencephalographic transients. Neurology 16: 59–66

6. Ktonas PY, Smith JR (1974) Quantitation of abnormal EEG spike characteristics. Comput Biol Med 4: 157–163

7. Lipton P, Whittingham TS (1984) Energy metabolism and brain slice function. In: Dingledine R (ed) Brain slices. Plenum Press, New York, London, pp 113–153

8. Lopez da Silva FH, Dijk A, Smiths H (1975) Detection of nonstationarities in EEGs using the autoregressive model—an application to the EEG of epileptics. In: Dolce G, Kunkel H (eds) CEAN-computerized EEG analysis. Fisher, Stuttgart, pp 180–199

9. Mars NJI (1982) Computer-augmented analysis of electroencephalograms in epilepsy. Dr. Eng. dissertation, Twente University of Technology, Enschede, the Netherlands

10. Nyström SHM, Eskelinen PH, Heikkinen ER (1984) Shape-factor intensity (SFI) analysis of EEG of patients treated surgically for epilepsy. Acta Neurochir (Wien) [Suppl] 33: 63–67

11. Salzberg B, Heath RG, Edwards RJ (1967) EEG spike detection in schizophrenia research, In: Dig 7th Int Conf Med Biol Eng, Stockholm, Sweden, pp 266

12. Salzberg B, Lustick LS, Heath RG (1971) Detection of focal depth spiking in the scalp EEG of monkeys. EEG Clin Neurophysiol 31: 327–333

Acta Neurochirurgica, Suppl. 39, 186–187 (1987)

Accessory to the Talairach's Apparatus for Orthogonal Approaches. Technical Note

M. Scerrati*, P. Pola, A. Fiorentino, and **M. Fiorentino**

Istituto di Neurochirurgia, Università Cattolica, Roma, Italy

Summary

A stereotactic device is described which, without modifying the main characteristics of the Talairach's stereotactic apparatus, increases its flexibility in orthogonal approaches. This accessory, which replaces the double-grid system, permits any orthogonal access on the X and Y coordinates and insertion of any type of electrode, catheter or probe, without the constraints of the double grid holes. Its use in different stereotactic operations (brain tumor biopsy or curietherapy, implantation of intracerebral depth electrodes for stereo-EEG recording) proved its wide working capability and great reliability.

Keywords: Stereotaxy; orthogonal coordinates; Talairach's apparatus.

Introduction

The main characteristic of the Talairach's stereotactic apparatus is to make accessible simultaneously different brain sites along multiple orthogonal trajectories by means of double grid system[3]. However, these same advantages have two main limitations: 1. the impossibility of using polar approaches; 2. the constraint of fixed access through the holes of the grids[1, 2].

As for the former we earlier described a special device permitting work with polar coordinates[2]. To overcome the latter we describe a new acccessory which, operating orthogonally as well, is more flexible than the classical double-grid system.

Technical Description

The device consists of two bars perpendicular to each other which reproduce and replace the double-grid system (Fig. 1): the horizontal bar (A, Fig. 1) can be fixed on either sides of the stereotactic frame; the vertical bar (B, Fig. 1) is mounted on the horizontal bar and sliding on it. The vertical bar holds a sliding carrier (C, Fig. 1) with two grooves (E, Fig. 1) perpendicular to the X and Y axes, thus replacing the holes of the double grid. The carrier can be moved in such a way as to place the grooves in exact correspondence with preselected double holes of the grid or in any position between the grid holes. Trephine and guides for inserting probes or catheters or electrodes can be fitted into the carrier groove. A graduated bar (Fig. 2) allows calculation outside the instrument of the distance of the target point from the midline (Z-axis coordinates), as indicated by the micrometric screw (A, Fig. 2). In addition a special set has been designed for inserting catheters to be used for brachycurietherapy (A, Fig. 3) and multilead electrodes (stainless steel rings mounted on similar catheters). Both of these have an outer diameter of 1.3 mm and ar closed with a pellet (B, Fig. 3) on the tip. The electrodes or catheters are introduced by means of a grooved guide (C, Fig. 3) narrowed at one end in such a way as to push on the pellet for the insertion. A lock (D, Fig. 3), which fitts on the guide and slides on it, and is set according to the desired depth along the Z axis, keeping the electrode or catheter exactly in place. Hollow conic nails (A, Fig. 4) of various length are fitted into the holes on the skull through which electrodes or catheters pass. The nails are provided with a spring (B, Fig. 4) acting as a lock after insertion. This set turned out to be very useful for chronic stereo-EEG explorations with intracerebral electrodes as well as for temporary brachycurietherapy with removable catheters.

Discussion

This device presumes that the repérage be made according to the well known technique of the Talairach stereotactic instrument. Compared to the classical double grid system, our device has two main advantages: 1. the possibility of selecting and reaching any point in the intracranial space along the X and Y

---

* Massimo Scerrati, M.D., Istituto di Neurochirurgia, Università Cattolica S. Cuore, Largo A. Gemelli 8, I-00168 Roma, Italy.

Fig. 1. The two perpendicular bars and the carrier of the accessory. See text for detailed description

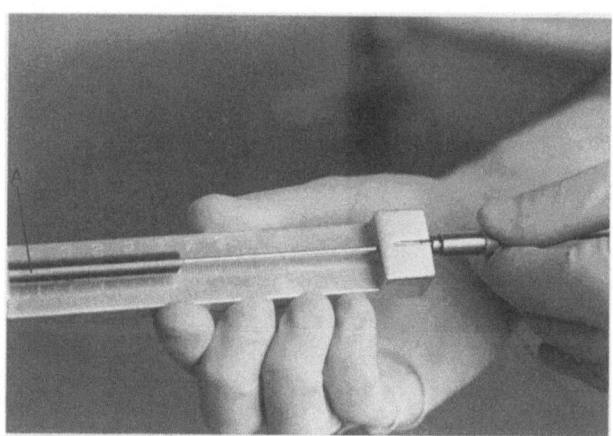

Fig. 2. The graduated bar with the micrometric screw (*A*)

Fig. 3. Insertion set for electrodes or catheters. See text for detailed description

Fig. 4. Hollow conic nails (*A*) provided with lock spring (*B*)

coordinates, without the constraints of fixed access cooresponding to the holes of the grid; 2. the possibility, when chronic SEEG or brachycurietherapy have to be performed, to free the electrodes or the catheters which have been left in place from the carrier groove in an easier way than from the double grid. The device we describe proved its usefulness and wide capability in many different stereotactic operations requiring orthogonal aproaches (tumor biopsy, permanent or temporary brachycurietherapy, implant of depth electrodes for stereo-EEG) showing great flexibility, accuracy and safety.

Instruments manifactured by Fiorentino A.M., S.r.l. via delle Benedettine, 48, I-00135 Roma, Italy.

Acknowledgements

This work is partially supported by the Ministry of Public Education.

References

1. Olivier A, Bertrand G (1982) Stereotactic device for percutaneous twistdrill insertion of depth electrodes and for brain biopsy. Technical note. J Neurosurg 56: 307–308
2. Scerrati M, Fiorentino A, Fiorentino M, Pola P (1984) Stereotactic device for polar approaches in orthogonal systems. Technical note. J Neurosurg 61: 1146–1147
3. Talairach J, David M, Tournoux P (1957) Atlas d'anatomie stéréotaxique. Masson, Paris

Pain

A Medical and Anthropological Challenge

Proceedings of the First Convention of the Academia Eurasiana Neurochirurgica, Bonn, September 25–28, 1985

Editors: **J. Brihaye,** Clinique Neuro-chirurgicale, Bruxelles, **F. Loew,** Neurochirurgische Universitäts-Klinik, Homburg/Saar, and **H. W. Pia**

1987. 111 partly coloured figures.
Approx. 200 pages.
Cloth DM 225,–, öS 1580,–
Reduced price for subscribers to
"Acta Neurochirurgica":
Cloth DM 202,50, öS 1422,–
ISBN 3-211-81990-8

(Acta Neurochirurgica/Supplementum 38)

The book gives a survey of the medical, philosophical and religious aspects of chronic pain and suffering. Experts in the fields of neurophysiology, neuropharmacology, anaesthesiology, psychology and psychotherapy, neurology and neurosurgery as well as representatives of the main world religions and of different philosophical directions were brought together during the First Convention of the Academia Eurasiana Neurochirurgica in September 1985, and discussed the various aspects of pain and suffering, including the possibilities for treatment. The combination of religious, philosophical and medical facets of pain means a new approach to a better understanding of the problems related to pain and suffering.

Stereotactic Techniques in Clinical Neurosurgery

By **D. A. Bosch,** Department of Neurosurgery, St. Elisabeth Hospital, Tilburg, The Netherlands

1986. With 216 partly colored figures. Drawings by D. Buiter. XII, 278 pages. Cloth DM 125,–, öS 875,–
ISBN 3-211-81878-2

Recent advances in stereotactics have made clear that in the surgical management of deep seated brain lesions not only diagnostic but also therapeutic procedures can be carried out in one session. Therefore, stereotactics will become an integrated part of general neurosurgery. Particularly with recent progress in high resolution CT, NMR, and PET imaging of the brain has it become feasible to study and localize any brain area of interest. With the concomitant advances in computer technology, three-dimensional reconstruction of deep seated lesions in stereotactic space is possible. Thus the way is open for combined surgery, with stereotactic precision and computer monitored open resection.
The use of stereotactic laser vaporization in debulking and subsequent adjuvant therapy at the tumor site will give glioma surgery a new impetus. Clinical research in neurooncology is reviewed, and future applications in stereotactics are discussed.
This monograph is a comprehensive guide to stereotactic principles, methodology, and possibilities. It offers the basis for all types of surgery which need exact positioning of instruments and precise localization of lesions and is the first to cover the whole field including functional, therapeutic, diagnostic, and combined stereotactic techniques.

Springer-Verlag Wien New York

Moelkerbastei 5, A-1010 Wien · Heidelberger Platz 3, D-1000 Berlin 33 · 175 Fifth Avenue, New York, NY 10010, USA
37-3, Hongo 3-chome, Bunkyo-ku, Tokyo 113, Japan

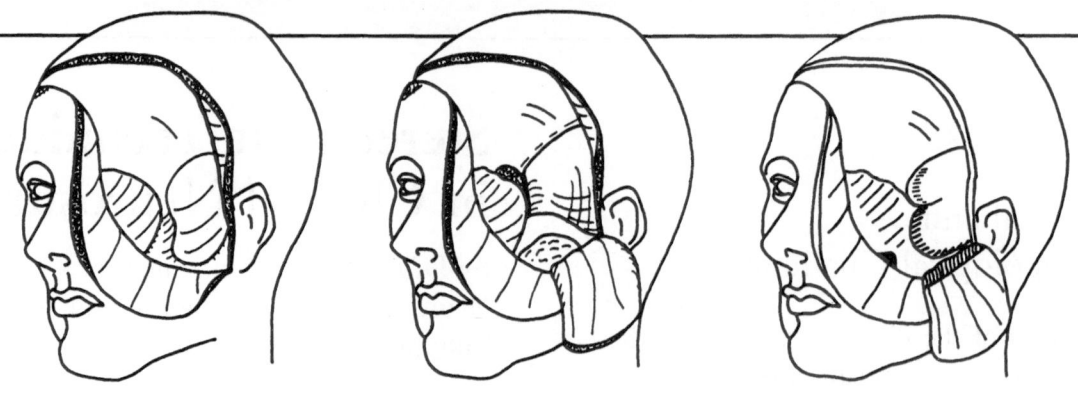

The Cavernous Sinus

V. V. Dolenc (ed.)

The management of vascular and tumorous lesions of the parasellar region still remains one of the most demanding tasks in neurosurgery. It is only a short time ago that the major concepts of the anatomy of the so-called cavernous sinus were described in detail.

Pioneer anatomical studies of the parasellar region done by Taptas, and the daring direct operative approach introduced by Parkinson promoted the development of modern neuroradiological intervention procedures and further refinement of neurosurgical techniques. Today, it is hard to imagine a successful management of vascular pathologies of this region without a complementary use of the two techniques.

Similar is the situation in the treatment of parasellar tumors. Modern neuroradiological diagnostic procedures and neurophysiological tests employed prior to surgery can supply many valuable data which are indispensable for the surgeon who wishes to devise optimal treatment strategies for each case separately. Preoperative embolization of the tumor vessels and peroperative monitoring of the function of cranial nerves III through VI enable the surgeon to plan a more radical excision and provide for greater safety of the procedure.

The book stresses that the improved understanding of normal structure and function of the cavernous sinus makes risk-free and effective operative treatment of intracavernous aneurysms, carotid-cavernous fistulas and tumors possible. It gives a view of our present understanding of the structure and function of the cavernous sinus and presents results and possibilities of the treatment of parasellar pathologies.

This stimulating book ist the first comprehensive and up-to-date text dealing with the cavernous sinus and is addressed to anyone who is concerned with the diagnosis and treatment of lesions of the skull base.

Contents: Historical Review and Pioneer Work. – Anatomy. – Diagnostic Procedures. – Occlusion Techniques. – Surgery of Vascular Lesions. – Tumor Surgery.

1987. 195 figures.
Approx. 400 pages.
Cloth approx. DM 240,–, öS 1700,–,
ISBN 3-211-82000-0

SPRINGER-VERLAG WIEN NEW YORK

Moelkerbastei 5, A-1010 Wien ● Heidelberger Platz 3, D-1000 Berlin 33 ● 175 Fifth Avenue, New York, NY 10010, USA ● 37-3, Hongo 3-chome, Bunkyo-ku, Tokyo 113, Japan